International Review of
Cytology

A Survey of
Cell Biology

VOLUME 164

International Review of Cytology

A Survey of

Cytology Cell Biology

Edited by

Kwang W. Jeon
Department of Zoology
University of Tennessee
Knoxville, Tennessee

Jonathan Jarvik
Department of Biological Sciences
Carnegie Mellon University
Pittsburgh, Pennsylvania

VOLUME 164

ACADEMIC PRESS
San Diego   New York   Boston   London   Sydney   Tokyo   Toronto

Copyright © 1996 by ACADEMIC PRESS, INC.

Academic Press, Inc.
A Division of Harcourt Brace & Company
525 B Street, Suite 1900, San Diego, California 92101-4495

United Kingdom Edition published by
Academic Press Limited
24-28 Oval Road, London NW1 7DX

International Standard Serial Number: 0074-7696

International Standard Book Number: 0-12-364568-9

PRINTED IN THE UNITED STATES OF AMERICA
96 97 98 99 00 01 BB 9 8 7 6 5 4 3 2 1

CONTENTS

Multifunctional Proteins in *Tetrahymena*: 14-nm Filament Protein/Citrate Synthase and Translation Elongation Factor-1α

Osamu Numata

Some Characteristics of Neoplastic Cell Transformation in Transgenic Mice

Irina N. Shvemberger and Alexander N. Ermilov

Integration of Intermediate Filaments into Cellular Organelles

Spyros D. Georgatos and Christèle Maison

Developmental Failure in Preimplantation Human Conceptuses

Nicola J. Winston

G-Protein-Coupled Receptors in Insect Cells

Jozef J. M. Vanden Broeck

Microtubules and Microtubule Motors: Mechanisms of Regulation

Catherine D. Thaler and Leah T. Haimo

CONTRIBUTORS

Numbers in parentheses indicate the pages on which the authors' contributions begin.

Alexander N. Ermilov (37), *Laboratory of Chromosome Stability and Cell Engineering, Institute of Cytology of Russian Academy of Sciences, 194064 St. Petersburg, Russia*

Spyros D. Georgatos (91), *Program of Cell Biology, European Molecular Biology Laboratory, 69117 Heidelberg, Germany, and Department of Basic Sciences, Faculty of Medicine, The University of Crete, 711 10 Heraklion, Greece*

Leah T. Haimo (269), *Department of Biology, University of California, Riverside, California 92521*

Christèle Maison (91), *Program of Cell Biology, European Molecular Biology Laboratory, 69117 Heidelberg, Germany*

Osamu Numata (1), *Institute of Biological Sciences, University of Tsukuba, Tsukuba, Ibaraki 305, Japan*

Irina N. Shvemberger (37), *Laboratory of Chromosome Stability and Cell Engineering, Institute of Cytology of Russian Academy of Sciences, 194064 St. Petersburg, Russia*

Catherine D. Thaler (269), *Department of Biology, University of California, Riverside, California 92521*

Jozef J. M. Vanden Broeck (189), *Laboratory for Developmental Physiology and Molecular Biology, Department of Biology, Zoological Institute K. U. Leuven, B-3000 Leuven, Belgium*

Nicola J. Winston (139), *Laboratoire de Physiologie du Developpement, Institut Jacques Monod, CNRS-Université Paris VII, 75005 Paris, France*

Multifunctional Proteins in *Tetrahymena:* 14-nm Filament Protein/ Citrate Synthase and Translation Elongation Factor-1 α

Osamu Numata

Institute of Biological Sciences, University of Tsukuba, Tsukuba, Ibaraki 305, Japan

One gene encoding a protein has been shown to have two entirely different functions. Such a phenomenon, which has been called "gene sharing," was first known in crystallins. We found two multifunctional proteins in the ciliated protozoan *Tetrahymena:* 14-nm filament protein and protein translation elongation factor 1-α (EF-1 α). The 14-nm filament protein has dual functions as a citrate synthase in mitochondria and as a cytoskeletal protein in cytoplasm. In cytoplasm, the 14-nm filament protein was involved in oral morphogenesis and in pronuclear behavior during conjugation. The observation that *Tetrahymena* intramitochondrial filamentous inclusions contain the 14-nm filament protein and that the citrate synthase activity of the 14-nm filament protein is decreased by polymerization and increased by depolymerization, suggests a possible modulating mechanism of citrate synthase activity by monomer–polymer conversion in mitochondria *in situ.* The EF-1 α functions as an F-actin-bundling protein and a 14-nm filament-associated protein as well as an elongation factor in protein synthesis. The F-actin-bundling activity of EF-1 α was regulated by Ca^{2+} and calmodulin. Here we review the properties and functions of two multifunctional proteins in *Tetrahymena.*

KEY WORDS: Multifunctional protein, 14-nm Filament protein, Elongation factor 1 α (EF-1 α), *Tetrahymena,* Citrate synthase, F-Actin-bundling protein, Ca^{2+}, Calmodulin, Conjugation, Mitochondria.

I. Introduction

The plethora of nucleotide sequence databases has brought about unexpected findings in present-day molecular biology. It was a great surprise to discover that duck ε-crystallin was found to be essentially identical to lactate dehydrogenase B4 (Wistow *et al.*, 1987), an ancient, highly conserved glycolytic enzyme. Sequence comparisons have shown relationships between eye lens crystallins and enzymes (Piatigorsky and Wistow, 1989, 1991; Wistow, 1993). In particular, taxon-specific crystallins turn out to be metabolic enzymes directly recruited to a new role by modification of gene expression without prior gene duplication (Piatigorsky and Wistow, 1989, 1991; Wistow, 1993). Because two distinct protein phenotypes are produced by the same transcriptional unit, this phenomenon has been called "gene sharing" (Piatigorsky *et al.*, 1988). This kind of gene recruitment contradicts the conventional view that gene duplication must precede the acquisition of new protein functions.

In the ciliated protozoan *Tetrahymena*, we found two multifunctional proteins: 14-nm filament protein and protein translation elongation factor-1 α (EF-1 α). In cytoplasm, the 14-nm filament protein has crucial roles in fertilization either alone or in association with some microtubules having specialized function (Numata *et al.*, 1985; Takagi *et al.*, 1991), whereas in mitochondria the protein functions as citrate synthase which is a key enzyme in the tricarboxylic acid (TCA) cycle (Numata *et al.*, 1991a, 1995; Kojima *et al.*, 1995). EF-1 α catalyzes the glucose triphosphate (GTP)-dependent binding of aminoacyl-tRNAs to their respective mRNA anticodons at the A site of ribosomes in the peptide elongation phase of protein synthesis (Riis *et al.*, 1990; Moldave, 1985). *Tetrahymena* EF-1 α binds to F-actin and induces the formation of the F-actin bundle; moreover, the bundling is modulated by the addition of both Ca^{2+} and calmodulin (CaM). In this chapter, we describe the properties and functions of these mulifunctional proteins and discuss their biological significance.

II. 14-nm Filament Protein

A. Properties

About 13 years ago, during a study of cytoskeletal proteins in *Tetrahymena*, we found that a 49-kDa filament-forming protein from an extract of *Tetrahymena* acetone powder could be purified by assembly into 14-nm filaments and subsequent disassembly (Numata and Watanabe, 1982). The 14-nm filament protein was purified from the $1M$ KCl extract of *Tetrahymena* acetone powder [Fig. 1(a), A] by ammonium sulfate fractionation [Fig. 1(a), B], followed by assembly [Fig. 1(a), C] and disassembly [Fig. 1(a), D] of 14-nm filaments. Finally, a pure 49-kDa protein was obtained by preparative sodium dodecyl sulfate (SDS)-polyacrylamide gel electro-

a   b

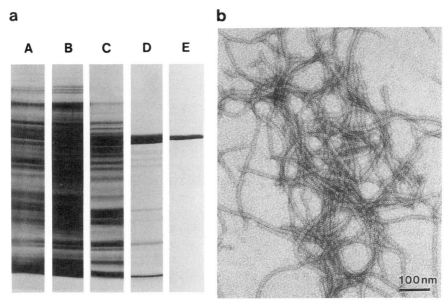

FIG. 1 (a) SDS/polyacrylamide gel analysis of fractions through the purification of 14-nm filament protein. A, KCl extract; B, crude 14-nm filament protein fraction after ammonium sulfate fractionation between 40% and 70% saturation; C, pellet from assembled 14-nm filament fraction after centrifugation; D, supernatant from disassembled 14-nm filament fraction after centrifugation; E, highly purified 14-nm filament protein by preparative gel electrophoresis. (b) Electron micrograph of negatively stained, reassembled 14-nm filaments. Disassembled 14-nm filament fraction was polymerized in the assembly conditions.

phoresis [Fig. 1(a), E] and was used to produce an antibody in rabbits (Numata *et al.*, 1983). The conditions for assembling a 14-nm filament protein are incubation of a protein sample (2 mg protein/ml) in 5 mM 2-(*N*-morpholino)ethanesulfonic acid (MES) buffer (pH 6.6) containing 0.1mM N^{α}-tosyl-L-lysyl-chloromethane hydrochloride (TLCK), 50 mM KCl, 0.6 mM ATP, and 1.2 mM CaCl$_2$ at 30°C for 30 min. Figure 1(b) is an electron micrograph of reassembled 14-nm filaments. The disassembly conditions are dialysis of the 14-nm filament suspension (3 mg protein/ml) against 10 mM Tris-acetate buffer (pH 8.2) containing 5 mM 2-mercaptoethanol, 1 mM ethylene glycol bis(β-aminoethyl ether)-*N,N*-tetraacetic acid (EGTA), and 0.05 mM TLCK at 4°C for 24 hr. The assembly and disassembly were repeatable, and resulted in the exclusive retention of the 49-kDa protein.

The assembly of the 14-nm filament is a two-step process, involving association of subunits to form short filaments, followed by their elongation and bundle formation. We think that assembly of the 14-nm filament might occur by a condensation polymerization mechanism consisting of nucleation and elongation steps like those in the assembly of actin filaments (Kasai *et al.*, 1962; Kasai, 1969; Oosawa and Asakura, 1975) and microtubules (Johnson and Borisy, 1977).

The 14-nm filaments appeared as flexible rods and showed no tendency to form thinner fibrils; they were composed of 7-nm globular subunits, as revealed by electron microscopy after negative staining (Numata *et al.*, 1980) [Fig. 1(b)]. We consider that these globular subunits interact with one another to form a complex fibrous structure directly as actin (Holmes *et al.*, 1990) and tubulin (Amos and Baker, 1979) do. Table I contains a comparison of actin, tubulin, and 14-nm filament protein polymers. Although intermediate filaments in mammalian cells also appear as flexible rods (Steinert and Roop, 1988; Traub, 1985), the final 10-nm diameter of the intermediate filament is thought to be composed of 8 protofilaments (32 polypeptide chains) joined end on end to neighbors by staggered overlaps to form the long, ropelike filament (Ip *et al.*, 1985). Thus, the 14-nm filament is different from three principal types of protein filament: actin filament, microtubules, and intermediate filaments. So we think that the 14-nm filament is a quite new filament.

B. Functions

1. Oral Morphogenesis during Cell Cycle

To study the biological function of the 14-nm filament protein, we used highly purified protein and preparative gel electrophoresis to prepare anti-

TABLE I

Comparison of Actin, Tubulin, and 14-nm Filament Protein Polymers

	Actin	Tubulin	14-nm Filament protein
Polypeptide molecular weight	42,000	50,000 (α-tubulin) 50,000 (β-tubulin)	49,000
pI	5.0–5.1	5.2–5.5	8.0
Unpolymerized form	Globular, monomer	Globular, dimer ($\alpha\,\beta$)	Globular, dimer (α_2)
Bound nucleotide	ATP (one/ monomer)	GTP (two/dimer)	ATP?
Form of polymer	Helical filament	Hollow tube composed of 13 protofilaments	Flexible rods
Diameter of polymer	7 nm	24 nm	14 nm

14-nm filament protein antiserum, and examined the immunofluorescence localization of the 14-nm filament protein within *Tetrahymena* cells. The results showed that the immunofluorescence was localized in the oral apparatus and mitochondria (Numata *et al.*, 1983). The oral apparatus is composed of three membranelles (M1, M2, and M3) and the undulating membrane (UM), all of which are joined by fibrous and tubular interconnections. These four membranes are the origin of the genera *Tetrahymena*. The oral apparatus in *Tetrahymena* has been described in detail by several investigators (Forer *et al.*, 1970; Sattler and Staehelin, 1979; Bakowska *et al.*, 1982a,b). The relationship between 14-nm filament protein and mitochondria is mentioned in the latter part of this review.

The fluorescence of the 14-nm filament protein in the oral apparatus was of great interest because of its unique region and role in the cell cycle: a τ-shaped region of the oral apparatus intensely fluoresced during interphase, but the fluorescence completely disappeared during the dividing phase (Numata *et al.*, 1983). The τ-shaped region corresponded to "posterior connectives" and the root part of the "deep fiber" to the conjunction parts of microtubule bundles. In those parts, there was electron-dense material in the microtubule bundles. Hence, it is conceivable that the 14-nm filament protein corresponds to the dense material and has the function of joining microtubule bundles.

On the other hand, the disappearance of immunofluorescence from the old oral apparatus of most dividing cells reflected the regression and remodeling of the oral apparatus which have been known as necessary sequential events in the cell cycle. We observed that oral fluorescence disappeared concurrently with the onset of oral regression and constriction of the division furrow, whereas at a late dividing stage, immunofluorescence began to appear simultaneously in both the new and old oral apparatus (Numata *et al.*, 1983). Some regression processes are known to occur at the initial period of the reorganization of the old oral apparatus. They include: (1) disappearance of the deep fiber bundle (Williams and Zeuthen, 1966); (2) regression of the cilia of UM and M1-3 (Frankel and Williams, 1973); and (3) loss of the buccal cavity (membranellar bases become situated on the same plane) (Frankel and Williams, 1973). We speculate that the disappearance of 14-nm filament protein from the τ-shaped region triggers the changes mentioned, since its disappearance coincides completely with the cessation of food vacuole formation and since the removal of 14-nm filament protein may induce some disintegration of the higher ordered oral structure.

At the late stage of oral primordium development or of old oral apparatus remodeling, 14-nm filament protein was integrated into both new and old oral apparatus (Numata *et al.*, 1983). There was an obvious lag period between the appearance of oral immunofluorescence and recommencement of food vacuole formation by the oral apparatus. The sequential events during the lag period should occur in the following order: (1) 14-nm filament protein is integrated into the posterior connectives; (2) the protein next appears on the root part of the deep fiber bundle; and (3) some time after that, oral morphogenesis is completed and the oral apparatus becomes functional (Numata *et al.*, 1983). Thus, the 14-nm filament protein may play a crucial role not only in the morphogenesis of oral primordia but also in transient morphogenesis in the old oral system.

2. Oral Replacement during Conjugation

During the cell cycle, the preexisting anterior oral apparatus is also known to show stage-specific regressive and remodeling changes (Frankel and Williams, 1973). Other than these division cycle-dependent oral changes, much more extensive regression and reformation of the anterior oral apparatus has been known to occur under certain physiological conditions without being accompanied by cell division (Frankel and Williams, 1973). This phenomenon is referred to as "oral replacement," and is observed, for example, in cells cultivated in an amino acid-free medium (Frankel, 1970) and in cells during conjugation (Tsunemoto *et al.*, 1988). In the former case, the process of oral replacement has been well studied (Frankel, 1970; Frankel and Williams, 1973) using the synchronous rounding (abortive division) system developed by Watanabe (1963, 1971; Tamura *et al.*, 1966).

Oral replacement during conjugation was found to proceed synchronously, so that the process could be classified into 12 stages, including 6 regressive and 6 reformation stages: Regression of the old oral apparatus occurred in the order of membranelles M1, M2, M3, and UM. After this, the paired cells approached a virtually astomatous state, and then, reformation of the oral apparatus at the preexisting area started from the right ends of each membranelle (Tsunemoto *et al.*, 1988). The 14-nm filament protein presumably plays a crucial role in oral morphogenesis in binary fission cells (Numata *et al.*, 1983). The same is true for oral morphological changes during conjugation (Tsunemoto *et al.*, 1988). The 14-nm filament protein is dissociated from the oral apparatus at an early phase of conjugation, which may trigger the regression of the old oral apparatus (Tsunemoto *et al.*, 1988). Then, the protein plays important roles in various germ nuclear events, such as meiosis, selection of functional meiotic products, transfer of pronucleus, and zygote formation (Numata *et al.*, 1985). After this, the protein reassociates with the immature oral apparatus and may cause the oral apparatus to become a functional one (Tsunemoto *et al.*, 1988). It follows from these findings that the 14-nm filament protein is indispensable for various processes during conjugation and is multifunctional.

3. Nuclear Events at Fertilization

In *Tetrahymena thermophila,* the nuclear events of conjugation start with meiosis of the micronucleus, resulting in four haploid nuclei. Three of them degenerate and only one remains functional in the cell–cell junction region. A mitotic division of this surviving nucleus generates the migratory and stationary pronuclei. The migratory pronucleus of each cell quickly transfers to its mate and fertilizes the resident stationary pronucleus. The diploid-fertilization nucleus divides mitotically twice and forms four diploid nuclei. The two in the anterior differentiate into macronuclei, while the other two in the posterior differentiate into micronuclei (Elliott and Hayes, 1953; Nanney, 1953; Ray, 1956; Martindale *et al.,* 1982; Orias, 1986).

During conjugation in *Tetrahymena* (Nanney, 1953) and *Paramecium* (Sonneborn, 1954; Mikami, 1980; Grandchamp and Beisson, 1981; Yanagi and Hiwatashi, 1985), there are two stages where the fates of nuclei are known to correlate with their position in the cell. One is the stage of selection and survival of the meiotic products (Yanagi and Hiwatashi, 1985). The other is the stage of macro- and micronuclear differentiation (Mikami, 1980; Grandchamp and Beisson, 1981). Nanney (1953) suggested that nuclear behavior in *Tetrahymena* is controlled by the cytoplasm, although the exact mechanism still remains to be elucidated. Based on immunofluorescence localization of the 14-nm filament protein in *Tetrahymena* during the early stages of conjugation, we suggest that the protein is involved in some nuclear events (Numata *et al.,* 1985): (1) the selection of one functional nucleus from four haploid nuclei generated by meiosis (Fig. 2,a); (2) the division of the functional nucleus to generate migratory and stationary pronuclei (Fig. 2,b); (3) the determination of the migratory pronucleus (Fig. 2,b); (4) the migratory pronuclear exchange (Fig. 2,c); (5) the selection of two gametic pronuclei from five nuclei (including three relic nuclei) to form the fertilization nucleus (Fig. 2,d), and (6) the fusion of gametic pronuclei (Fig. 2,d). Thus we concluded that the 14-nm filament protein may control germ nuclear behavior at fertilization.

It is known that gametic pronuclear exchange is blocked by inhibitors of microtubule assembly (Orias, 1986; Hamilton *et al.,* 1988). Orias *et al.* (1983) showed that each migratory pronucleus is surrounded by a microtubular meshwork. We speculated that, in addition to the microtubular meshwork, the cytoskeletal 14-nm filament protein may have a crucial role in the transfer of gametic pronuclei across the cell junction, and that the 14-nm filament protein interacts with the microtubules. Such an association does not, however, hold true for all the cases in which microtubules are present in large numbers: ciliary axonemes, most parts of the oral apparatus, cortical microtubular bands, and the micronucleus at the crescent stage of meiotic prophase showed no anti-14-nm filament protein immunofluores-

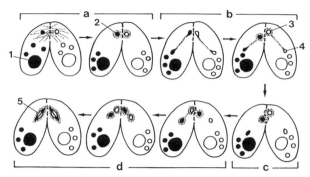

FIG. 2 Schematic diagram of the change in the distribution of 14-nm filament protein at conjugation in *Tetrahymena:* a–d represent the stages of conjugation. a, Only one functional nucleus selected from four meiotic products approaches the junction of the two cells. b, The functional nucleus divides mitotically to yield the migratory and stationary pronuclei. c, The migratory pronucleus moves across the cell junction. d, The stationary pronucleus moves toward the incoming migratory pronucleus and then two gametic pronuclei fuse to generate a fertilization nucleus. 1, macronucleus; 2, functional nucleus; 3, migratory pronucleus; 4, stationary pronucleus; 5, fertilization nucleus. The fine black dots show the localization of the 14-nm filament protein.

cence (Numata *et al.,* 1983, 1985; Tsunemoto *et al.,* 1988). So we studied the localization of the 14-nm filament protein in relation to tubulin during conjugation by double-staining immunofluorescence, and also studied the dependencies of those localizations on the presence of germinal nuclei by using conjugating pairs with star strains (B* and C*) having defective micronuclei (Allen, 1967; Doerder and Shabatura, 1980; Bruns, 1986).

Detailed studies of immunofluorescence localizations of *Tetrahymena* 14-nm filament protein in relation to tubulin in conjugating wild-type *Tetrahymena thermophila* (B strain) pairs and in pairs between B strain and star strain suggested that germ nuclear behavior may involve four successive cytoskeletal structures during conjugation (Takagi *et al.,* 1991): (1) during meiosis, the microtubular system is involved in micronuclear elongation and meiotic division; (2) at the postmeiotic stage, the 14-nm filament protein network structures that are formed independently of the existence of pronuclei are involved in the selection and survival of one of the four meiotic products (Fig. 2,a); (3) during the third prezygotic division–gametic pronucleus transfer and zygote formation–a cytoskeletal structure in which the 14-nm filament protein colocalizes with microtubules and which is dependent on the existence of normal gametic pronucleus, is involved in gametic pronuclear behavior (Fig. 2,b, c, d); and (4) during postzygotic division, the microtubular system is involved in nuclear behavior. Among four succeeding cytoskeletal structures during conjugation of *Tetrahymena,* the

14-nm filament protein network structures and colocalization of microtubules with 14-nm filament protein are biologically noteworthy with respect to important nuclear events related to meiosis and fertilization.

Experimental results using conjugating pairs between the B strain and star strain showed that the appearance of the respective cytoskeletal structures was dependent on the existence of germ nuclei, except for the 14-nm filament protein network structure at the postmeiotic stage (Takagi *et al.*, 1991). Orias (1986) observed that the fertilization basket was not formed in a star conjugant, and suggested that the assembly of this basket requires the presence of a functional postmeiotic nucleus. The 14-nm filament protein network structures appeared on the star-strain sides as well as the B-strain side in pairs between the B strain and star strain, and also appeared in pairs between star strains (Fig. 3A). Therefore, the appearance of this structure must be programmed in the timetable of conjugation, independently of the successful completion of meiosis.

As for the possible function of the network structures at the postmeiotic stage, it is inferred that the dynamic function of the network structures might cause one of the micronuclei to migrate into the cell–cell junction region in pairs between B strains. In pairs between B* strains, the macronuclei often approached the cell–cell junction as a substitute for the meiotic products (Fig. 3D and F). Even in pairs between B strains, the macronucleus is known to sometimes approach the cell–cell junction (Nanney and Nagel, 1964). This suggests that the network structures normally select one of the four meiotic products and bring it to the cell–cell junction region, but the structures presumably distinguish the meiotic product from macronuclei with relatively low specificity.

C. 14-nm Filament Protein Gene

In order to further understand the molecular structure of the 14-nm filament protein and for comparison with other cytoskeletal proteins, a full-length cDNA encoding the *Tetrahymena* 14-nm filament protein was cloned and sequenced (Numata *et al.*, 1991a). From the amino-terminal 19-amino acid sequence of 14-nm filament protein, a 35-mer oligonucleotide coding for amino acid residues 1–12 and a 30-mer oligonucleotide for residues 10–19 were synthesized. Using these synthetic probes, the cDNA for the 14-nm filament protein was isolated from a *Tetrahymena thermophila* cDNA library which was prepared by Takemasa *et al.* (1989). A 1497-bp cDNA clone was sequenced completely and the amino acid sequence was deduced. An open reading frame codes for a 462-amino acid polypeptide with a predicted relative molecular weight of 52,574. Though there are 15 TAA triplets within this open reading frame, which are usually recognized as one

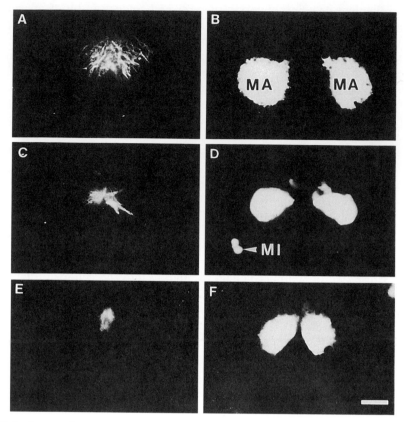

FIG. 3 Immunofluorescence micrographs showing the localization of 14-nm filament protein network structures in pairs between B* strains at postmeiotic stage. Mating cells at postmeiotic stage were extracted with 0.5% Nonidet P-40 and fixed with 2.5% paraformaldehyde. They were subjected to staining with anti-14-nm filament protein antibody followed by rhodamine-conjugated anti-rabbit IgG (A, C, E). The macro- (MA) and micronuclei (MI) were stained with 1 μg/ml 4,6,diamidino-2-phenylindole dihydrochloride (DAPI) (B, D, F). Bar = 10 μm.

of the universal stop codons, they have been proved to code for glutamine in *Paramecium* and *Tetrahymena* (Caron and Meyer, 1985; Preer *et al.,* 1985; Horowitz and Gorovsky, 1985; Takemasa *et al.,* 1989).

The amino-terminal of the amino acid sequence deduced from the 14-nm filament protein gene extends 21 residues beyond the amino-terminal of the purified 14-nm filament protein (Fig. 4). The 21-residue amino-terminal extension has some features similar to many mitochondrial targeting sequences, such as an abundance of scattered basic residues, several serine residues, and a lack of acidic residues (Hurt and van Loon, 1986;

```
                                              -20            -10
T49K                                        M RSINQLLKQA SLSQKSQYNF-
YCS1                         MSAILSTTS KSFLSRGSTR QCQ*MQKALF ALLNARH*SSA
YCS2                                      * TVPYLNSNRN VA*YLQSNSS-
PHCS                                                              A


          1        10        20        30        40        50        60
T49K    SQTNLKKVIA EIIPQKQAEL KEVKEKYGDK VVGQYTV-KQV IGGMRGMKGL MSD-LSRCDPY
YCS1    *EQT**ERF* ****A*AQ*I *KF*KEH*KT *I*EVLLEE*A Y***#*I*** VWE-G*VL**E
YCS2    QEKT**ERFS **Y*IHAQDV RQFVKEH*KT KISDVLL-E** Y***#*IP*S VWE-G*VL**E
PHCS    *S****DIL* DL**KE**RI *TFRQQH*NT ****I**-DMM Y****#***** VYE-T*VL**D


          70        80        90       100       110       120
T49K    QGIIFRGYTI PQLKEFLPKA DPKAADQANQ EPLPEGIFWL LMTGQLPTHA QVDALKHEWQ
YCS1    F**R***R** *EIQRE**** E------GST ****#*AL*** *L**EI**D* **K**SADLA
YCS2    D**R***R** ADIQKD**** --*GSS*--- -*##*AL*** *L**EV**Q* **EN*SADLM
PHCS    E**R***S* *ECQKM**** --*GGEE--- -*****L*** *V***I**EE **SW*SK**A


          130       140       150       160       170       180
T49K    NRGTVNQDCV NFILNLPKDL HSMTMLSMAL LYLQKDSKFA KLYDEGKISK KDYWEPFYED
YCS1    A*SEIPEHVI QLLDS***** *P*AQF*I*V TA*ESE**** *A*AQ*-V*# *E**SYTF**
YCS2    S*SELPSHV* QLLD****** *P*AQF*I*V TA*ESE**** *A*AQG-*#* Q***SYTF**
PHCS    K*AALPSHV* TMLD*F*TN* *P*SQ**A*I TA*NSE*N** RA*A**-*HR TK***LI***


          190       200       210       220       230       240
T49K    SMDLIAKIPR VAAIIYRHKY RDSKLIDS-DS KLDWAGNYAH MMGFEQHVVK ECIRGYLSIH
YCS1    *L**LG*L*V I*SK***NVF K*G*ITST-*P NA*YGK*L*Q LL*Y*NKDFI DLM*L**T**
YCS2    *L**LG*L*V I**K***NVF K*G*MGEV-*P NA*Y*K*LVN LI*SKDEDFV DLM*L**T**
PHCS    C******L*C ***K***NL* *EGSS*GAI** ****SH*FTN *L*YIDAQFT *LM*L**T**


          250       260       270       280       290       300
T49K    CDFEGGNVSA HTTHLVGSAL SDPYLSYSAG VNGLAGPLHG LANQEVLKWL LQFIEEKGTK
YCS1    S*#****** ****#**** *S****LA** L********# R******E** FKLR**VKGD
YCS2    S*#****** **S****** *S****LAS* L********# R******E** FALK**VNDD
PHCS    S*#****** **S****** ******FA*A M********# ******V** T*LQK*V*KD


          310       320       330       340       350       360
T49K    VSDKDIEDYV DHVISSGRVV PGYGHAVLRD TDPRFHHQVD FSKFHLKDDQ MIKLLHQCAD
YCS1    Y*KET**K*L WDTLNA**## *#*******K ****YTA*RE *ALK*FP*YE LF**VSTIYE
YCS2    Y*KDI**K*L WDTLN****I *#*******K ****YMA*RK *AMD*FP*YE LF**VSSIYE
PHCS    ***EKLR**I WNTLN****# *#*******K ****YTC*RE *ALK**PH*P *F**VA*LYK


          370       380       390       400       410       420
T49K    VIPKKLLTYK KIANPYPNVD CHSGVLLYSL GLTEYQYYTV VFAVSRALGC MANLIWSRAF
YCS1    *A*GV*TKHG *TK**W**** S******QYY **#**ASF*** L*G*A#*I*V LPQ**ID**V
YCS2    *A*GV*TEHG *TK**W**** A******QYY **K*SSF*** L*G*#*F*I L*Q**TD**I
PHCS    IV*NV**EQG *AK**W**#* A******QYY *M**MN**** L*G*#***V L*Q******L


          430       440
T49K    GLPIERPGSA DLKWFHDKYR E
YCS1    *A***#*K*F STEKYKELVK KIESKN
YCS2    *AS**#*K*Y STEKYKELVK NIESKL
PHCS    *F*L*#*KSM STDGLIKLVD SK
```

FIG. 4 Comparison of amino acid sequence of the *Tetrahymena* 14-nm filament protein with those of citrate synthases from *Saccharomyces cerevisiae* (YCS1 = mitochondrial form; YCS2 = nonmitochondrial form) (Rosenkrantz *et al.*, 1986) and from porcine heart (PHCS) (Bloxham *et al.*, 1982). Residue No. 1 indicates the serine (S) at the amino terminal of the purified 14-nm filament protein. Residues identical with those of the 14-nm filament protein are indicated by asterisks. Gaps introduced to maximize homology are indicated by dashes. Residues implicated in substrate or product binding by X-ray crystallography of porcine heart citrate synthase (Remington *et al.*, 1982) are indicated by boxes.

Roise and Schatz, 1988). If the amino-terminal 21 residues are removed, a relative molecular weight of 50,107 is predicted, which is in agreement with the molecular weight (49,000) of the 14-nm filament protein.

From the result of a computer search (EMBL protein database) for sequence homology with other proteins, it was a great surprise to discover that the 14-nm filament protein gene product exhibited high homology with "citrate synthases" (EC 4.1.3.7) of various organisms, such as 51.5% identity with that of porcine heart citrate synthase (Bloxham *et al.,* 1982; Evans *et al.,* 1988) (Fig. 4), 45–47% identity with yeast mitochondrial citrate synthase and yeast nonmitochondrial citrate synthase (Rosenkrantz *et al.,* 1986) (Fig. 4) and 21.2% identity with *Escherichia coli* citrate synthase (Ner *et al.,* 1983). Citrate synthase is a key enzyme of the Krebs tricarboxylic acid cycle and catalyzes the stereospecific synthesis of citrate from acetyl coenzyme A and oxalacetate (Srere, 1975; Weitzman and Danson, 1976). X-Ray crystallographic studies on porcine citrate synthase have highlighted 18 residues thought to interact with its substrate (acetyl CoA) or product (citrate) (Remington *et al.,* 1982). In this regard, there is one divergence in the 14-nm filament protein, but the other 17 residues required for the interaction are conserved in the 14-nm filament protein as in other citrate synthases (Fig. 4). We therefore assayed the 14-nm filament protein purified by assembly and disassembly for citrate synthase activity, and found that it has significant enzyme activity (Numata *et al.,* 1991a). Next, we determined the distribution of the 14-nm filament protein in *Tetrahymena* cells, using monoclonal (49K15F6) and polyclonal antibodies (49K I) against the 14-nm filament protein, which have been shown to be highly specific in immunoblots. Both antibodies stained mitochondria (Numata *et al.,* 1983, 1991a, 1995) and cytoskeletal structures in cytoplasm, the network structures spreading from a cell–cell junction at fertilization (Numata *et al.,* 1985, 1991a; Takagi *et al.,* 1991) and the posterior connectives of the oral apparatus in interphase cells (Numata *et al.,* 1983, 1991a). These results strongly suggest that the 14-nm filament protein is simultaneously a mitochondrial enzyme and a structural protein in the cytoskeleton.

To determine whether the 14-nm filament protein gene exists as a single-type gene in *Tetrahymena* macronucleus, Southern blotting analysis was performed. For this purpose, three probes were used. Probe A corresponded to full-length cDNA (nucleotides +1 to +1498); probe B corresponded to 63 bases (nucleotides +8 to +70) of the coding region of the mitochondrial targeting sequence. Probe C corresponded to 1205 bases (nucleotides +294 to +1498) of the coding region of the mature protein. A discrepancy between the results of Southern blotting analysis and the replication map of the 14-nm filament protein cDNA suggested that macronuclear DNA of the 14-nm filament protein possessed a few introns. So the macronuclear DNA encoding the 14-nm filament protein was amplified by polymerase chain reaction from *Tetrahymena* macronuclear DNA. The amplified product was cloned and its sequence was verified. The amplified DNA definitely possessed two introns (Fig. 5). A correspondence between the patterns of the Southern blotting analysis and the patterns deduced from the restriction maps of the amplified product indicates that *Tetrahymena* possesses only a single type of

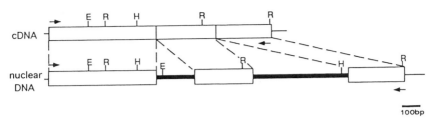

FIG. 5   Structural organization of 14-nm filament protein cDNA and macronuclear DNA (nuclear DNA). Macronuclear DNA encoding 14-nm filament protein was amplified by polymerase chain reaction from *Tetrahymena* macronuclear DNA. Arrows represent position of primers. Open reading frames and untranslated regions are shown by boxes and lines, respectively. Solid lines correspond to introns. Symbols for restriction enzymes: E, *Eco*RI; H, *Hind*III; R, *Rsa*I.

14-nm filament protein gene in the macronucleus (Numata *et al.*, 1996). Using the three probes, Northern hybridizations were performed. A single band of 1.7 kb was intensely hybridized with each of the three probes, suggesting that the transcription of the 14-nm filament protein gene is a single type (Numata *et al.*, 1996). These results indicate that one protein encoded from a single gene may have two functions—as the citrate synthase in mitochondria and as the 14-nm filament protein in the cytoplasm.

D. Identification of 14-nm Filament Protein as Citrate Synthase

To ascertain whether *Tetrahymena's* mitochondrial citrate synthase is identical with the 14-nm filament protein, *Tetrahymena* citrate synthase was purified from mitochondria by monitoring its enzymatic activity. The following evidence was obtained from a comparison between purified mitochondrial citrate synthase and the 14-nm filament protein purified by a polymerization and depolymerization procedure (Kojima *et al.*, 1995). (1) The apparent molecular weight of mitochondrial citrate synthase is 49 kDa, which is the same as the 14-nm filament protein. (2) Mitochondrial citrate synthase reacts as strongly to an anti-14-nm filament protein antibody as the 14-nm filament protein. (3) Both mitochondrial citrate synthase and the 14-nm filament protein possess nearly the same enzymatic properties in terms of optimum pH, optimum ionic strength, effects of substrate concentrations, and inhibitory effect by ATP (Fig. 6). (4) Mitochondrial citrate synthase activity is specifically inhibited with an anti-14-nm filament protein monoclonal antibody, down to less than one-fifth of its activity. In addition to these results, we confirmed that a portion of the amino acid sequence of

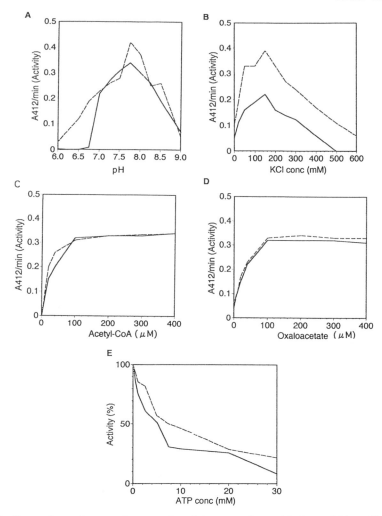

FIG. 6 Comparison of enzymatic properties of citrate synthase with those of 14-nm filament protein. pH optimum (A) on the citrate synthase activity was measured by the DTNB method (Srere *et al.,* 1963; Kispal and Srere, 1991) at 37°C. Effects of concentration of KCl (B), acetyl-CoA (C), and oxaloacetate (D) on enzyme activity, and inhibition of enzyme activity by ATP (E) were measured at pH 7.75. A solid line and a broken line represent *Tetrahymena* citrate synthase and 14-nm filament protein, respectively.

mitochondrial citrate synthase completely matched with the corresponding sequence of the 14-nm filament protein (Kojima *et al.,* 1996). Thus, we conclude that mitochondrial citrate synthase and the 14-nm filament protein are identical, and that the 14-nm filament protein has dual functions as a cytoskeletal element in cytoplasm and as a citrate synthase in mitochondria.

A similar phenomenon which has been called "gene sharing" is known in crystallins. Wistow and Piatigorsky recently reported that the major taxon-specific crystallins of vertebrates and invertebrates are either several kinds of enzymes, or closely related to enzymes (Wistow and Piatigorsky, 1988; Piatigorsky and Wistow, 1989, 1991; Wistow, 1993). Therefore, a gene encoding a single protein may acquire and maintain a second function without duplication and without loss of the primary function. In the case of the *Tetrahymena* citrate synthase/14-nm filament protein, the gene for the 14-nm filament protein seems to maintain sequences necessary for both enzyme activity and filament formation. This surprising discovery is the first indication of gene sharing in ciliated protozoa and implies that gene sharing may occur in a wide variety of organisms.

The fact that the identical protein is localized in quite different sites and plays quite different functions in mitochondria or cytoplasm led to the following possibilities: (1) the existence of a proteolytic activity which specifically cleaves between the mitochondrial targeting sequence and the mature protein in cytoplasm, and (2) post-translational modification as a mechanism for the compartmentalization. The former possibility was suggested in the case of uracil–DNA glycosylase (UDG). UDG catalyzes the first step in a DNA repair pathway which removes uracil from DNA (Dianov *et al.,* 1992), and has been found both in the cell nucleus and in the mitochondria (Anderson and Friedberg, 1980). Slupphaug *et al.* (1993) indicated that the signals for mitochondrial translocation reside in the presequence, whereas signals for nuclear import are within the mature protein. It is therefore likely that the mature nuclear UDG is localized to the nucleus after the presequence is processed by a protease in cytoplasm.

On the other hand, the significance of post-translational modification as a mechanism for the compartmentalization of cytoplasmic tubulin and axonemal tubulin has now been appreciated. For α-tubulin, lysine-ε-amino acetylation is associated with flagellar morphogenesis in *Chlamydomonas* (L'Hernault and Rosenbaum, 1983, 1985a,b; Piperno and Fuller, 1985). The acetylated form (α_3-tubulin) was shown to be present in flagellar microtubules, but not in soluble cytoplasmic tubulins.

To study the compartmentalization of cytoplasmic 14-nm filament protein and mitochondrial citrate synthase, an investigation into the post-translational modification of these proteins is being undertaken in our laboratory. Recently a very small difference between mitochondrial citrate synthase and the 14-nm filament protein has been validated (Kojima *et al.,* 1996). The difference may be based on some post-translational modification of the 14-nm filament protein. We speculate that a protease in cytoplasm may recognize the modification in the 14-nm filament protein and then cleave between the mitochondrial targeting sequence and the mature protein. In the case of *Tetrahymena* 14-nm filament protein/citrate synthase, the 14-nm filament protein may remain in the cytoplasm after processing of the mitochondrial targeting sequence by a protease in cytoplasm, while the citrate synthase may translocate to mitochondria and then the mitochondrial targeting sequence may be processed by a protease in mitochondria.

E. Intramitochondrial Filamentous Inclusions

The occasional occurrence of unusual intramitochondrial inclusions has been reported in a number of cell types from various species of organisms (Mugnaini, 1964; Fujita and Machino, 1964; Behnke, 1965; Wills, 1965; Suzuki and Mostofi, 1967; Jessen, 1968; Nagano et al., 1982; Sasaki and Suzuki, 1989). Of these inclusions, the most frequent are those showing fibrillar or filamentous subunits. In ciliated protozoa, intramitochondrial inclusions were also occasionally found within the matrices of some mitocondria in Tetrahymena (Elliott and Bak, 1964) and Pseudourostyla (Suganuma and Yamamoto, 1980). There are several possible explanations for the formation of intramitochondrial inclusions: (1) the inclusions are formed for in vivo storage of mitochondrial enzyme or (2) the formation of inclusions is a result of aging of cytopathological change. However, the entity involved in these intramitochondrial inclusions has not yet been identified.

Besides possible involvement of the 14-nm filament protein as a cytoskeletal element, we have always observed strong immunofluorescence localized in mitochondria (Numata et al., 1983, 1985). The following results were obtained (Numata et al., 1995) in a recent study. (1) Immunofluorescence studies of the 14-nm filament protein/citrate synthase in mitochondria found that intense mitochondrial fluorescence remained unchanged in Tetrahymena cells taken from logarithmic growth phase to the stationary phase. (2) Electron microscopic studies showed that electron-dense rod-shaped structures found in mitochondrial matrices tended to increase in Tetrahymena cells in the growth decline phase (Fig. 7). (3) The rods were composed of side-by-side straight filaments with diameters of approximately 14 nm (Fig. 7). (4) Serial sections revealed that one to four rods occurred in the matrix of every mitochondrion in Tetrahymena during the growth decline phase. (5) Immunoelectron microscopy using an anti-14-nm filament protein antibody clearly showed that a filament bundle in the electron-dense rod was the bundle of polymerized filaments of 14-nm filament protein/citrate synthase. These results supported the hypothesis that Tetrahymena citrate synthase can sometimes form rod-shaped filamentous structures in the mitochondrial matrix.

Elliott and Bak (1964) reported mitochondria with both a spindle- and a spherical-shaped intramitochondrial mass in stationary phase Tetrahymena and suggested that the intramitochondrial mass was formed as a result of aging. The spindle-shaped intramitochondrial mass that they described is

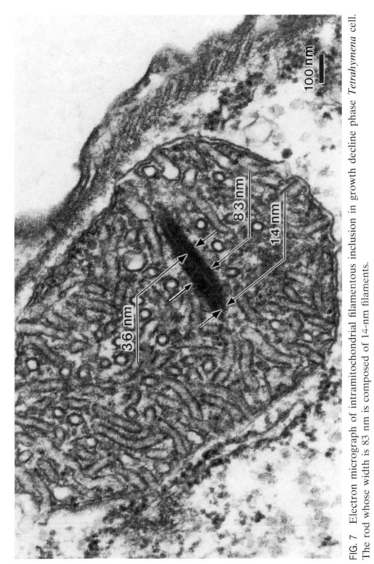

FIG. 7 Electron micrograph of intramitochondrial filamentous inclusion in growth decline phase *Tetrahymena* cell. The rod whose width is 83 nm is composed of 14-nm filaments.

similar to the rod-shaped structures that we reported (Nuamta *et al.*, 1995). However, the intramitochondrial rod-shaped structures tended to form during the growth decline phase (5.65×10^5 cells/ml), so the formation of rod-shaped structures was not a result of senescence or aging of the cells.

F. Modulation Mechanism of Citrate Synthase Activity

The intramitochondrial filament bundles which are composed of 14-nm filament protein/citrate synthase have been observed (Numata *et al.*, 1995) in almost all the mitochondria of *Tetrahymena* in the growth decline phase. Therefore, 14-nm filament protein/citrate synthase appears to exist in both monomeric and polymeric forms in mitochondria. The questions then arise of the meaning of filament formation by a metabolic enzyme, and whether filament formation modulates enzymatic activity. At present it is very difficult to examine directly how filaments of 14-nm filament protein/citrate synthase function *in vivo*, but an important clue for understanding the significance of the occurrence of 14-nm filaments can be gained by examining the change in citrate synthase activity accompanying polymerization and depolymerization of the 14-nm filament protein *in vitro*. Recently Takeda *et al.* (1995) presented evidence that citrate synthase activity of the 14-nm filament protein decreases upon formation of the filament, but increases upon depolymerization of the 14-nm filaments (Fig. 8). The mechanism by which the citrate synthase activity decreased may be the disturbance of the catalytic sites of the enzyme by the adjacent 14-nm filament protein molecule or conformational changes of 14-nm filament protein molecules under polymerization conditions.

As indicated by the results of the *in vitro* experiments, the citrate synthase activity of the 14-nm filament protein/citrate synthase is presumably regulated by monomer–polymer conversion. This may reflect the *in vivo* modulation mechanism of citrate synthase activity. Citrate synthase activity in general is known to be inhibited by ATP (Weitzman and Danson, 1976), and the same is true for citrate synthase of *Tetrahymena* (Kojima *et al.*, 1995). ATP is also an indispensable factor for *in vitro* 14-nm filament formation (Numata and Watanabe, 1982). So we propose that citrate synthase activity is regulated by monomer–polymer conversion, and ATP accumulation in mitochondria may promote filament formation and result in a decrease of the enzymatic activity. This is consistent with the fact that the appearance of filamentous structures in mitochondria increases in the growth decline phase and during the stationary phase compared with that during the log phase. That is, ATP consumption is very active in the log phase, but becomes inactive in the stationary phase, and this inactive ATP

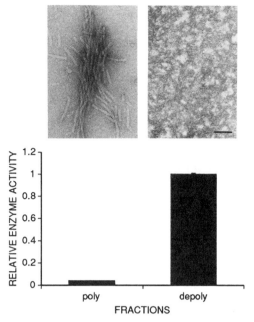

FIG. 8   Decrease and increase of citrate synthase activity by polymerization and depolymeriza-tion of 14 nm filament protein. Citrate synthase activities of polymerized 14-nm filaments (poly) formed with purified 14-nm filament protein and depolymerized fraction (deploy) of 14-nm filaments were measured by DTNB method. The enzyme activity/mg protein of the depolymerized fraction was taken as 1.0, and that of the polymerized fraction (poly) was expressed as the relative ratio. In each fraction, the presence or absence of 14-nm filaments was checked electron microscopically (photographs above). Bar = 100 nm.

consumption may result in transient overaccumulation on ATP and forma-tion of filamentous structures.

Wallimann's group reported that adult rat cardiomyocytes cultured in creatine-deficient medium display large mitochondria with paracrystalline inclusions, and they demonstrated that the paracrystalline inclusions were labeled with anticreatine kinase antibody (Eppenberger-Eberhardt *et al.*, 1991). The addition of creatine to creatine-free culture medium caused the disappearance of the giant mitochondria as well as the paracrystalline inclusions and was accompanied by an increase in the intracellular level of total creatine (Eppenberger-Eberhardt *et al.*, 1991). They suggested that the accumulation of creatine kinase within the inclusions represents a com-pensatory effort of the cardiomyocytes to cope with a metabolic stress situation caused by low intracellular total creatine level.

In starved *Tetrahymena* cells, electron-dense rod-shaped structures were found in mitochondria matrices and immunogold particles were also deposited on the rod-shaped structures (Numata *et al.*, 1995). So we speculated that, during the growth decline phase, the accumulation of citrate synthase within the rod-shaped structures may be induced by a compensatory effort of *Tetrahymena* cells to deal with low intracellular oxaloacetate and acetyl-coenzyme A levels and/or by overaccumulation on ATP as a result of inactive ATP consumption. These observations strongly suggest that the dynamic monomer–polymer conversion of the 14-nm filament protein/citrate synthase in mitochondria depends upon the physiological conditions of *Tetrahymena* cells. Thus we conclude that the accumulation of 14-nm filament protein/citrate synthase within rod-shaped structures may be a regulatory system for citrate synthase activity in mitochondria.

III. Elongation Factor-1α

A. 14-nm Filament-Associated Protein and EF-1 α

Immunoblotting using monoclonal and polyclonal antibodies following two-dimensional gel electrophoresis showed that the 14-nm filament protein fraction contained two 49-kDa proteins whose isoelectric points were 8.0 and 9.0; a monoclonal antibody (MAb), 26B4, and a polyclonal antibody, 49K I, reacted only to a pI 8.0 protein, while two other MAbs, 11B6 and 11B8, reacted only to a pI 9.0 protein (Kurasawa *et al.*, 1992). From the N-terminal amino acid sequences, the pI 8.0 protein was identified as the 14-nm filament protein/citrate synthase, but the pI 9.0 protein N-terminal sequence had no similarity with that of the pI 8.0 protein. The pI 9.0 protein is considered to be a 14-nm filament-associated protein since the pI 9.0 protein copurified with the pI 8.0 protein during two cycles of an assembly and disassembly purification protocol (Kurasawa *et al.*, 1992).

To ascertain the properties of the 14-nm filament-associated protein and for comparison with other proteins, a full-length cDNA encoding the 14-nm filament-associated protein was cloned and sequenced. A 1474-bp cDNA clone was sequenced completely and the amino acid sequence deduced. An open reading frame codes for a polypeptide consisting of 435 amino acids with a predicted molecular weight of 48,297 (Kurasawa *et al.*, 1992). Southern blotting analysis showed that *Tetrahymena pyriformis* possesses only a single type of 14-nm filament-associated protein gene in the macronucleus (Kurasawa *et al.*, 1992).

A computer search of the EMBL and NBRF protein databases demonstrated that the 14-nm filament-associated protein shared 73–76% sequence identity with polypeptide elongation factor 1 α: those from human (MOLT-

4 lymphoid cell line; Brands *et al.*, 1986), 75.12%; *Xenopus,* 76.04% (Coppard *et al.*, 1991); *Drosophila,* 73.04% (Hovemann *et al.*, 1988); *Dictyostelium,* 73.56% (Yang *et al.*, 1990), and *Saccharomyces,* 75.41% (Cottrelle *et al.*, 1985). EF-1 α catalyzes the GTP-dependent binding of aminoacyl-tRNA to ribosomes in the peptide elongation phase of eukaryotic protein synthesis (Moldave, 1985). All the primary structures of the EF-1 α selected for comparison share common characteristics; about one-third of the way along the sequence from the N-terminal there are three GTP binding regions (Jurnak, 1985; Linz *et al.*, 1986; Woolley and Clark, 1989), a tRNA binding region is included in the middle region of the primary sequence (Bosch *et al.*, 1985; Linz *et al.*, 1986), and these sequences are highly conserved (Fig. 9). In addition, the 14-nm filament-associated protein purified from *Tetrahymena pyriformis* had GTP-binding properties (Kurasawa *et al.*, 1996). The molecular weight and isoelectric point of the 14-nm filament-associated protein are very similar to those of EF-1 α. These EF-1 α-specific characteristics hold true for *Tetrahymena* 14-nm filament-associated protein. Thus we conclude that the 14-nm filament-associated protein is identified as *Tetrahymena* EF-1 α.

It is possible that EF-1 α may play a role in formation of the 14-nm filament because EF-1 α copurifies with the 14-nm filament protein upon polymerization and depolymerization. To determine whether the 14-nm filament alone or EF-1 α alone can form the 14-nm filaments, or whether both proteins are needed for filament formation, using the purified 14-nm filament protein and/or purified EF-1 α, the interaction between 14-nm filament protein and EF-1 α in filament formation was investigated electron microscopically. The results demonstrated that the purified 14-nm filament protein was capable of forming 14-nm filaments without EF-1 α (Takeda *et al.*, 1995). Therefore, the relationship between 14-nm filament protein and EF-1 α is quite different from that of tubulin and MAPs in microtubule formation; tubulin and MAPs are copurified through polymerization and depolymerization steps during purification, and MAPs are necessary for the formation of microtubules (Olmsted, 1986; Vallee *et al.*, 1984).

Recently, an actin bundling protein from *Dictyostelium* was identified as EF-1 α (Yang *et al.*, 1990). More recently, Itano and Hatano (1991) reported that in *Physarum polycephalum,* a 50-kDa protein cross-reacting to ant-EF-1 α antiserum could bundle not only actin filaments but also microtubules. Furthermore, it was shown that *Tetrahymena* EF-1 α could also bundle actin filaments (Kurasawa *et al.*, 1996). Thus we speculated that EF-1 α might take part in the interaction between 14-nm filament protein and other cytoskeletal elements, such as microtubules which colocalize with 14-nm filament protein during the conjugation process in *Tetrahymena* (Takagi *et al.*, 1991).

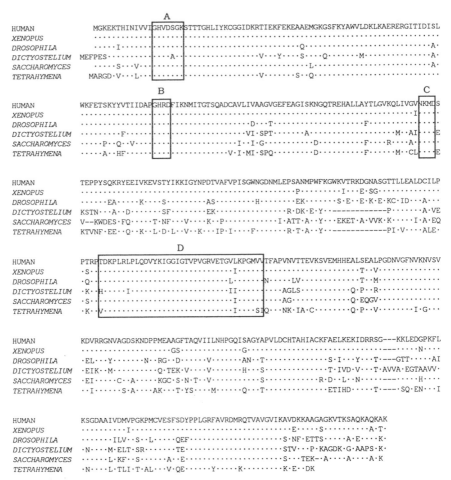

FIG. 9   Comparison of *Tetrahymena* 14-nm filament-associated protein with EF-1 α from other species. Comparison of the predicted amino acid sequence of the 14-nm filament-associated protein with that of EF-1 α from human (MOLT-4 lymphoid cell line), *Xenopus (X. laevis)*, *Drosophila (D. melanogaster)*, *Dictyostelium (D. discoideum)*, and *Saccharomyces (S. cerevisiae)*. The alignment was maximized by allowing gaps (bars). The amino acid sequence of human EF-1 α is shown in one-letter codes at the top. A dot in place of a residue indicates a residue identical with that of human EF-1 α. Consensus sequences GXXXXGK (region A), GXXD (region B), and NKXD (region C) corresponding to the three GTP binding domains (Jurnak, 1985; Woolley and Clark, 1989) are indicated. Region D corresponds to the tRNA binding domain of *Escherichia coli* EF-Tu (Bosch *et al.*, 1985; Linz *et al.*, 1986).

B. Presence of EF-1 α within the Replication Band of *Euplotes*

Using a monoclonal antibody, MAb 11B6, which reacted only to EF-1 α, immunofluorescence experiments have shown that EF-1 α was localized in the replication band of *Euplotes eurystomus* (Numata *et al.*, 1991b). As is well known in the hypotrichous ciliate *Euplotes,* the chromatin is replicated within a specialized macronuclear structure called the replication band during the S-phase (Gall, 1959; Kimball and Prescott, 1962). Our result implies that EF-1 α might be involved in DNA replication in *Euplotes.*

In this regard, studies of the replicase of the single-stranded RNA coliphage Qβ suggested that Qβ replicase was a tetramer consisting of a phage-encoded RNA polymerase and three host proteins, that is, the elongation factors Tu and Ts and ribosomal protein S1 (Blumenthal and Carmichael, 1979). Since both plus and minus strands of the Qβ genome have a 3'-terminal CCA which is a typical 3'-terminal tRNA-like structure, Weiner and Maizels (1987) suggested that the EF-Tu and EF-Ts subunits of Qβ replicase might enable the enzyme to bind to 3'-terminal tRNA-like structures on the viral RNAs. The *Euplotes* macronucleus is transcriptionally active and composed of short gene-sized DNA molecules. Weiner and Maizels (1987) proposed that 3'-terminal tRNA-like structures of single-stranded RNA viruses are in fact molecular fossils of a putative original RNA world and suggested that modern DNA telomeres with the repetitive structure CmAn could be direct descendants of the CCA terminal of tRNA. Because EF-1 α is the eukaryotic analog of prokaryotic EF-Tu, it is plausible that the translation elongation factors may be used for initiation of DNA replication in *Euplotes* as well as RNA replication in Qβ.

C. F-Actin-Bundling Activity of EF-1 α

Recently direct and indirect evidence has indicated a relationship between actin bundling protein and EF-1 α. One example is ABP-50, which is an abundant actin filament bundling protein of 50 kDa from *Dictyostelium discoideum* (Demma *et al.*, 1990). Yang *et al.* (1990) identified ABP-50 as *D. discoideum* EF-1 α from the cDNA sequencing data. Moreover, 52-kDa F-actin bundling protein isolated from *Physarum polycephalum* can cross-react with an antibody against yeast EF-1 α (Itano and Hatano, 1991). This protein is also capable of bundling microtubules (Itano and Hatano, 1991). In this way, it cross-links actin filaments and microtubules to form cobundles of actin filaments and microtubules. The 52-kDa protein cross-reacts with a monoclonal antibody against an HeLa F-actin-bundling protein of 55 kDa (Yamashiro-Matsumura and Matsumura, 1985).

To investigate whether *Tetrahymena* EF-1 α could interact with F-actin, the interaction between EF-1 α and F-actin was examined with a cosedimentation experiment. It showed that EF-1 α bound to *Tetrahymena* F-actin. Electron microscopic observation demonstrated that *Tetrahymena* EF-1 α possesses F-actin bundling activity (Fig. 10). F-Actin bundles formed by *Tetrahymena* EF-1 α looked as tight as the bundles formed by *Physarum* EF-1 α or *Dictyostelium* EF-1 α (Kurasawa *et al.*, 1996).

The primary structure of *Tetrahymena* actin shows only 74.6% homology with that of skeletal muscle actin (Hirono *et al.*, 1987). In *Tetrahymena* actin, essential properties such as actin–actin interaction and actin–myosin interaction are highly conserved, but some other properties common to ubiquitous actins, such as interactions with tropomyosin, α-actinin, phalloidine, and DNase I, are lost due to the deflected evolution of actin in the ciliated protozoa (Hirono *et al.*, 1989, 1990). Therefore, it is necessary to determine whether *Tetrahymena* EF-1 α can bind to *Tetrahymena* actin alone, or whether *Tetrahymena* EF-1 α can bind to skeletal muscle actin as well as *Tetrahymena* actin. The results indicated that *Tetrahymena* EF-1 α possesses the ability to bundle skeletal muscle actin filaments (Kurasawa *et al.*, 1996). It is suggested that binding sites for *Tetrahymena* EF-1 α are essentially conserved in *Tetrahymena* actin and skeletal muscle actin.

D. Regulation of Bundle Formation by Ca²⁺/CaM

It came as a great surprise when EF-1 α was found to bind calmodulin as well as F-actin. Durso and Cyr (1994a,b) reported that a carrot EF-1 α homolog bundled microtubules *in vitro,* and moreover, this bundling was modulated by the addition of Ca^{2+} and CaM together (Ca^{2+}/CaM). Kaur and Ruben (1994) demonstrated a direct interaction between CaM and the translation elongation factor EF-1 α from *Trypanosoma brucei*. Therefore, it is necessary to examine whether Ca^{2+}/CaM can regulate the interaction between EF-1 α and F-actin.

Using EF-1 α, actin and CaM separately purified from *Tetrahymena,* the interactions among these three proteins were investigated with cosedimentation and electron microscopy. It was demonstrated that Ca^{2+}/CaM bound to the EF-1 α/F-actin complex. Although the formation of the EF-1 α/F-actin complex and F-actin bundling by EF-1 α were Ca^{2+}-insensitive, a binding between CaM and the EF-1 α/F-actin complex was sensitive to Ca^{2+}. Cosedimentation showed that Ca^{2+}/CaM directly bound to EF-1 α and did not affect the formation of the EF-1 α/F-actin complex. In addition, Ca^{2+}/CaM clearly inhibited the formation of F-actin bundles by EF-1 α. Based on these findings, Kurasawa *et al.* (1996) proposed that the EF-1 α mediates the formation of F-actin bundles by a Ca^{2+}/CaM-sensitive mechanism.

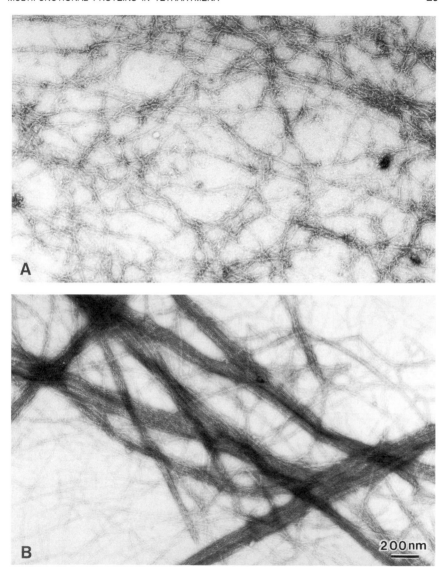

FIG. 10 Electron micrographs of *Tetrahymena* F-actin bundles formed by *Tetrahymena* EF-1 α. *Tetrahymena* F-actin was polymerzied alone (A) or with equimolar amounts of *Tetrahymena* EF-1 α (B) in the presence of 40 mM KCl and 2 mM MgCl₂.

E. Multifunction of EF-1 α

Several recent reports have linked EF-1 α with the microtubule cytoskeleton. The mitotic apparatus-associated 51-kDa protein from sea urchin eggs was closely related to the nucleation of astral microtubules *in vitro* (Ohta *et al.,* 1988, 1990). The 51-kDa protein was shown to have both GTP-binding activity and EF-1 α activity (Ohta *et al.,* 1990). Kuriyama and Borisy (1985) reported that a monoclonal antibody SU5 preferentially stained centrosomes in sea urchin eggs and recognized a 50-kDa protein on immunoblots. They further cloned and sequenced a cDNA encoding the 50-kDa protein and showed that its deduced amino acid sequence bore a striking similarity to those of EF-1 α from several species (Kuriyama *et al.,* 1990). These reports indicated that the EF-1 α family of proteins functions in the formation of the mitotic apparatus of sea urchin eggs. More recently, Shiina *et al.* (1994) reported that a 48-kDa microtubule-severing protein purified from *Xenopus* eggs was identical with EF-1 α. An activity that severs stable microtubules is thought to be involved in the microtubule reorganization that occurs between interphase and mitosis. Thus, EF-1 α appears to have a second role as a regulator of microtubule rearrangements.

On the other hand, from direct and indirect evidence, it appears that actin bundling proteins are identified as an EF-1 α or an EF-1 α homolog. ABP-50 is a 50-kDa actin-binding protein that bundles F-actin *in vitro* and is localized with bundled F-actin structures in *Dictyostelium discoideum* (Demma *et al.,* 1990). Yang *et al.,* (1990) identified ABP-50 as *D. discoideum* EF-1 α. A 52-kDa F-actin bundling protein from *Physarum polycephalum* was reported to react with antibodies raised against yeast EF-1 α (Itano and Hatano, 1991). The 52-kDa protein also reacted with antibodies raised against an actin-bundling protein from HeLa cells, indirectly suggesting that an EF-1 α/ actin association occurs in mammalian cells as well. The 52-kDa protein has two additional activities: microtubule bundling and microtubule/F-actin co-bundling. *Tetrahymena* EF-1 α can also bind to F-actin and form F-actin bundles (Kurasawa *et al.,* 1996). Thus, the EF-1 α family of proteins are regarded as regulators in the organization of microfilaments as well as microtubules.

Kaur and Ruben (1994) isolated a cytosolic CaM-binding protein from *Trypanosoma brucei* and identified it as EF-1 α. Durso and Cyr (1994a) found a CaM-sensitive interaction between microtubules and a carrot EF-1 α homolog and proposed that an EF-1 α homolog mediates the lateral association of microtubules in plant cells by a Ca^{2+}/CaM-sensitive mechanism. In addition, the formation of F-actin bundles induced by EF-1 α is regulated by Ca^{2+}/CaM (Kurasawa *et al.,* 1996). Based on these findings, we speculate that the functions of the EF-1 α family of proteins as regulators in the organization of microtubules as well as microfilaments are modulated by Ca^{2+}/CaM. Possible functions of EF-1 α are represented in Table II.

TABLE II

Possible Functions of EF-1 α

Functions	EF-1 α or its homologous proteins	Reference
Polypeptide elongation	EF-1 α	
F-actin bundling	*Dictyostelium* EF-1 α	(Demma *et al.*, 1990)
	Physarum 52-kDa protein	(Itano and Hatano, 1991)
	Tetrahymena EF-1 α	(Kurasawa *et al.*, 1996)
Microtubule (MT) bundling	*Physarum* 52-kDa protein	(Itano and Hatano, 1991)
	carrot EF-1 α homolog	(Durso and Cyr, 1994a)
MT/F-actin cobundling	*Physarum* 52-kDa protein	(Itano and Hatano, 1991)
Microtubule-organizing center (MTOC)	sea urchin (*Stronglycentrotus*) egg 50-kDa protein	(Kuriyama *et al.*, 1990)
	sea urchin (*Pseudocentrotus*) egg 51-kDa protein	(Ohta *et al.*, 1990)
MT severing	*Xenopus* egg EF-1 α	(Shiina *et al.*, 1994)
Calmodulin binding	*Trypanosoma brucei* EF-1 α	(Kaur and Ruben, 1994)
	carrot EF-1 α homolog	(Durso and Cyr, 1994a)
	Tetrahymena EF-1 α	(Kurasawa *et al.*, 1996)
Initiation of Qβ phage replication	EF-Tu	(Blumenthal and Carmichael, 1979)
DNA replication in *Euplotes* (?)	EF-1 α	(Numata *et al.*, 1991b)

IV. Concluding Remarks

In this chapter we reviewed two multifunctional proteins—14-nm filament protein–citrate synthase and EF-1 α in *Tetrahymena*. Here we would like to point out several important projects along this line in the future.

Tetrahymena 14-nm filament protein–citrate synthase is a crucial protein which is related to essential biological metabolism and phenomena, such as the Krebs tricarboxylic acid cycle and sexual reproduction. Each function is displayed in separate compartments. Nevertheless, the 14-nm filament protein–citrate synthase has been shown to be encoded by the gene of a single type and to be produced by the same transcriptional unit. After translation, the 14-nm filament protein remaining in the cytoplasm is involved in oral morphogenesis during vegetative binary fission and in germ nuclear fertilization during conjugation, while the protein transported into the mitochondria functions as citrate synthase. Therefore, solving the mechanisms of compartmentalization of cytoplasmic 14-nm filament protein and mitochondrial citrate synthase is one of the important projects in the future.

Concerning the filament formation of citrate synthase and its activity in mitochondria, we have shown the following two facts: (1) *Tetrahymena* citrate synthase can form 14-nm filament structures in mitochondrial matrix in the growth decline phase (Numata *et al.*, 1995), and (2) the citrate synthase activity of the 14-nm filament protein–citrate synthase is drastically changed in monomer–polymer conversion *in vitro* (Takeda *et al.*, 1995). This strongly suggests that the accumulation of 14-nm filament protein–citrate synthase within rod-shaped filamentous structures consisting of 14-nm filaments is a regulatory system for citrate synthase activity in mitochondria. Site-directed mutagenesis (Alter *et al.*, 1990; Zhi *et al.*, 1991) and X-ray crystallographic studies (Remington *et al.*, 1982) on porcine citrate synthase have highlighted the residues thought to interact with its substrate and product. The relationship between catalytic residues and residues responsible for filament assembly may hold the key for solving the regulatory system for citrate synthase activity by monomer–polymer conversion. Therefore, it is necessary to determine which residues are related to the assembly of the 14-nm filament protein–citrate synthase.

We have stressed that *Tetrahymena* citrate synthase is a multifunctional protein. Then the questions arise: Do citrate synthases in organisms other than *Tetrahymena* have several functions and can citrate synthases in other organisms form 14-nm filaments? In this regard, our preliminary experiment has shown that commercial pig and chicken citrate synthases failed to form 14-nm filaments. However, it is necessary to determine whether citrate synthase purified from fresh pig heart has the ability to form filaments.

The EF-1 α or EF-1 α homolog is a highly conserved protein ubiquitous in the cytoplasm of eukaryotes and has several functions, such as GTP-dependent binding of aminoacyl-tRNA to ribosomes and interactions with cytoskeletal elements; possibly it also functions in the initiation of DNA replication.

The EF-1 α or EF-1 α homolog induces the formation of F-actin bundles as well as microtubule bundles, and moreover, these bundlings are modulated by Ca^{2+}/CaM. Condeelis's group reported that in *Dictyostelium* the incorporation of EF-1 α into the cytoskeleton is regulated during chemotaxis (Dharmawardhane *et al.*, 1991) and suggested that the reversible association of EF-1 α with actin might provide a mechanism for temporal and spatial regulation of protein synthesis in eukaryotic cells (Yang *et al.*, 1990). In this regard, some reports have suggested that actin synthesis is feedback-regulated by the status and/or level of existing actin (Serpinskaya *et al.*, 1990; Reuner *et al.*, 1991; Lloyd *et al.*, 1992). In addition, Theodorakis and Cleveland (1993) reported that unassembled tubulin levels autoregulate *de novo* tubulin synthesis through cotranslational effects on tubulin mRNA stability. EF-1 α may function as a communicator between the cytoskeletons and protein synthetic machinery. Therefore, it is very important to investigate the relationships between Ca^{2+}/CaM and this communication.

Finally, we stress the notion that both citrate synthase and EF-1 α are essential proteins and are highly conserved during evolution. *Tetrahymena* EF-1 α and *Dictyostelium* EF-1 α have F-actin bundling activity, so that EF-1 α must have acquired the F-actin bundling activity in the first stage of molecular evolution and has maintained the activity to the present. Also, an ancient progenitor of *Tetrahymena* must have acquired sexual reproduction and the TCA cycle, and then citrate synthase of the progenitor must have taken part in a control system of fertilization during conjugation. During evolution, *Tetrahymena* citrate synthase as well as EF-1 α has been maintaining a second role without gene duplication. Thus, we insist that both *Tetrahymena* citrate synthase and EF-1 α are conservative multifunctional proteins.

Acknowledgments

I thank Professor Yoshio Watanabe (Joubu University) for helpful discussion and his critical reading of the manuscript. This work was supported in part by grants from the Ministry of Education, Science, and Culture, Japan.

References

Allen, S. L. (1967). Genomic exclusion: A rapid means of inducing homozygous diploid lines in *Tetrahymena pyriformis,* syngen 1. *Science* **155**, 575–577.

Alter, G. M., Casazza, J. P., Zhi, W., Nemeth, P., Srere, P. A., and Evans, C. T. (1990). Mutation of essential catalytic residues in pig citrate synthase. *Biochemistry* **29**, 7557–7563.

Amos, L. A., and Baker, T. S. (1979). The three dimensional structure of tubulin protofilaments. *Nature (London)* **279**, 607–612.

Anderson, C. T. M., and Friedberg, E. C. (1980). The presence of nuclear and mitochondrial uracil-DNA glycosylase in extracts of human KB cells. *Nucleic Acids Res.* **8**, 875–888.

Bakowska, J., Frankel, J., and Nelsen, E. M. (1982a). Regulation of the pattern of basal bodies within the oral apparatus of *Tetrahymena thermophila. J. Embryol. Exp. Morphol.* **69**, 83–105.

Bakowska, J., Nelsen, E. M., and Frankel, J. (1982b). Development of the ciliary pattern of the oral apparatus of *Tetrahymena thermophila. J. Protozool.* **29**, 366–382.

Behnke, O. (1965). Helical filaments in rat liver mitochondria. *Exp. Cell Res.* **37**, 687–689.

Bloxham, D. P., Parmelee, D. C., Kumar, S., Walsh, K. A., and Titani, K. (1982). Complete amino acid sequence of protein heart citrate synthase. *Biochemistry* **21**, 2028–2036.

Blumenthal, T., and Carmichael, G. G. (1979). RNA replication: Function and structure of Qβ-replicase. *Annu. Rev. Biochem.* **48**, 525–548.

Bosch, L., Kraal, B., van Noort, J. M., van Delft, J., Talens, A., and Vijenboom, E. (1985). Novel RNA interactions with the elongation factor EF-Tu: Consequences for protein synthesis and *tuf* gene expression. *Trends Biochem. Sci.* **10,** 313–316.

Brands, J. H. G. M., Maassen, J. A., van Hemert, F. J., Amons, R., and Moller, W. (1986). The primary structure of the α subunit of human elongation factor 1. *Eur. J. Biochem.* **155,** 167–171.

Bruns, P. J. (1986). Genetic organization of *Tetrahymena. In* "The Molecular Biology of Ciliated Protozoa" (J. G. Gall, ed.), pp. 27–44. Academic Press, Orlando, FL.

Caron, F., and Meyer, E. (1985). Does *Paramecium primaurelia* use a different genetic code in its macronucleus? *Nature (London)* **314,** 185–188.

Coppard, N. J., Poulsen, K., Madsen, H. O., Frydenberg, J., and Clark, B. F. C. (1991). 42Sp48 in previtellogenic *Xenopus* oocytes is structurally homologous to EF-1 α and may be a stage-specific elongation factor. *J. Cell Biol.* **112,** 237–243.

Cottrelle, P., Thiele, D., Price, V. L., Memet, S., Micouin, J. Y., Marck, C., Buhler, J. M., Sentenac, A., and Fromageot, P. (1985). Cloning, nucleotide sequence, and expression of one of two genes coding for yeast elongation factor 1 α. *J. Biol. Chem.* **260,** 3090–3096.

Demma, M., Warren, V., Hock, R., Dharmawardhane, S., and Condeelis, J. (1990). Isolation of an abundant 50,000-dalton actin filament bundling protein from *Dictyostelium* amoebae. *J. Biol. Chem.* **265,** 2286–2291.

Dharmawardhane, S., Demma, M., Yang, F., and Condeelis, J. (1991). Compartmentalization and actin binding properties of ABP-50: The elongation factor-1 alpha of *Dictyostelium. Cell Motil. Cytoskel.* **20,** 279–288.

Dianov, G., Price, A., and Lindahl, T. (1992). Generation of signal-nucleotide repair patches following excision of uracil residues from DNA. *Mol. Cell. Biol.* **12,** 1605–1612.

Doerder, F. P., and Shabatura, S. K. (1980). Genomic exclusion in *Tetrahymena thermophila:* A cytogenetic and cytofluorimetric study. *Dev. Genet.* **1,** 205–218.

Durso, N. A., and Cyr, R. J. (1994a). A calmodulin-sensitive interaction between microtubules and a higher plant homolog of elongation factor-1 α. *Plant Cell* **6,** 893–905.

Durso, N. A., and Cyr, R. J. (1994b). Beyond translation: Elongation factor-1 α and the cytoskeleton. *Protoplasma* **180,** 99–105.

Elliott, A. M., and Bak, I. J. (1964). The fate of mitochondria during aging in *Tetrahymena pyriformis. J. Cell Biol.* **20,** 113–129.

Elliott, A. M., and Hayes, R. E. (1953). Mating types in *Tetrahymena. Biol. Bull. (Woods Hole, Mass.)* **105,** 269–284.

Eppenberger-Eberhardt, M., Riesinger, I., Messerli, M., Schwarb, P., Müller, M., Eppenberger, H. M., and Wallimann, T. (1991). Adult rat cardiomyocytes cultured in creatine-deficient medium display large mitochondria with paracrystallin inclusions, enriched for creatine kinase. *J. Cell Biol.* **113,** 289–302.

Evans, C. T., Owens, D. D., Sumegi, B., Kispal, G., and Srere, P. A. (1988). Isolation nucleotide sequence, and expression of a cDNA encoding pig citrate synthase. *Biochemistry* **27,** 4680–4686.

Forer, A., Nilsson, J. R., and Zeuthen, E. (1970). Studies on the oral apparatus of *Tetrahymena pyriformis* GL. *C. R. Trav. Lab. Carlsberg* **38,** 67–86.

Frankel, J. (1970). The synchronization of oral development without cell division in *Tetrahymena pyriformis* GL-C. *J. Exp. Zool.* **173,** 79–100.

Frankel, J., and Williams, N. E. (1973). Cortical development in *Tetrahymena. In* "The Biology of *Tetrahymena*" (A. M. Elliott, ed.), pp. 375–409. Dowden, Hutchinson & Ross, Stroudsburg, PA.

Fujita, H., and Machino, M. (1964). Fine structure of intramitochondrial crystals in rat thyroid follicular cell. *J. Cell Biol.* **23,** 383–385.

Gall, J. G. (1959). Macronuclear duplication in the ciliated protozoan *Euplotes. J. Biophys. Biochem. Cytol.* **5,** 295–308.

Grandchamp, S., and Beisson, J. (1981). Positional control of nuclear differentiation in *Paramecium. Dev. Biol.* **81,** 336–341.

Hamilton, E. P., Suhr-Jessen, P. B., and Orias, E. (1988). Pronuclear fusion failure: An alternate conjugational pathway in *Tetrahymena thermophila,* induced by vinblastine. *Genetics* **118,** 627–636.

Hirono, M., Endoh, H., Okada, N., Numata, O., and Watanbe, Y. (1987). *Tetrahymena* actin. Cloning and sequencing of the *Tetrahymena* actin gene and identification of its gene product. *J. Mol. Biol.* **194,** 181–192.

Hirono, M., Kumagai, Y., Numata, O., and Watanabe, Y. (1989). Purification of *Tetrahymena* actin reveals some unusual properties. *Proc. Natl. Acad. Sci. USA* **86,** 75–79.

Hirono, M., Tanaka, R., and Watanabe, Y. (1990). *Tetrahymena* actin: Copolymerization with skeletal muscle actin and interactions with muscle actin-binding proteins. *J. Biochem. (Tokyo)* **107,** 32–36.

Holmes, K. C., Popp, D., Gebhard, W., and Kabsch, W. (1990). Atomic model of the actin filament. *Nature (London)* **347,** 44–49.

Horowitz, S., and Gorovsky, M. A. (1985). An unusual genetic code in nuclear genes of *Tetrahymena. Proc. Natl. Acad. Sci. USA* **82,** 2452–2455.

Hovemann, B., Richter, S., Walldorf, U., and Cziepluch, C. (1988). Two genes encode related cytoplasmic elongation factors 1 α (EF-1 α) in *Drosophila melanogaster* with continuous and stage specific expression. *Nucleic Acids Res.* **16,** 3175–3194.

Hurt, E. C., and van Loon, A. P. G. M. (1986). How proteins find mitochondria and intramitochondrial compartments. *Trends Biochem. Sci.* **11,** 204–207.

Ip, W., Hartzer, M. K., Pang, S. Y. -Y., and Robson, R. M. (1985). Assembly of vimentin *in vitro* and its implications concerning the structure of intermediate filaments. *J. Mol. Biol.* **183,** 365–375.

Itano, N., and Hatano, S. (1991). F-actin bundling protein from *Physarum polycephalum:* Purification and its capacity for co-bundling of actin filaments and microtubules. *Cell Motil. Cytoskel.* **19,** 244–254.

Jessen, H. (1968). The morphology and distribution of mitochondria in ameloblasts with special reference to a helix-containing type. *J. Ultrastruct. Res.* **22,** 120–135.

Johnson, K. A., and Borisy, G. G. (1977). Kinetic analysis of microtubule self-assembly in vivo. *J. Mol. Biol.* **117,** 1–31.

Jurnak, F. (1985). Structure of the GDP domain of EF-Tu and location of the amino acids homologous to *ras* oncogene proteins. *Science* **230,** 32–36.

Kasai, M. (1969). Thermodynamical aspect of G-F transformations of actin. *Biochim. Biophys. Acta* **180,** 399–409.

Kasai, M., Asakura, S., and Oosawa, F. (1962). The G-F equilibrium in actin solutions under various conditions. *Biochim. Biophys. Acta* **57,** 13–21.

Kimball, R. F., and Prescott, D. M. (1962). Deoxyribonucleic acid synthesis and distribution during growth and amitosis of the macronucleus of *Euplotes. J. Protozool.* **9,** 88–92.

Kaur, K. J., and Ruben, L. (1994). Protein translation elongation factor-1 α from *Trypanosoma brucei* binds calmodulin. *J. Biol. Chem.* **269,** 23045–23050.

Kispal, G., and Srere, P. A. (1991). Studies on yeast peroxisomal citrate synthase. *Arch. Biochem. Biophys.* **286,** 132–137.

Kojima, H., Chiba, J., Watanabe, Y., and Numata, O. (1995). Citrate synthase purified from *Tetrahymena* mitochondria is identical to *Tetrahymena* 14-nm filament protein. *J. Biochem. (Tokyo)* **118,** 189–195.

Kojima, H. et al. (1996). In preparation.

Kurasawa, Y., Numata, O., Katoh, M., Hirano, H., Chiba, J., and Watanabe, Y. (1992). Identification of *Tetrahymena* 14-nm filament-associated protein as elongation factor 1 α. *Exp. Cell Res.* **203,** 251–258.

Kurasawa, Y. *et al.* (1996). F-actin bundling activity of *Tetrahymena* elongation factor 1α is regulated by Ca^{2+}/calmodulin. Submitted.

Kuriyama, R., and Borisy, G. G. (1985). Identification of molecular components of the centrosphere in the mitotic spindle of sea urchin eggs. *J. Cell Biol.* **101,** 524–530.

Kuriyama, R., Savereide, P., Lefebvre, P., and Dasgupta, S. (1990). The predicted amino acid sequence of a centrosphere protein in dividing sea urchin eggs is similar to elongation factor (EF-1 α). *J. Cell Sci.* **95,** 231–236.

L'Hernault, S. W., and Rosenbaum, J. L. (1983). *Chlamydomonas* α-tubulin is posttranslationally modified in the flagella during flagellar assembly. *J. Cell Biol.* **97,** 258–263.

L'Hernault, S. W., and Rosenbaum, J. L. (1985a). *Chlamydomonas* α-tubulin is postranslationally modified by acetylation on the ε-amino group of lysine. *Biochemistry* **24,** 473–478.

L'Hernault, S. W., and Rosenbaum, J. L. (1985b). Reversal of the posttranslational modification on *Chlamydomonas* flagellar α-tubulin occurs during flagellar resorption. *J. Cell Biol.* **100,** 457–462.

Linz, J. E., Lira, L. M., and Sypherd, P. S. (1986). The primary structure and the functional domains of an elongation factor-1 α from *Mucor racemosus. J. Biol. Chem.* **261,** 15022–15029.

Lloyd, C., Schevzov, G., and Gunning, P. (1992). Transfection of nonmuscle β- and γ-actin genes into myoblasts elicits different feedback regulatory responses from endogenous actin genes. *J. Cell Biol.* **117,** 787–797.

Martindale, D. W., Allis, C. D., and Bruns, P. J. (1982). Conjugation in *Tetrahymena thermophila:* A temporal analysis of cytological stages. *Exp. Cell Res.* **140,** 227–236.

Mikami, K. (1980). Differentiation of somatic and germinal nuclei correlated with intracellular localization in *Paramecium caudatum* exconjugants. *Dev. Biol.* **80,** 46–55.

Moldave, K. (1985). Eukaryotic protein synthesis. *Annu. Rev. Biochem.* **54,** 1109–1119.

Mugnaini, E. (1964). Helical filaments in astrocytic mitochondria of the corpus striatum in the rat. *J. Cell Biol.* **23,** 173–182.

Nagano, M., Kodama, K., Baba, N., Kanaya, K., and Osumi, M. (1982). Filamentous structure in yeast mitochondria by freeze-etching of replica combined with rapid freezing. *J. Electron Microsc.* **31,** 268–272.

Nanney, D. L. (1953). Nucleo-cytoplasmic interaction during conjugation in *Tetrahymena. Biol. Bull. (Woods Hole, Mass.)* **105,** 133–148.

Nanney, D. L., and Nagel, M. J. (1964). Nuclear misbehavior in an aberrant inbred *Tetrahymena. J. Protozool.* **11,** 465–473.

Ner, S. S., Bhayana, V., Bell, A. W., Giles, I. G., Duckworth, H. W., and Bloxham, D. P. (1983). Complete sequence of the *glt* A gene encoding citrate synthase in *Escherichia coli. Biochemistry* **22,** 5243–5249.

Numata, O., and Watanabe, Y. (1982). In vitro assembly and disassembly of 14-nm filament from *Tetrahymena pyriformis.* The protein component of 14-nm filament is a 49,000-dalton protein. *J. Biochem. (Tokyo)* **91,** 1563–1573.

Numata, O., Yasuda, T., Ohnishi, K., and Watanabe, Y. (1980). *In vitro* filament formation of a new fiber-forming protein from *Tetrahymena pyriformis. J. Biochem. (Tokyo)* **88,** 1505–1514.

Numata, O., Hirono, M., and Watanabe, Y. (1983). Involvement of *Tetrahymena* intermediate filament protein, a 49K protein, in the oral morphogenesis. *Exp. Cell Res.* **148,** 207–220.

Numata, O., Sugai, T., and Watanabe, Y. (1985). Control of germ cell nuclear behavior at fertilization by *Tetrahymena* intermediate filament protein. *Nature (London)* **314,** 192–194.

Numata, O., Takemasa, T., Takagi, I., Hirono, M., Hirano, H., Chiba, J., and Watanabe, Y. (1991a). *Tetrahymena* 14-nm filament-forming protein has citrate synthase activity. *Biochem. Biophys. Res. Commun.* **174,** 1028–1034.

Numata, O., Tomiyoshi, T., Kurasawa, Y., Zhen-xing, F., Takahashi, M., Kosaka, T., Chiba, J., and Watanabe, Y. (1991b). Antibodies against *Tetrahymena* 14-nm filament-forming protein recognize the replication band in *Euplotes. Exp. Cell Res.* **193,** 183–189.

Numata, O., Yasuda, T., and Watanabe, Y. (1995). *Tetrahymena* intramitochondrial filamentous inclusions contain 14-nm filament protein/citrate synthase. *Exp. Cell Res.* **218**, 123–131.

Numata, O. et al. (1996). *Tetrahymena* 14-nm filament protein (49k protein) and citrate synthase are encoded from a single gene. Submitted.

Ohta, K., Toriyama, M., Endo, S., and Sakai, H. (1988). Mitotic apparatus-associated 51-kD protein in mitosis of sea urchin eggs. *Zool. Sci.* **5**, 613–621.

Ohta, K., Toriyama, M., Miyazaki, M., Murofushi, H., Hosoda, S., Endo, S., and Sakai, H. (1990). The mitotic apparatus-associated 51-kDa protein from sea urchin eggs is a GTP-binding protein and is immunologically related to yeast polypeptide elongation factor 1 α. *J. Biol. Chem.* **265**, 3240–3247.

Olmsted, J. B. (1986). Microtubule-associated proteins. *Annu. Rev. Cell Biol.* **2**, 421–457.

Oosawa, F., and Asakura, S. (1975). "Thermodynamics of the Polymerization of Protein." Academic Press, New York and London.

Orias, E. (1986). Ciliate conjugation. *In* "The Molecular Biology of Ciliated Protozoa" (J. G. Gall, ed.), pp. 45–85. Academic Press, Orlando, FL.

Orias, J. D., Hamilton, E. P., and Orias, E. (1983). A microtubule meshwork associated with gametic pronuclear transfer across a cell-cell junction. *Science* **222**, 181–184.

Piatigorsky, J., and Wistow, G. J. (1989). Enzyme/crystallins: Gene sharing as an evolutionary strategy. *Cell (Cambridge, Mass.)* **57**, 197–199.

Piatigorsky, J., and Wistow, G. J. (1991). The recruitment of crystallins: New functions precede gene duplication. *Science* **252**, 1078–1079.

Piatigorsky, J., O'Brien, W. E., Norman, B. L., Kalumuck, K., Wistow, G. J., Borras, T., Nickerson, J. M., and Wawrousek, E. F. (1988). Gene sharing by δ-crystallin and argininosuccinate lyase. *Proc. Natl. Acad. Sci. U.S.A.* **85**, 3479–3483.

Piperno, G., and Fuller, M. T. (1985). Monoclonal antibodies specific for an acetylated form of α-tubulin recognize the antigen in cilia and flagella from a variety of organisms. *J. Cell Biol.* **101**, 2086–2094.

Preer, J. R., Jr., Preer, L. B., Rudman, B. M., and Barnett, A. J. (1985). Deviation from the universal code shown by the gene for surface protein 51A in *Paramecium*. *Nature (London)* **314**, 188–190.

Ray, C., Jr. (1956). Meiosis and nuclear behavior in *Tetrahymena pyriformis*. *J. Protozool.* **3**, 88–96.

Remington, S., Wiegand, G., and Huber, R. (1982). Crystallographic refinement and atomic models of two different forms of citrate synthase at 2.7 and 1.7 Å resolution. *J. Mol. Biol.* **158**, 111–152.

Reuner, K. H., Schlegel, K., Just, I., Aktories, K., and Katz, N. (1991). Autoregulatory control of actin synthesis in cultured rat hepatocytes. *FEBS Lett.* **286**, 100–104.

Riis, B., Rattan, S. I. S., Clark, B. F. C., and Merrick, W. C. (1990). Eukaryotic protein elongation factors. *Trends Biochem. Sci.* **15**, 420–424.

Roise, D., and Schatz, G. (1988). Mitochondrial presequences. *J. Biol. Chem.* **263**, 4509–4511.

Rosenkrantz, M., Alam, T., Kim, K.-S., Clark, B. J., Srere, P. A., and Guarente, L. P. (1986). Mitochondrial and nonmitochondrial citrate synthase in *Saccharomyces cerevisiae* are encoded by distinct homologous genes. *Mol. Cell. Biol.* **6**, 4509–4515.

Sasaki, H., and Suzuki, T. (1989). Intramitochondrial helical filaments in medullary tubules of the rat kidney. *J. Ultrastruct. Mol. Struct. Res.* **102**, 229–239.

Sattler, C. A., and Staehelin, L. A. (1979). Oral cavity of *Tetrahymena pyriformis*. A freeze-fracture and high-voltage electron microscopy study of the oral ribs, cytostome, and forming food vacuole. *J. Ultrastruct. Res.* **66**, 132–150.

Serpinskaya, A. S., Denisenko, O. N., Gelfand, D. I., and Bershadsky, A. D. (1990). Stimulation of actin synthesis in phalloidin-treated cells: Evidence for autoregulatory control. *FEBS Lett.* **277**, 11–14.

Shiina, N., Gotoh, Y., Kubomura, N., Iwamatsu, A., and Nishida, E. (1994). Microtubule severing by elongation factor 1 α. Science 266, 282–285.

Slupphaug, G., Markussen, F.-H., Olsen, L. C., Aasland, R., Aarsaether, N., Bakke, O., Krokan, H. E., and Helland, D. E. (1993). Nuclear and mitochondrial forms of human uracil-DNA glycosylase are encoded by the same gene. Nucleic Acids Res. 21, 2579–2584.

Sonneborn, T. M. (1954). Patterns of nucleo-cytoplasmic integration in Paramecium. Caryologia, Suppl. 1, 307–325.

Srere, P. A. (1975). The enzymology of the formation and breakdown of citrate. Adv. Enzymol. 43, 57–101.

Srere, P. A., Brazil, H., and Gordon, L. (1963). The citrate condensing enzyme of pigeon breast muscle and moth flight muscle. Acta Chem. Scand. 17, S129–S134.

Steinert, P. M., and Roop, D. R. (1988). Molecular and cellular biology of intermediate filaments. Annu. Rev. Biochem. 57, 593–626.

Suganuma, Y., and Yamamoto, H. (1980). Occurrence, composition, and structure of mitochondrial crystals in a hypotrichous ciliate. J. Ultrastruct. Res. 70, 21–36.

Suzuki, T., and Mostofi, F. K. (1967). Intramitochondrial filamentous bodies in the thick limb of henle of the rat kidney. J. Cell Biol. 33, 605–623.

Takagi, I., Numata, O., and Watanabe, Y. (1991). Involvement of 14-nm filament-forming protein and tubulin in gametic pronuclear behavior during conjugation in Tetrahymena. J. Protozool. 38, 345–351.

Takeda, T., Kurasawa, Y., Watanabe, Y., and Numata, O. (1995). Polymerization of highly purified Tetrahymena 14-nm filament protein/citrate synthase into filament and its possible role in regulation of enzymatic activity. J. Biochem. (Tokyo) 117, 869–874.

Takemasa, T., Ohnishi, K., Kobayashi, T., Takagi, T., Konishi, K., and Watanabe, Y. (1989). Cloning and sequencing of the gene for Tetrahymena calcium-binding 25-kDa protein (TCBP-25). J. Biol. Chem. 264, 19293–19301.

Tamura, S., Toyoshima, Y., and Watanabe, Y. (1966). Mechanism of temperature-induced synchrony in Tetrahymena pyriformis. Analysis of the leading cause of synchronization. Jpn. J. Med. Sci. Biol. 19, 85–96.

Theodorakis, N. G., and Cleveland, D. W. (1993). Translationally coupled degradation of tubulin mRNA. In "Control of the mRNA Stability" (G. Brawerman and J. Balesco, eds.), pp. 219–238. Academic Press, San Diego, CA.

Traub, P. (1985). "Intermediate Filaments: A Review." Springer-Verlag, New York.

Tsunemoto, M., Numata, O., Sugai, T., and Watanabe, Y. (1988). Analysis of oral replacement by scanning electron microscopy and immunofluorescence microscopy in Tetrahymena thermophila during conjugation. Zool. Sci. 5, 119–131.

Vallee, R. B., Bloom, G. S., and Theurkauf, W. E. (1984). Microtubule-associated proteins: Subunits of the cytomatrix. J. Cell Biol. 99, 38s–44s.

Watanabe, Y. (1963). Some factors necessary to produce division conditions in Tetrahymena pyriformis. Jpn. J. Med. Sci. Biol. 16, 107–124.

Watanabe, Y. (1971). Mechanism of synchrony induction. I. Some features of synchronous rounding in Tetrahymena pyriformis. Exp. Cell Res. 68, 431–436.

Weiner, A. M., and Maizels, N. (1987). tRNA-like structures tag the 3' ends of genomic RNA molecules for replication: Implications for the origin of protein synthesis. Proc. Natl. Acad. Sci. U.S.A. 84, 7383–7387.

Weitzman, P. D. J., and Danson, M. J. (1976). Citrate synthase. Curr. Top. Cell. Regul. 10, 161–204.

Williams, N. E., and Zeuthen, E. (1966). The development of oral fibers in relation to oral morphogenesis and induced division synchrony in Tetrahymena. C. R. Trav. Lab. Carlsberg 35, 101–118.

Wills, E. J. (1965). Crystalline structures in the mitochondria of normal human liver parenchymal cells. J. Cell Biol. 25, 511–514.

Wistow, G. J. (1993). Lens crystallins: gene recruitment and evolutionary dynamism. *Trends Biochem. Sci.* **18,** 301–306.

Wistow, G. J., and Piatigorsky, J. (1988). Lens crystallins: Evolution and expression of proteins for a highly specialized tissue. *Annu. Rev. Biochem.* **57,** 479–504.

Wistow, G. J., Mulders, J. W. M., and de Jong, W. W. (1987). The enzyme lactate dehydrogenase as a structural protein in avian and crocodilian lenses. *Nature (London)* **326,** 622–624.

Woolley, P., and Clark, B. F. C. (1989). Homologies in the structures of GTP-binding proteins: An analysis based on elongation factor EF-Tu. *Bio/Technology* **7,** 913–920.

Yamashiro-Matsumura, S., and Matsumura, F. (1985). Purification and characterization of an F-actin-bundling 55-kilodalton protein from HeLa cells. *J. Biol. Chem.* **260,** 5087–5097.

Yanagi, A., and Hiwatashi, K. (1985). Intracellular positional control of survival or degeneration of nuclei during conjugation in *Paramecium caudatum. J. Cell Sci.* **79,** 237–246.

Yang, F., Demma, M., Warren, V., Dharmawardhane, S., and Condeelis, J. (1990). Identification of an actin-binding protein from *Dictyostelium* as elongation factor 1 α. *Nature (London)* **347,** 494–496.

Zhi, W., Srere, P. A., and Evans, C. T. (1991). Conformational stability of pig citrate synthase and some active-site mutants. *Biochemistry* **30,** 9281–9286.

Some Characteristics of Neoplastic Cell Transformation in Transgenic Mice

Irina N. Shvemberger and Alexander N. Ermilov
Laboratory of Chromosome Stability and Cell Engineering, Institute of Cytology
of Russian Academy of Sciences, 194064 St. Petersburg, Russia

The role of the expression of different cellular genes and viral oncogenes in malignant cell transformation is discussed. We pay special attention to the role of the genes for growth factors and their receptors and homeobox genes in oncogenesis. Based on both the literature and our own data, specific features of tumors developed in transgenic mice are discussed. All of these data are used to analyze current theories of multistep oncogenesis and the stochastic component in this process. We suggest that all known evidence about the mechanisms of oncogenesis be used in studying the problem at various structural and functional levels in an organism. The chapter shows that transgenic mice are a most suitable model for studying various aspects of malignant transformation from the molecular to the organismal and populational levels.
KEY WORDS: Transgenic mice, Carcinogenesis, Oncogenes.

I. Introduction

Understanding such a complex process as oncogenesis requires constant development and improvement of experimental methods and adequate models for further study of malignant growth. Several stages can be distinguished in the development of the science of oncology, each marked by particular methods of investigation.

There is no doubt about the importance of numerous studies carried out early in this century and dealing with spontaneous and transplanted tumors in laboratory animals. During this period, investigators had to concentrate on only a few features of malignant growth because there were only a few models that could be used to evaluate specific features of tumors. Despite these limitations, by the second half of this century a lot of data on tumor development and growth had been accumulated. An important achievement of clinical and experimental oncology in this period was the detection of a number of characteristic features of tumor development: autonomous growth of the tumor, invasive growth and metastatic spread of some types of tumors, a higher variability of tumor cells during tumor growth, a detailed characterization of cytological and histological peculiarities of many tumors in man and animals (Pimentel, 1989). These theoretical and experimental data resulted in a new step in the investigation of malignant growth. This step was based on studies of oncogenesis at all levels, from molecules to a whole organism.

The information obtained on tumors induced by numerous carcinogenic substances is extremely valuable. It includes the isolation and characterization of particular histological types of tumors, analysis of their histogenesis and biochemical properties, the tumor-organism relationship, and the role of regeneratory, inflammatory and hyperplastic changes in tumor growth. These studies were essential in understanding many specific features of oncogenesis, such as a high incidence of tumor development after ionizing radiation, the hereditary character of some tumors, a higher incidence of particular tumors induced by some infectious diseases, and the disturbance of some chromosomes and karyotypes revealed by cytogenetic methods (Burck *et al.*, 1988).

The development of molecular-biological methods has opened a new stage in oncology. One of the major achievements on oncological science is the discovery of viral oncogenes and their cellular analogs—proto-oncogenes—as well as of viral and eukaryotic genes inducing and/or promoting the malignant transformation of cells (Burck *et al.*, 1988).

The results of molecular-biological studies on tumors developed in the intact organism as well as model experiments allowed a number of conclusions to be drawn. The most important of them seems to be that oncogenesis is accompanied by structural and functional disturbances of one or several proto-oncogenes. Three types are found in tumor cells: disturbances in the expression of normal proto-oncogenes, point mutations, and amplification of these genes. These disturbances are considered to be sufficient to cause serial events leading to the appearance of malignant cells and tumors. Moreover, studies of different models have made it possible to detect viral and eukaryotic genes, the expression of which in certain cell types induces malignant transformation. These genes are classed as cancer genes.

Until recently, a principal model for such studies was cell cultures, which allowed elucidation of the role and significance of normal and altered products of cancer genes. In addition, investigation of the gene products expressed in cell cultures allows detection of the cell regulatory networks characteristic of phenotypically normal cells as well as of malignant ones. However, it should be emphasized that studies of malignant cells *in vitro* do not always agree with data obtained by experiments *in vivo* and studies of tumors in man. Use of the cell culture alone is insufficient for analyzing such complex biological phenomena, which inevitably include a number of discrete events connected by complex links. Although an important role of oncogenes in tumor development is quite evident, oncogenesis cannot be considered just a simple consequence of the expression of one oncogene or even of several oncogenes. Studies on oncogenesis can succeed only if this process is analyzed simultaneously at all structural and functional levels of the organism. The widespread detection of oncogene expression both in experimental tumors and in tumors developed in man coincided with detection of oncogene expression in cell cultures.

During the early stages of oncogene studies, it was found that there was no direct connection between the expression of oncogenes and development of tumors. Moreover, evidence indicated a hereditary nature for numerous tumors in men and animals (Knudson, 1973, 1991). It became necessary to search for a more suitable model of oncogenesis which would include all modern information about malignant growth.

The next period of studies on oncogenesis is associated with the use of transgenic mice carrying different proto-oncogenes, oncogenes, and cancer genes in their genomes. Transgenic mice allow disturbances in the expression of particular genes in malignant transformation to be studied at a certain level of differentiation, taking into account various functional states, disturbances in cell interactions caused by changes in cell structure or function under conditions of heterogeneity of the surrounding cells that are inherent in any tissue or organ structures, and the effects of humoral and immune systems on the development and progression of tumors. This feature of the model makes it more suitable for studying the key problems of malignant growth in a system approaching the natural one. At the same time, such a model is characterized by quite significant difficulties in interpretation of the results.

Transgenesis as a model is used for different purposes by numerous scientists. It would be difficult to review the entire literature on this topic. Therefore we will concentrate on cytological aspects of oncogenesis in transgenic mice carrying different cellular or viral genes.

Several problems seem especially important. One of them is the role that the expression of foreign DNA sequences in the genome of transgenic mice plays in tumor development. Of particular importance is whether the disturbance of oncogene expression is the cause of malignant transformation. Most of the data on oncogenesis in transgenic mice are accounted for by the theory of multistep carcinogenesis. However, it is not clear what is regarded as a step in multistep oncogenesis in transgenic mice: an individual event or a cascade of events. The balance of genetic and homeostatic disturbances necessary for a single injured cell to develp malignant features and a possible stochastic character of the malignant transformation is under discussion.

II. Characteristics of the Method

Transgenic mice seem to be one of the most complex models ever used. The results obtained should be interpreted in compliance with specific features of transgenesis. Following the early reports on transgenic mice produced from fertilized eggs injected with foreign genetic material (Gordon et al., 1980; Gordon and Ruddle, 1981; Brinster et al., 1981; Costantini and Lacy, 1981), an intensive study of the phenomenon of transgenesis in mice was started. The results of this work are summarized in many reports (Gordon, 1983; Gordon and Ruddle, 1983; Brinster et al., 1984; Palmiter and Brinster, 1985, 1986; Hogan et al., 1986; Jaenish, 1988; Babinet et al., 1989; Went and Stranzinger, 1991), and many specific properties of transgenesis were characterized in detail (Westphal, 1987; Ermilov and Shvemberger, 1988; Hanahan, 1989; Rusconi, 1991; Stein, 1991).

Recombinant DNA techniques have made it possible to create various artificial genetic constructions with genes of interest (Byrne et al., 1991; Palmiter et al., 1991; Ornitz et al., 1991). It has become evident that the expression of the transgene is primarily determined by regulatory elements rather than by its integration into a particular site on chromosomes (Brinster and Palmiter, 1986; Palmiter and Brinster, 1986; Wilkie et al., 1986). A set of regulatory sequences may provide tissue specificity, cytospecificity, and stage specificity of transgene expression, as well as its inducibility. However, the level of foreign gene expression is apparently determined by a larger number of factors than regulatory elements. Thus, it has been assumed that the functional activity of a transgene may depend on the DNA sequences flanking it and on specific features of organization of the chromosome that affect the imprinting of the foreign sequence and the pattern of its methylation (Palmiter and Brinster, 1986; Sapienza et al., 1987; Al-Shawi et al., 1990; Feriotto et al., 1991; Lettmann et al., 1991).

The possibilities for making models to study oncogenesis increased greatly after the binary system for regulation of expression was proposed (Ornitz et al., 1991). According to this theory, the expression of a foreign gene can only occur in hybrids of two independent lines of transgenic mice. This allows the founder lines to be maintained in cases where the pathology induced by the expression of the foreign sequence develops at early stages of ontogenesis and can result in the death of the animal prior to its sexual maturity.

The introduction of foreign genetic information into one-cell mouse embryos allowed the integration of foreign DNA fragments as large as 50 kb into the mouse genome (Brinster *et al.*, 1985; Hogan *et al.*, 1986). Although primary transgenic-F_0 mice may be cell or even tissue mosaics, each somatic cell starting with F_1 usually contains a foreign sequence in its genome. Analysis of the hereditability of transgenes reveals that the foreign DNA sequence, being involved in the structure of chromosomal DNA, follows a Mendelian distribution (Gordon, 1983; Wilkie *et al.*, 1986). Study of transgenic offspring has shown that the foreign genes maintain their ability for expression through several generations, although the intensity of the expression varies significantly and may be different from that in the parents.

All the evidence on transgenic mice as a model for studying different diseases, including tumors, clearly indicates that transgenic mice represent a stable model in which the activity of a foreign sequence can be reproduced for a number of generations. Nevertheless, the investigator deals with a genetically mixed population of transgenic animals, with differences in homo- and heterozygosity, in the number of copies per genome, and in the site of integration on a chromosome when several lines carrying the same genetic construction are being studied. There are also differences in insertional mutations, and in unique genetic properties conditioned by peculiarities of the inbred animals. The latter is of great importance because the fertilized eggs used for the injection of foreign genes are usually obtained from F_1 hybrids, and hence are more viable than eggs from mice of inbred lines (Gordon, 1983; Palmiter and Brinster, 1985, 1986; Hogan *et al.*, 1986; Krulewski *et al.*, 1989; Beddington, 1992). Therefore, the expression of a foreign gene in cells of the same type but in different lines of mice characterized by different genetic backgrounds may have different consequences. Such a nontraditional approach to an experimental model is fully justified because it allows unique material to be obtained for studying different pathologies associated with particular genetic disturbances and for clarifying general interrelations in the phenomena studied.

III. Tumor Development in Transgenic Mice

A. Tumors in Transgenic Mice Harboring Some Cell Genes

Cellular proto-oncogenes were the first genes to be investigated for tumorigenesis in transgenic mice. Other genes that had been unknown up to now but were closely related to cell proliferation and differentiation were also studied (Table I).

TABLE I

Pathologic Changes in Transgenic Mice Bearing Cell Proto-Oncogenes on Their Mutant Analogs

Regulatory region/gene	Expression	Disturbances	References
Eμ/c-myc	Lymphoid tissue	B-cell lymphoma	Adams et al., 1985; Schmidt et al., 1988; Haupt et al., 1991; Verbeek et al., 1991; Sidman et al., 1993
MMTV/c-myc	Mammary gland	Well-differentiated adenocarcinoma	Stewart et al., 1984
	Testis, kidney, salivary gland, brain, thymus, pancreas, lung, spleen	No lesions	
CGRP/c-myc	Neurons, hind-brain, spinal cord, dorsal root ganglia, thyroid gland	No lesions	Baetscher et al., 1991
alb/c-myc	Liver	Hepatocellular dysplasia, adenoma	Murakami et al., 1993
Eμ/N-myc	Lymphoid tissue	B-cell lymphoma	Rosenbaum et al., 1989;
Eμ/N-myc	Lymphoid tissue	B- and T-cell lymphomas	Dildrop et al., 1989; Y. Wang et al., 1992
c-Ha-ras	Spleen	Angiosarcoma	Saitoh et al., 1990
	Mesentery	Angiosarcoma	
	Kidney	Angiosarcoma	
	Skin	Angiosarcoma, papilloma	
	Lung	Adenocarcinoma	
	Harderian gland	Adenocarcinoma	
	Lymph node	Lymphoma	
	Thymus	Lymphoma	
	In all other tissues	No lesions	
Wap/c-Ha-ras	Mammary gland	Carcinomas	Andres et al., 1987, 1988; Nielsen et al., 1991, 1992
	Salivary gland	Carcinomas	

Construct	Organ	Lesion	Reference
MMTV/*Hα-ras* (mutation in 61st codon)	Harderian gland	Hypertrophy, adenocarcinoma	Sinn et al., 1987
	Mammary gland	Adenocarcinoma	
	Salivary gland	Adenocarcinoma	
	Seminal vesicle, thymus, lung, spleen	No lesions	
MMTV/*v-Hα-ras*	Salivary gland	Adenocarcinoma	Tremblay et al., 1989
	Mammary gland	Adenocarcinoma	
	Lungs	Adenocarcinoma	
	Harderian gland	Hyperplasia, adenocarcinoma	
	Spleen	Hyperplasia	
	Epididymis, seminal vesicle, thymus, testis, bone marrow	No lesions	
MMTV/*N-ras*	Salivary gland	Hyperplasia, carcinoma	Mangues et al., 1992
	Mammary gland	Carcinoma	
	Harderian gland	Hyperplasia, benign tumor	
	Small intestine	Malignant neoplasms	
	Thymus	Hyperplasia	
	Lymph nodes	Hyperplasia	
	Lung	Carcinoma	
	Testis, ovary, prostate	No lesions	
MMTV/*N-ras* (mutation in 61st codon)	Lymph nodes	Lymphoma	
	Salivary gland, mammary gland, thymus	No lesions	
H2-K^b/*c-fos*	Bones	Osteosarcoma	Grigoriadis et al., 1993
	Salivary gland, thymus, heart, skeletal muscle, lung, brain	No lesions	
H2-K^b/*c-fos*	Spleen	Enlargement	Rüther et al., 1988
	Thymus	Hypertrophy	
	Almost in all organs	No lesions	

(continued)

TABLE I (*continued*)

Regulatory region/gene	Expression	Disturbances	References
MT I/c-fos	Bones	Lesions in long bones	Rüther et al., 1987, 1988, 1989
	Kidney, brain, sceletal muscle, testis, pancreas	No lesions	
H2-K^b/fos B	Spleen, liver, thymus, bones, intestine	No lesions	Grigoriadis et al., 1993
Keratin gene/v-fos	Epidermis	Hyperplasia benign tumor	Greenhalgh et al., 1993
H2-K^b/c-jun	Almost in all organs	No lesions	Grigoriadis et al., 1993
H2-K^b/v-jun	Skin	Hyperplasia	Schuh et al., 1990, 1992
	Skeletal muscle	Sarcomas	
	Thymus, spleen, heart, lung, liver, kidney, brain, bone, testis, ovary, uterus	No lesions	

1. Role of Cellular Proto-Oncogenes and Their Mutant Analogs in the Development of Tumors

a. c-myc *Transgenic Mice*   The expression of *c-myc* regulated by a foreign promoter can facilitate a high frequency of tumor development, whereas tumors do not develop if *c-myc* is regulated by its own promoter (Adams, 1985; Schmidt, 1988). Thus, highly malignant lymphomas from pre-B lymphocytes, mature B lymphocytes, and their intermediate forms developed in almost all transgenic mice bearing the mouse *c-myc* proto-oncogene controlled by an enhancer from the immunoglobulin heavy (μ) chain and in a high percentage of transgenic mice bearing the human *c-myc* gene controlled by the same enhancer. Further study of these transgenic mice has shown that the expression of a foreign gene in B lymphocytes starts long before development of tumors in the mice. In such mice, the proliferative B-lymphocyte activity increased markedly to result in a four- to five-fold increase of the pre-B-cell population compared with the norm (Langdon *et al.*, 1986). A study of immunoglobulin rearrangements in malignant lymphoma cells allowed us to conclude that in spite of the tremendous proliferation and the differentiation block, developing tumors in these transgenic mice are of a monoclonal origin (Alexander *et al.*, 1989; Suda *et al.*, 1987).

At the same time, investigations of homozygous and hemizygous Eμ-*myc* transgenic mice showed that the life span of homozygous Eμ-*myc* mice is about one-fourth that of hemizygous mice that develop polyclonal, nontransplantable tumors opposite to the monoclonal, highly transplantable malignancies in hemizygous animals (Sidman *et al.*, 1993).

A high level of development of tumors of different histogenesis in *c-myc* transgenic mice was also observed in cases where the foreign sequence was regulated by a long terminal repeat (LTR) of the mouse mammary gland tumor virus (MMTV) (Stewart *et al.*, 1984; Leder *et al.*, 1986). Transgene expression in primary MTV-*myc* transgenic mice and in their offspring was seen in different organs and tissues. Histologic analysis of tumors in these animals made it possible to reveal malignant lymphomas of an immunoblastic type and a follicular center cell type, as well as malignant mast cell neoplasms. In addition, well-differentiated breast adenocarcinomas appeared in females, and stromal cell neoplasms of testicular Sertoli cells were observed in males. Although expression of foreign *c-myc* in various organs and tissues was revealed sufficiently early, tumors developed in the 10–16-month-old mice. Moreover, the neoplasms did not appear in all tissues that expressed the transgene and some transgenic mice developed only one of the types of tumors mentioned here.

Analysis of different models of *c-myc* transgenic mice shows that disturbance of expression of *c-myc* proto-oncogene does not always result in cellular injuries and the appearance of pathology. Thus, a low level of tumor development has been observed in transgenic mice bearing the foreign *c-myc* proto-oncogene controlled by the whey acidic protein (WAP) that is usually expressed in the mammary gland (Andres *et al.*, 1987; Schoenenberger *et al.*, 1988) or by the promoter gene albumin that is expressed in the liver (Sandgren *et al.*, 1989). Meanwhile there was no tumor development in transgenic mice bearing the foreign *c-myc* gene regulated by the promoter of the rat calcitonin gene (CGRP) in spite of its expression in the thyroid, spinal cord, and hindbrain (Baetscher *et al.*, 1991).

b. **c-fos** *Transgenic Mice*   To study role of the cellular proto-oncogene *c-fos* in the development of different types of pathologies, several experimental models in transgenic mice have been created. One series of papers deals with transgenic mice obtained with the mouse *c-fos* gene controlled by the human metallothionein 1 (MT) promoter and osteosarcoma virus long terminal repeat 3' region (FBJ-LTR). The MT/*c-fos*-LTR mice expressed the foreign gene in various tissues but developed specific lesions only in the long bones starting at 2–3 postnatal weeks. These focal changes of bones were multiple and developed at the background of the normal osseous tissue (Rüther *et al.*, 1987, 1988, 1989). At the same time, tumors in these mice occurred rarely and only after a long (9–10 months) latent period. Histologic analysis of the tumors characterized them as osteosarcomas.

A high efficiency (100%) in the occurrence of bone tumors has been noted by Grigoriadis and coauthors (1993) in transgenic mice bearing *c-fos*-LTR (FBJ-LTR) fused with the promoter of a class 1 major histocompatibility (MHC) gene (H2-K^b). The H2-*c-fos*-LTR mice developed chondroblastic osteosarcomas at the age of 10–12 weeks. The *fos* family genes are known to encode transactivating (or repressing) DNA-binding proteins that form homo- or heterodimeric complexes, the complexes with the proteins of the *jun* family, in particular. In this connection it would be interesting to know what particular transactivating complexes change the activity of the osseous cell genome so dramatically. In the same work, no bone pathology was revealed when H2-*c-fos*-LTR and H2-*c-jun*-LTR transgenic mice were crossed, although each foreign gene was expressed at a high level. Based on these results, the authors suggested that high levels of *fos* disturb the normal growth control of osteoblastic cells and specifically affect expression of the osteoblast phenotype.

Use of genetic constructions with control elements determining peculiarities of the foreign sequence in the transgenic mouse genome often results in essential differences in the development of a pathology bound to a disturbance in the functioning of a certain gene. For example, the murine *c-fos* gene was regulated by the promoter of a class 1 MHC gene H2-K^b but without the FBJ-LTR (Rüther *et al.*, 1988). In the transgenic mice, the transgene expression occurred in almost all organs and tissues, including the osseous tissue, but pathology consisted only of an enlarged spleen and hyperplastic thymus (increased number of thymic epithelial cells).

c. ras *Transgenic Mice*   Transgenic mice containing the normal murine *N-ras* gene regulated by the promoter and enhancer LTR MMTV developed hyperplasia and tumors in almost all tissues and organs where the foreign gene was expressed; however, this occurred only at a high level of expression after a long latent period and with a moderate frequency (Mangues *et al.,* 1990, 1992). Sequestration of the foreign *N-ras* and its control sequences cloned from tumor cells did not reveal any mutations. The same group studied the mutant (61 codon) murine *N-ras* gene in transgenic mice, in which the foreign sequence was also being controlled by LTR MMTV. In spite of the fact that transgene expression in transgenic mice with the mutant *N-ras* gene was observed in the same organs and tissues as in those in the mice with the normal *N-ras* gene, only nodular lymphomas developed in this stock of transgenic mice.

A low frequency of tumor formation after a long latent period from the start of foreign sequence expression was also seen in transgenic mice bearing the normal human *c-Ha-ras* gene regulated by the promoter WAP gene (Andres *et al.,* 1987, 1988; Nielson *et al.,* 1991, 1992). The WAP/*c-Ha-ras* transgenic mice developed adenosquamous carcinomas of the salivary and mammary glands. Only tumors of the same localization were revealed in the MMTV/*v-Ha-ras* transgenic mice (the 12 and 59 codons are activated), in spite of foreign gene expression in many organs and tissues (Tremblay *et al.,* 1989).

Use of the promoter and enhancer of the transcription of the murine elastase 1 gene (El) fused with the normal or mutant *H-ras* human gene (12-condon) enabled the creation of a model for studies of a highly tissue-specific tumor development in transgenic mice (Quaife *et al.,* 1987). In mice with the mutant El/*H-ras* oncogene, pancreatic tumors appeared at once after birth or during the first postnatal months. During prenatal development, cytotypical and histotypical pancreas differentiation was disturbed to delay the cytodifferentiation. Retardation of the acinar structure formation started at the 14–16th prenatal day to coincide with the start of cell and tissue differentiation in normal mouse development. Study of the foreign El/*H-ras* oncogene did not reveal any correlation between the mRNA transgene level and structural injuries such as acinar cell hypertrophy, acinar structure hyperplasia, cellular and nuclear pleiomorphism, stromal proliferation, and other kinds of pathology in the pancreas.

d. bcr/abl *Transgenic Mice*   As a result of chromosomal reconstructions, two types of disturbances bound to cell proto-oncogenes can develop: a change in the character of individual gene expression if the reconstructions occurred in the regulatory areas, or formation of the fused genes, the products of which possess abnormal functional activity. A classic example of such disturbances is the discovery of the Philadelphia chromosome in some types of human leukemias. The Philadelphia chromosome results from a reciprocal translocation t(9;22) (q34;q11) in which the *c-abl* proto-oncogene located at 9q34 is translocated to the chromosome 22q11, where it is fused with the *bcr* gene (breakdown claster region). As a result of such reconstruction, a fused gene is formed, with regulatory sequences and a certain coded part of the *bcr* gene as well as a coded part of the *c-abl* gene. The *bcr/abl* fusion protein has an abnormal tyrosine kinase activity. To elucidate the direct role of the *bcr/abl* fusion protein in pathogenesis of leukemias, several different models of transgenic mice have been created (Hariharan *et al.*, 1989; Heisterkamp *et al.*, 1990; Cleary, 1991).

Since the expression of the *bcr/abl* construction under control of the *bcr* gene promoter is lethal during embryogenesis of the transgenic mouse, various *bcr/abl* constructions regulated by foreign promoters were used to create the models. The transgenic mice bearing the human *bcr/abl* fused gene controlled by the metallothionein promoter developed acute lymphoblastic leukemia and chronic myelogenous leukemia in a very high percentage of the cases. In one of the mice, two neoplasias developed simultaneously—acute lymphoblastic leukemia and lymphoblastic lymphoma. Neoplastic disturbances in these transgenic mice occurred immediately after birth, had a very aggressive character, and progressed rapidly to result in death of the animals at the 10–58th postnatal day. Molecular-biological and cell-populational analyses of the neoplasias have shown the peripheral blood cell population of the acute lympoblastic leukemic mice to be consistent with the presence of a polyclonal population of neoplastic pre-B lymphoblasts (Voncken *et al.*, 1992).

The transgenic mice bearing the *bcr/v-abl* fused gene under control either of the gene promoter immunoglobulin (IH) and enhancer Eμ or of LTR myeloproliferative sarcoma virus (MPSV) also developed lymphomas, although with a low frequency, the lymphomas being revealed only several months after birth (Fujita *et al.*, 1993). In these transgenic mice, a model of chronic myelogenous leukemia was reproduced. The tumors were of monoclonal character and developed as a rule from T lymphocytes. Hence, the *bcr/abl* transgenic mice turned out to be an adequate and efficient model of pathogenesis of human leukemias in which the Philadelphia chromosome is revealed in leukemia cells.

The results of these and some other works confirm, once more, the correctness of the concepts put forward earlier about an essential role for disturbances in proto-oncogene expression as well as disturbances of the oncoprotein functional activity in malignancies. At the same time, these disturbances usually are not sufficient for cells to become malignant or for tumors to develop. This is substantiated by numerous observations about transgenic mice with various oncogenes; in these animals, in spite of the oncogene being expressed in practically all organs and tissues, tumors developed with a quite different probability. This peculiarity is probably due, on one hand, to unique capabilities of the regulatory sequences to determine foreign gene expression only in certain periods of the cell cycle, and on the other hand, to the state of the cell functional activity as determined by differentiation stage, the activity of separate regulatory networks, sensitivity to external influences such as hormonal balance of the organism, growth factors, and so on.

An increase in proto-oncogene expression can not only stimulate cell proliferative activity and disturbances of differentiation but also, as a consequence, can result in changes in the structure of cell populations of several tissues.

e. Tumors in Transgenic Mice Bearing Growth Factor Receptor Genes and Growth Factor Genes Growth factor receptors and growth factors have a specific role in the functioning of the organism owing to their unique properties. The growth factor receptor and growth factor complex is an initiating step in the signal transduction system. Activation of the receptor by a growth factor results in a change in a most important cell regulatory networks via the second messengers involved in the signal transduction pathway. These messengers give rise to a multifold enhancement of the initial signal and are able to modify (for instance, to phosphorylate) various cell proteins. Such changes in the functional activity of quite different proteins, from enzymes to transcription factors, ultimately result in fundamental changes in the cell's functional activity. The most important of these changes for malignancies are changes in proliferation and differentiation. It is interesting that at different stages of the organism's development and in different tissues, the growth factors and growth factor receptors provide for different cell functional states (Parker and Katan, 1990).

Marked disturbances discovered so far in the functional activity of growth factors and growth factor receptors in ontogenesis have stimulated more detailed studies, including the use of transgenic mice, on the role of these changes in malignant transformation and progression of tumors.

Transgenic mice bearing the rat *neu* gene or its homolog, the human *c-erbB*-2 gene, provided a comprehensive and interesting material for studying the peculiarities of oncogenesis caused by malfunction of the cell receptor apparatus. Unique and rather unexpected results were reported in the paper dealing with the activated rat *c-neu* gene regulated by the MMTV promoter/enhancer and expressed in the mammary and salivary glands, epididymis, and Harderian glands as well as in multiple mammary tumors

(Muller *et al.*, 1988). A high expression of the MMTV/*c-neu* fusion gene in the mammary gland tissues brought about dramatic consequences both in males and females. In females, as early as at the 8th postnatal week, multiple hyperplastic and dysplastic nodules of mammary gland epithelium seemed to infiltrate the entire mammary fat pad. By the 95th postnatal day, multiple mammary gland tumors had already developed in all transgenic females. Mammary gland tumors were also observed in males, albeit somewhat later. Their earliest tumors developed at the age of 100 days while by 200 days not a single male was free of the tumor. The tumors found in both males and in females were classified histologically as multiple mammary adenocarcinomas. Histological analysis of the tumors showed an interesting peculiarity: in the *neu* transgenic mice, the entire mammary gland tissue was involved in the tumor so that no areas of normal tissue were found in the mammary glands in any of the animals.

Expression of the MMTV/*c-neu* gene in other organs and tissues of transgenic mice has been shown to result in the development of pathologic changes (Muller *et al.*, 1988). In the MMTV/*c-neu* transgenic mice, a uniform bilateral enlargement of the parotid glands, epididymis, and salivary glands was observed. It is interesting that in the salivary gland, mucous glandular cells were selectively hypertrophic while serous cells were selectively hyperplastic. In epididymis, papillary hyperplasia and hypertrophy with intralaminar engorgement were revealed. In the transgenic males with these disturbances, a decreased fertility and sometimes sterility was found. The Harderian and lacrimal glands, unlike mammary glands, salivary glands, and epididymis, never showed any pathologic changes, even at the hyperexpression of the foreign *c-neu* gene.

No other model for studying the role of the *neu* gene in oncogenesis has demonstrated such characteristic features as bilaterality and uniformity of the tumor and nontumor pathology at certain stages of ontogenesis in all transgenic mice bearing the activated rat *c-neu* gene.

At present transgenic mice have been obtained that contain the nonactivated rat *neu* gene (Guy *et al.*, 1992), the activated human *c-erbB*-2 gene (Stöcklin *et al.*, 1993) as well as the activated rat *neu* gene (Bouchard *et al.*, 1989) with small changes in the genetic construction and with mice of other lines compared with those used by Muller and coauthors (Muller *et al.*, 1988). All these studies used the LTR MMTV promoter-enhancer area as a regulatory sequence. In all the transgenic mouse lines obtained, tumor and nontumor pathology developed, with rather common characteristic features. In almost all the organs and tissues in which the foreign gene expression was revealed there were hypertrophy and hyperplasia. Development of the mammary gland tumors in females was of a stochastic character although the percentage of their development was very high. In cases when the same animal developed several tumors, they were of a different size and occurred independently and asynchronously (Bouchard *et al.*, 1989). In addition, in the transgenic mice bearing the nonactivated rat *c-neu* gene (Guy *et al.*, 1992) and the activated rat *c-neu* gene (Bouchard *et al.*, 1989), the mammary gland tumors sometimes metastasized to the lungs. Apart from the mammary gland tumors in the mice expressing the activated *c-neu* gene, tumors of salivary and Harderian glands were found (Bouchard *et al.*, 1989), while in the transgenic mice expressing the activated *c-erbB-2* gene, diffuse and focal malignant mastocytosis was revealed (Stöcklin *et al.*, 1993).

In general, studies on transgenic mice with the *neu* gene have shown that malfunction of the cell receptor apparatus can give rise to various pathological cellular and tissue changes, including malignant transformation, which not uncommonly involves a substantial part of the tissue. The character of the pathology revealed in these studies seems to depend on peculiarities both of the gene constructs and of the genetic background.

Disturbances of the expression of the growth factor genes as well as of the growth factor receptor genes in transgenic mice also can result in various kinds of pathology. As shown in a number of papers (Matsui *et al.*, 1990; Muller *et al.*, 1990; Sandgren *et al.*, 1990; S. J. Kim *et al.*, 1991; G.-H. Lee *et al.*, 1992; Takagi *et al.*, 1992), transgenic mice expressing growth factors at a high level developed hyperplasias, dysplasias, and neoplasias similar to those in transgenic mice expressing, at a high level, activated or nonactivated genes of growth factor receptors. This may be illustrated by a series of studies on transforming growth factor-α (TGFα).

TGFα is a member of the epidermal growth factor (EGF) family of proteins. It is a ligand for the EGF receptor as well as possibly for other members of this receptor subfamily. The activated TGFα-receptor complex brings about signal transduction in a variety of cells (Parker and Katan, 1990). To elucidate the direct role of TGFα in oncogenesis, several models of transgenic mice have been created in a number of laboratories. Thus, transgenic mice bearing the human TGFα gene under regulation of the promoter/enhancer region of the MT gene (Jhappan *et al.*, 1990; G.-H. Lee *et al.*, 1992; Takagi *et al.*, 1992) developed, starting from 4–5 postnatal weeks, centrilobular hypertrophy of the liver. The foci of the hepatocytic hypertrophy were of a progressive character, and subsequently foci of hyperplasia were also observed. In 74% of the TGFα transgenic mice, hepatic adenomas and carcinomas developed by the end of the first year. In several cases these tumors were multifocal.

Single hepatocellular carcinomas in a low percentage of the foci were also found in transgenic mice bearing the rat TGFα gene under regulation of the promoter/enhancer MT (Sandgren *et al.*, 1990). In addition, these animals developed multifocal pseudoductal acinar metaplasia of the pancreas and adenocarcinomas of the postlactational mammary glands. Mammary gland adenomas and carcinomas were also revealed in transgenic mice bearing the human TGFα gene fused with the regulatory sequences LTR/MMTV (Matsui *et al.*, 1990).

Hyperplasia of the mammary glands in females and hyperplasia of the prostatic epithelium developed in transgenic mice bearing the growth factor *int*-2 gene under regulation of the MMTV promoter/enhancer (Muller *et al.*, 1990). In transgenic mice bearing the human insulin-like growth factor I gene under regulatory elements of the gene MT, there was an enlargement of multiple tissues, including muscle, fat, and connective tissue, as well as a significant elevation of the total body weight of adult animals (Mathews *et al.*, 1988).

In all the TGFα transgenic mice lines described, the development of tumors was preceded by hypertrophic or hyperplastic changes in the tissues. These changes most commonly were diffuse, which resulted in an alteration of the tissue architecture, depended on the expression level of the foreign sequence, and occurred at a certain stage of ontogenesis. Thus, studies on the TGFα transgenic mice allowed all the authors cited to conclude that TGFα may have an extremely important role in progression of tumors at later stages of their development.

Experiments on TGFα transgenic mice (Jhappan *et al.*, 1990; Matsui *et al.*, 1990; Sandgren *et al.*, 1990; G.-H. Lee *et al.*, 1992; Takagi *et al.*, 1992) as well as on transgenic mice with the growth factor receptor *neu* gene (Muller *et al.*, 1988; Bouchard *et al.*, 1989; Guy *et al.*, 1992; Stöcklin *et al.*, 1993) are of particular interest in understanding oncogenesis. These experiments have made it possible to evaluate the role and importance of genetic background in the appearance and progression of tumors. This evaluation is based on the observation that different pathological changes, such as hypertrophy, hyperplasia, dysplasia, and tumors of different degrees of malignancy occurred in transgenic mice with the same substrate genes in the genetic constructions used. At the same time, overexpression of the transgene in some organ or tissue does not necessarily result in development of a tumor or pretumor pathology.

f. Oncogenesis in Transgenic Mice Bearing the Chimeric Homeobox Gene E2A-PBX1 At present it has become quite evident that the transformation of a normal cell into a malignant one is accompanied by disturbances in the intracellular regulation that determines the functional state of the cell. A most important role in the functional activity of the genome is played by a specific class of nuclear proteins and protein complexes with a DNA-binding activity. In this part of the chapter, we will deal with studies on the role of transcriptional factors in oncogenesis using transgenic mice. Of a particular interest is the paper by Dedera and coauthors (1993), who studied transgenic mice bearing the chimeric homeobox gene E2A-PBX1, because the results of their work convincingly demonstrate a connection between such biological phenomena as the malignant transformation of cells and programmed cell death in the entire organism.

Owing to screening of the genes specifically expressed in the tumor cells, a great many genes coding the DNA-binding proteins have been cloned at present. To elucidate the direct role of these genes in oncogenesis *in vivo*, transgenic mice bearing some of them have been obtained. Changes in the transcriptional activity of different genes affect the behavior of the cell in different ways in the entire organism. For instance, no tumor pathology has been revealed in the transgenic mice expressing *c-ski* (Sutrave *et al.*, 1990). In these animals, only an increase in muscle mass occurred, at the expense of increased myofibril volume. This muscle tissue hypertrophy was not accompanied by any changes in the functional, physiological activity of the tissue. Fisch and coauthors (1992) observed the development of T-cell lymphoblastoses in RBTN1 and RBTN2 transgenic mice. These tumors developed in the mice after a long latent period and only in a small percentage of the animals. In addition, there were no prelymphomatous changes in these animals.

To elucidate the role of the chimeric homeobox protein E2A-PBX1 in oncogenesis, transgenic mice have been obtained in which the E2A-PBX1 gene was regulated by the Eμ enhancer (Dedera *et al.*, 1993). This protein is of special interest because the chimeric E2A-PBX1 gene is found only in the genome of the leukemic cells in children and is formed as a result of the t(1:19) (q23:p13) translocation (Mellentin *et al.*, 1990). The chimeric protein E2A-PBX1 is composed of the transactivating domain E2A and the carboxy-terminal portion of PBX1. The E2A gene codes one of the proteins of the basic-helix-loop-helix family. This protein has a nonspecific transactivating domain in its amino-terminal portion (Quong *et al.*, 1993). The PBX1 protein is one of the members of the homeobox family and has a DNA-binding homeodomain. The PBX1 gene is expressed in many fetal and adult tissues but not in the lymphoid cell lines (Monica *et al.*, 1991).

Lymphoma developed in all transgenic mice expressing E2A-PBX1, and these mice died within the first 5 months of their life. The E2A-PBX1 gene was expressed in transgenic mice in both T cells and B cells, and at a higher level in the former. In the E2A-PBX1 transgenic mice, as in the transgenic mice expressing the foreign Eμ/*c-myc* gene (Section III,A,1,a), proliferative activity of the lymphoid cells increased. However, in contrast to the Eμ/*c-myc* transgenic mice in which an increase in lymphoid cell proliferative activity resulted in the development of pretumor hyperplasia, no hyperplastic changes have been revealed in the E2A-PBX1 transgenic mice. On the contrary, in these mice, prior to development of the tumors, the number of thymocytes and B cells fell to 20% of the normal level. The revealed lymphopenia was not bound to any lymphocytic fraction because, according to studies of molecular markers specific for the early and late stages of the lymphocytic differentiation, the T and B cells were reduced similarly at all differentiation stages.

The fact that the E2A-PBX1 transgenic mice, prior to tumor development, showed such a complex of disturbances as increase of lymphoid cell proliferative activity and lymphopenia with an equivalent reduction of the number of cells at all stages of differentiation has allowed the authors to conclude that the chimeric E2A-PBX1 transcriptional factor can activate processes resulting in programmed cell death. In addition, the data obtained have suggested a direct connection between the oncogene-induced cell cycle progression and programmed cell death. The E2A-PBX1 transgenic mice created by the authors (Dedera *et al.*, 1993) can be considered a convenient model for studying this phenomenon.

B. Tumors in Transgenic Mice Carrying Different Virus Genes

Transgenic mice bearing viral genes are of particular interest because the data about these animals are extremely important both in elucidating the role of individual viral genes in malignant cell transformation in the course of virologic investigations in oncology and in discovering the etiopathology involved in oncogenesis. For most viral genes, there are no direct analogs in the eukaryote genome; at the same time, cellular disturbances induced by viral genes are quite distinct and facilitate the development of various pathologic states.

1. Transgenic Mice Bearing Simian Virus 40 Genes

A great many interesting models for studying the oncogenesis of quite different organs, tissues, and cells (Table II) have been created using transgenic mice bearing in their genome an early transcription region of the simian virus 40 or the gene of the large tumor antigen regulated by various promoters. Numerous studies have made it possible to conclude that the large tumor antigen of the simian virus 40 is a multifunctional transforming protein. This seems to be partly connected with the fact that the large T antigen of the SV40 binds both the retinoblastoma gene product Rb and the cellular nuclear protein p53. The Rb and p53 genes are referred to as "tumor repressors," "recessive oncogenes," "anti-oncogenes," or "checkpoint genes" to imply that tumor formation may occur when these genes are inactivated. The mechanisms of functioning of the Rb and p53 proteins are not properly understood; however, these proteins are known to be involved in the cell cycle and to determine the expression of some genes.

The T antigen SV40 is able to disturb the most important cell regulatory networks, which probably accounts for the diversity of tumor pathology in transgenic mice: diffuse or focal hyperplasia, adenomatous formations, nonmetastatic and metastatic malignant tumors. We will review in more detail some models using SV40 that were created to study oncogenesis in transgenic mice (Parker and Katan, 1990).

One of the first experimental models was obtained in transgenic mice which had introduced into their embryonic line the early region SV40 (SV40 ER) or the entire genome SV40, both fused with the metallothionein gene regulatory elements (Brinster et al., 1984; Small et al., 1985). These transgenic mice develop characteristic brain tumors: papillomas and carcinomas of the choroid plexus. Further use of genetic constructions consisting of the SV40 ER, including regulatory sequences, human and/or murine growth hormone gene, and promoter of the metallothionein gene to obtain transgenic mice has shown that the multiple pathology is due to the presence of an SV40 sequence in a transgene (Palmiter et al., 1985; Messing et al., 1985). Apart from the characteristic choroid plexus papillomas revealed earlier, renal pathology, neuropathies, thymus hyperplasia, and adenomas of the liver and pancreas islets have been found in these transgenic mice. A certain pathology depended on changes in the construction regulatory sequences and structural genes. Development of the choroid plexus tumors has been shown in the same studies to be associated with the presence of a particular sequence of 21 pairs of nucleotides in the enhancer of SV40. Replacement of the regulatory sequences of the metallothionein gene in genetic constructions by the promoter and enhancer of the transcriptional unit of the type 5 adenovirus E1A, has made it possible to reveal the development of glioblastomas in transgenic mice (Kelly et al., 1986).

The enormous possibilities for studying brain tumor pathology in transgenic mice bearing the SV40 T antigen gene (Tag) suggested by the initial work encouraged the investigators to develop this model further. Thus, primitive neuroectodermal tumors turned out to occur in transgenic mice bearing the T antigen SV40 gene under control of the promoter, human cystic fibrosis transmembrane conductance regulator (CFTR) (Perraud et al., 1992), mouse opsin promoter (Al-Ubaidi et al., 1992b), and human interphotoreceptor retinoid-binding protein (IRBP) promoter (Al-Ubaidi et al., 1992a). Malignant Schwannomas were revealed in transgenic mice expressing the T antigen gene under regulation of the mouse myelin basic protein promoter (Jensen et al., 1993). We would like to review one of these papers in more detail because, in our opinion, it illustrates very well the importance of disturbances of cell differentiation in the development of neoplasias.

Al-Ubaidi and co-workers (1992a) have shown that the expression of the SV40 Tag under control of the human IRBP promoter in transgenic

TABLE II
Tumors in Transgenic Mice Bearing Early Region Transcription Genes SV40

Organ, tissue	Type of tumor	Description of transgene	References
Central and peripheral nervous system	Carcinomas choroid plexus	MT/ER	Brinster et al., 1984
	Carcinomas choroid plexus	CFTR/Tag	Perraud et al., 1992
	Papilloma choroid plexus	MT/ER	Brinster et al., 1984; Messing et al., 1985; Palmiter et al., 1985; Small et al., 1985
	Papilloma choroid plexus	SV40	Faas et al., 1987
	Glioblastoma	E1A/ER	Kelly et al., 1986
	Neuroectodermal tumors	IRBP/Tag	Al-Ubaidi et al., 1992a
	Neuroectodermal tumors	MOP/Tag	Al-Ubaidi et al., 1992b
	Schwannomas	MBP/Tag	Jensen et al., 1993
Pancreas	β-cell adenoma	Insulin/Tag	Hanahan, 1985; Adams et al., 1987
	Islet adenoma	MT/ER	Palmiter et al., 1985; Messing et al., 1985
	Islet adenoma	MSV/Tag	Theuring et al., 1990
	Acinar cell carcinoma	EL/ER	Ornitz et al., 1987; Glasner et al., 1992
	Acinar cell carcinoma	Amy-2.2/Tag	Ceci et al., 1991
Stomach	Carcinoma	Amy-2.2/Tag	Ceci et al., 1991
Liver	Carcinomas	MT/ER	Messing et al., 1985
	Carcinomas	MUP/Tag	Held et al., 1989
	Carcinomas	Alb/Tag	H.-G. Lee et al., 1990; Hino et al., 1991
	Carcinomas	Gastrin/Tag	Montag et al., 1993
	Adenoma	MT/ER	Palmiter et al., 1985
	Bile duct carcinoma	Gastrin/Tag	Montag et al., 1993

56

Tissue	Tumor type	Transgene	Reference
Lung	Bronchio-alveolar tumor	Uteroglobin/Tag	De Mayo et al., 1991
	Adenocarcinoma	MMTV/ER	Choi et al., 1988
Urogenital tract	Undifferentiated tumor, ovarian glandular tumor	Uteroglobin/Tag	De Mayo et al., 1991
	Ependymomas	CFTR/Tag	Perraud et al., 1992
	Renal adenocarcinoma	MMTV/ER	Choi et al., 1988
	Renal adenocarcinoma	MSV/Tag	Theuring et al., 1990
Bone	Osteosarcoma	P1/ER	Benringer et al., 1988
	Osteosarcoma	MBP/Tag	Jensen et al., 1993
	Osteosarcoma	Amy-1[a]/Tag	Knowles et al., 1990
Muscle tissue	Rhabdomyosarcoma	ß-globin/Tag	Teitz et al., 1993
Heart	Rhabdomyosarcoma	P1/ER	Benringer et al., 1988
Mammary gland	Adenocarcinoma	MMTV/ER	Choi et al., 1988
Harderian gland	Adenocarcinoma	MMTV/ER	Choi et al., 1988
Thyroid gland	Carcinomas	CGRP/Tag	Baetscher et al., 1991
Pineal gland	Adenoma	MSV/Tag	Theuring et al., 1990
Skin	Melanoma	Tyr/ER	Bradl et al., 1991; Klein-Szanto et al., 1991; Larue et al., 1993
Eye	Neuroectodermal tumors	IRBP/Tag	Al-Ubaidi et al., 1992a
	Melanoma	Tyr/ER	Bradl et al., 1991; Klein-Szanto et al., 1991; Larue et al., 1993
Connective tissue	Fibrosarcoma	MSV/Tag	Theuring et al., 1990
Lymphoid tissue	Lymphoma	MMTV/ER	Choi et al., 1988
Yellow adipose tissue	Metastazing hybernoma	Amy-1[a]/Tag	Fox et al., 1989

mice results in the formation of trilateral tumors: bilateral retinal and cerebral tumors. The tumors were undifferentiated and appeared to be initiated at the embryonic or early postnatal stages. According to preliminary studies, in the transgenic mice bearing the bacterial reporter CAT gene (chloramphenicol acetyltransferase) under regulation of the same IBRP promoter, expression of the foreign sequence starts at embryonic day 13, that is, when the developing retinal cells just begin their initial differentiation into rod and cone photoreceptor cells.

It has been shown immunohistochemically that the retinoblastoma cells express neuron-specific enolase, S antigen, and Leu 7 typical for normal retinal photoreceptor cells, but do not express opsin and phosphodiesterase characteristic of the differentiated rod and cone photoreceptors. These results have allowed the authors to suggest that the disturbances resulting in the malignant transformation take place in the stem cells prior to their final differentiation. The disturbances due to T antigen at this early differentiation stage probably are able to lead to a malignant transformation because expression of the T antigen at later differentiation stages due to the mouse opsin promoter (Al-Ubaidi *et al.*, 1992b), and located only in rod photoreceptor cells, did not result in tumor development.

One of the characteristic features of the transgenic mice bearing the early transcription region SV40 is that coexpression of the large T and small t antigens not uncommonly changes the character of the cellular pathology compared with that developed at expression of the T antigen only, although the small t antigen itself has not been shown to lead to the malignant transformation of cells.

This peculiarity allows the experimentators to enlarge the possibilities of creating various models of tumor pathology in transgenic mice. Changes in the neoplasm character that depend only on the expression of the early region SV40 or of the T antigen gene have been convincingly demonstrated by Choi and coauthors (1988). These authors have shown that the MMTV regulatory sequences determine transgene expression in the ductal epithelium of several organs such as mammary glands, lungs, and kidney as well as lymphoid cells. Pulmonary and renal tumors develop only in transgenic mice harboring the genetic constructions containing both small t and large T antigen genes. This has enabled the authors to conclude that it is only a combination of the SV40 small and large antigens that would result in slowly proliferating pulmonary and renal epithelial cells.

Characteristic histologic disturbances such as the hyperplasia and dysplasia found in the transgenic mice with SV40 ER as well as cytologic, karyologic, and molecular-genetic disturbances revealed in the cells expressing T antigen are typical for the changes found in the development of human and animal tumors. Thus, these transgenic mice make it possible to further follow similar changes in certain tissues. The mice modelling the development of pancreatic tumors are an example.

Hanahan (1985) was the first to study the genetic constructions that in the transgenic mouse genome would lead to the development of tumors from a particular pancreatic cell type. These constructions consisted of the SV40 T antigen gene regulated by the promoter and enhancer of the rat insulin gene. The two constructions differed in the promoter orientation. The transgenic mice obtained with both these constructions and living for 10 to 20 weeks developed tumors of Langerhans' islet β cells.

Expression of the SV40 Tag in the insulin-producing β cells induced a high cell proliferation index as early as 4–6 weeks which resulted in a progressive proliferation in the Langerhans' islet. Karyologic and cytologic studies of the development of pancreatic tumors in the transgenic mice expressing the SV40 Tag under control of the elastase promoter have made it possible to follow the sequence of characteristic disturbances (Ornitz et al., 1987; Levine et al., 1991). The development of hyperplasia and tumors turned out to correlate well with karyologic disturbances in the form of the development of aneuploidy and cytologic disturbances such as multipolar mitosis which is probably due to the failure of centriole formation.

Another similarity of tumor development in transgenic mice bearing the SV40 T antigen gene in their genome to tumor development in humans consists in the ability of the developed tumors in some models to metastasize (Knowles et al., 1990; Fox et al., 1989; Cartier et al., 1992). Thus, in transgenic mice bearing T and t antigen genes regulated by tyrosinase promoter (Tyr-SV40ER), ocular and cutaneous melanomas developed several weeks after birth. The ocular tumors often were bilateral, highly malignant, and highly invasive (Bradl et al., 1991). Histologically, these tumors, like human ocular and cutaneous melanomas, were pleomorphic, with an epithelioid, spindle-shaped, small-cell, glanduliform or, less commonly, myxoid pattern. In addition, these tumors rapidly acquired the ability for local and distant metastases. The pattern of metastasizing in these transgenic mice had many features similar to those in human tumors. Human melanomas metastasize as a rule to the liver, lung, bone, kidney, brain; in the transgenic mice Tyr-SV40E, metastases also were present in the lung, bone, brain, and, in addition, in lymph nodes as well as occasionally in the salivary glands, thymus, and subcutaneous tissue. However, unlike human melanomas, no metastases were revealed in the kidney and liver.

Changes in the structural-functional state of the genome, cell, and tissue as a whole, which are due to expression of SV40 Tag in transgenic mice, are so fundamental that they can bring about oncogenesis. However, some papers report cases in which expression of the T antigen did not result in tumor development. The study of such transgenic mice is of interest because it allows a better understanding of the malignant transformation of the cell and tumor development. We would like to review some of these models in greater detail.

There were several attempts at obtaining tumors from neurons in transgenic mice using SV40. Thus, transgenic mice were obtained which contained the SV40 Tag regulated by neuron-specific expressing promoters such as the calcitonin gene (CGRP), the proglucagon gene, the phenylethanolamine N-methyltransferase (PNMT) gene (Baetscher *et al.*, 1991). This work has shown that in many cases Tag expression can be traced in the neurons of adult mice. In some experiments such transgenic mice developed brain tumors of an unexplained histogenesis. However, the authors could not be certain that the tumors actually developed from neurons because the tumor cells did not contain neuron-specific molecular markers. In this connection the paper describing the development of neuroblastomas in transgenic mice coexpressing the *c-myc* and transcriptional factor *tax* at a high level (Benvenisty *et al.*, 1992) is of great significance. Special molecular-biological analysis has confirmed that these tumors originate from neurons. Probably some neurons continue to divide up to the time of birth in the mouse, and certain molecular-biological disturbances brought about by products of the *c-myc* and *tax* gene expression in these cells can produce their malignant transformation. It is evident that these disturbances are most likely of a different nature than those induced by SV40 Tag expression.

In transgenic mice bearing the early transcription region SV40 regulated by the promoter intestinal fatty acid binding protein gene (I-FABP), no tumors developed up to 9 months of age (Hauft *et al.*, 1992). Expression of the T antigen gene has been revealed in the epithelial cells in the uppermost portion of small intestine crypts. Study of the proliferative activity and cell cycle of the villus-associated enterocytes has shown that in the I-FABP/Tag transgenic mice, disturbances characteristic of the cells expressing SV40 Tag take place, such as an increase of the proliferative activity and a longer S phase or block at the G2/M boundary. In spite of the high proliferative activity of the cells, there was no increase in the villus area, which probably is connected with "squeezing." Moreover, as shown immunohistochemically by using many cell-specific molecular markers, no disturbance of differentiation of the villus-associated enterocytes took place. The authors think that the absence of intestinal neoplasms in the I-FABP/Tag animals might reflect one of the peculiarities of small intestine tissue organization such as the rapid migration of enterocytes along the crypt-to-villus axis in the process of proliferation and differentiation. Thus, such a fundamental property of the stem cell as anchorage is necessary for a tumor to develop from the initial cell.

2. Tumor Formation in Transgenic Mice Harboring Some Viruses or Virus Genes

Numerous studies of tumor development in transgenic mice carrying various sequences of SV40 have already been discussed (Section III,B,1). These studies have allowed a better understanding of the role of different virus genes in the malignant transformation of cells and progression of turmors. At the same time, the role of tissue-specific promoters was also revealed in the papers cited. We believe it will be useful to discuss here several studies dealing with virus genes or viruses of which introduction *in toto* produced the development of tumors.

a. Human Hepatitis B Virus Human hepatitis B virus (HBV) was used to obtain transgenic mice as a suitable mouse model not only to study hepatitis. It is well known that in the past humans suffering from hepatitis B had a tendency to develop hepatocarcinomas. Therefore an attempt has been made to obtain a model for studying tumor development in transgenic mice carrying HBV or some genes of this virus. At first, only hepatocellular injury in the HBV transgenic mice was reported (Moriyama *et al.*, 1990; Kapitskaya *et al.*, 1990; Shvemberger *et al.*, 1990). In these works the HBV envelope gene was used to obtain transgenic mice, and expression of the surface antigen was detected at the hepatocyte surface. Both the degeneration of hepatocytes and hypoplasia of the lymphoid system in HBsAg transgenic mice revealed in our experiments have suggested a role for changes in the immunological state of the human organism when hepatocarcinoma developed after hepatitis B (Shvemberger *et al.*, 1990).

Moriyama and coauthors (1990) have found envelope-specific antibodies in transgenic mice and also have drawn a conclusion about the immunologically mediated character of the hepatocyte injury. At the same time, this group of researchers reported the development of hepatocellular carcinoma in transgenic mice that expressed large envelope polypeptides under the control of albumin promoter (Chisari *et al.*, 1989; Dunsford *et al.*, 1990) and suggested a scheme of phenotypic changes and molecular pathogenesis of hepatocellular carcinoma in HBV transgenic mice. These data have allowed the authors to discuss the general mechanisms of hepatocarcinogenesis regardless of etiology: viral, toxic, etc. The hepatitis B virus transactivator X protein (HBx) was not tumorigenic in one series of experiments (T. H. Lee *et al.*, 1990), but it did produce tumors in another series in which hepatocarcinomas developed in the HBx transgenic mice (C. M. Kim *et al.*, 1991). The authors suggested a possible direct involvement of the HBx protein acting as a transcriptional transactivator of viral genes into the host gene expression (C. M. Kim, *et al.*, 1991). The authors believe that such disturbances in cell regulatory systems can lead to the development of hepatocellular carcinoma. So the HBx transgenic mice may be used for studying molecular events responsible for the development of liver cancer.

b. Polyomavirus   Polyomavirus is known to be a virus which is able to develop numerous tumors if it is inoculated into newborn mice. The tumorigenic activity of this virus is connected with early region genes coding large T antigen (LT), middle T antigen (MT), and small T antigen (ST). Several studies of oncogenesis in transgenic mice were performed with all of these antigens (Bautch *et al.,* 1987; Bautch, 1989). In the first group of experiments, the LT, MT, and ST genes of polyoma virus were under their own regulatory sequencies. Pituitary adenoma, hypoplasia of the adrenal medulla and changes in male reproductive organs were observed in the lineages of the LT transgenic mice. Hemangiomas were found only in the MT transgenic mice while the transgenic mice carrying the ST gene were free of any tumors. These results have allowed the authors to draw some conclusions about the tumorigenecity of the LT and MT polypeptides of polyoma virus.

Experiments with the polyoma virus antigens in transgenic mice were also made with LT, MT, and ST genes fused with insulin promoter (Bautch, 1989). The β-cell tumors in the pancreas were observed only in the transgenic mice bearing the LT gene.

Of particular interest is the work by Wang and Bautch (1991) in which transgenic mice carrying the entire polyomavirus early region consistently developed vascular and bone tumors and sometimes other tumors of mesenchymal origin, as is observed in mice infected with polyomavirus. The authors have found that the number of transgene expression sites in mice of a given lineage correlates with the severity and latency of the tumor phenotype in these animals. In spite of the transgene expression in various tissues, most of them were refractory to tumorigenesis. However, in another series of experiments, large bilateral testicular tumors were observed in every old transgenic male harboring the polyoma LT (Paquis-Flucklinger *et al.,* 1993). In these experiments, a high level of LT expression was detected in Sertoli and germ cells in transgenic mice during their whole life without any pathological consequences, but the cell lines could be established not only from the tumors but also from the "normal" testis.

The discovery that the polyomavirus MT antigen mediates malignant transformation via an interaction with $pp60^{c-src}$ and that the complementary viral oncogenes are found complexed with other cellular proteins (p53 or/ and p105-RB) which were identified as tumor supressors has allowed Bastin (1992) to propose a new model in viral tumorigenesis investigations.

c. Papillomavirus   Bovine papilloma/virus type 1 (BPV-1) was used to obtain transgenic mice to study the pathogenesis of cutaneous papilloma

(Lacey *et al.*, 1986). The BPV-1 transgenic mice developed skin papillomas beginning at 8 months of life, but they appeared only after scratching or wounding of the skin. It is possible that the cells disintegrated from tissue and losing their intratissue interactions, were more predisposed to malignant transformation. Although the BPV-1 was integrated in all examined tissues, its extrachromosomal fraction was found only in the tumors. So it may be concluded that normal tissues were nonpermissive for the BPV-1.

Chromosomal abnormalities were studied in the benign dermal fibromatosis and malignant fibrosarcomas in one of the lines of the transgenic mice carrying BPV-1 (Lindgren *et al.*, 1989). The fibrosarcomas showed consistent abnormalities of chromosome 8 (trisomy or duplication) and chromosome 14 (monosomy or translocation). These chromosomal abnormalities were not connected directly with integration of BPV-1, because it was mapped at chromosome 15. Since such chromosomal abnormalities were absent in the fibromatosis cells, a secondary origin of these genomic disturbancies accompanying the tumor progression was considered (Howley, 1991).

Similar results were obtained in the study of oncogenesis in transgenic mice harboring human papillomaviruses 16 E6 and 16 E7 (Arbeit *et al.*, 1993; Griep *et al.*, 1993). Development of neuroepithelial carcinomas in such transgenic mice was also observed (Arbeit *et al.*, 1993).

d. Human T-lymphotropic Virus   Human T-lymphotropic virus type 1 (HTLV-1) might have produced human T-cell leukemia. It is natural that the *tat* and *tax* genes of this virus were detected in the leukemiagenesis in transgenic mice (Nerenberg *et al.*, 1987; Green *et al.*, 1989).

Transgenic mice carrying the *tat* gene under the control of its own LTR did not develop any leukemias but did develop mesenchymal tumors, such as spindle cell and oval cell tumors (Nerenberg *et al.*, 1987). Extensive thymic depletion and growth retardation were also observed in several mice.

The other gene of HTLV-1, the *tax* gene, also was tested in transgenic mice (Green *et al.*, 1989). Three founder lines of transgenic mice carrying the *tax* gene under the control of the viral LTR developed neurofibromas and exocrinopathy involving salivary and lacrimal glands. No hemoblastosis has been found in these transgenic animals.

e. Human Immunodeficiency Virus   The *tat* gene of the human immunodeficiency virus (HIV) was introduced into a germ line under its own regulatory sequences to obtain transgenic mice as a model for studying acquired immunodeficiency syndrome (Vogel *et al.*, 1991). The *tax* transgenic mice obtained showed a high incidence of hepatocellular carcinoma after a long latency. The authors have predicted that with prolongation of the life of patients with acquired immunodeficiency syndrome by modern treatment, the occurence of malignancies caused by HIV will be increased.

C. Multiple Tumors

The development of multiple tumors in the genome in transgenic mice with different genetic constructions is of special interest since these results show cancerogenesis in a new light that is a little different from the traditional concept of cancerogenesis as a complex of several consecutive molecular-biological and morphological disturbances in the cell (Table III). It is the models with multiple tumors and hyperplasia in quite different organs and tissues and, not infrequently, in the same mouse that enable us to consider the possibility of pathological changes in different cells and tissues of the same mouse during disturbances associated with the activity of some genes, rather than certain molecular-biological disturbances in a single cell. In addition, the reproducibility of typical pathological changes in transgenic mice suggests the possibility of several different routes to some pathologies.

Owing to the stochastic character of tumor development in most cases in transgenic mice with viral and cellular cancer genes, it is possible to select and estimate the importance of each of the structural and functional changes occurring at different levels. Their unique combinations not only induce the malignant transformation of a single cell, which cannot be detected in the context of the whole organism, although undoubtedly it occurs much more often than tumor development, but they also lead to the ability of a transformed cell to escape the control of multilevel homeostatic mechanisms and to develop into a malignant tumor. In emphasizing this stage of oncogenesis, we would like to stress the necessity of studying malignant growth as a disease of the organism and of searching for models and specific methods for these investigations.

D. Co-Oncogenic Factors of Tumor Development in Transgenic Mice

The creation of more complex models for the study of oncogenesis in transgenic mice with viral oncogens or cellular proto-oncogenes was theoretically based on the now classical idea of a multistep mechanism of carcinogenesis (Hanahan, 1988). If oncogene expression or an insertional mutation caused by integration of a transgene into the genome can be regarded as the first step toward malignant transformation of the cell, it would be quite logical to expand this system by one or several events which would increase the probability of the completion of the malignant cell transforma-

TABLE III

Multiple Hyperplastic and Neoplastic Changes in Transgenic Mice

Regulatory region/gene	Pathology	Organs, tissues	References
Δ gag-myc	Hyperplasia	Thymus, lungs, lymphoid tissue, kidney, adrenal, liver, parathyroid, thyroid and salivary glands	Ermilov et al., 1988; Shvemberger et al., 1993; Shvemberger and Ermilov, 1993
	Benign tumors	Brain, adrenal, pancreas, pituitary, stomach, lymphoid tissue, uterus, liver, soft tissues, ovary, kidney, prostate, thyroid, parathyroid, Harderian, and salivary glands	
	Malignant tumors	Thymus, stomach, lung, skin, liver, uterus, lymphoid tissue, large intestine, soft tissues, thyroid, salivary, Harderian and mammary glands	
H-2Kᵇ/SV40 Tag	Hyperplasia	Brain, thymus, spleen, lymph nodes, pancreas	Reynolds et al., 1988
	Benign tumors	Brain, heart, kidney, skeletal muscle, testis, thyroid gland, thymus	
	Malignant tumors	Brain, pituitary, thyroid gland, adrenal, pancreas, testis	
MT/TGFα or EL/TGFα	Hyperplasia	Liver, pancreas, stomach, small and large intestine, coagulation gland	Jhappan et al., 1990; Sandren et al., 1990
	Malignant tumors	Mammary gland, liver	
Tyr/SV40 ER	Melanosis	Nasal mucosa, endocardium, lung, meninges, peripheral nervous system, dermis, lymph nodes, central nervous system, genital organs and accessory glands, skeletal muscle, oral mucosa, choroid plexus, larynx, pineal, mammary and salivary glands	Bradl et al., 1991; Kline-Szanto et al., 1991
	Malignant tumors	Eye, skin, choroid plexus, cochlear, heart, peripheral nervous system, pineal and salivary glands	

tion and the development of a benign or malignant tumor. These theoretical considerations have been used as the basis for a series of experimental studies in several different directions, with the use of different co-oncogenic factors.

One of the first experiments dealing with the influence of some somatic events on the malignant transformation of cells was performed in transgenic mice carrying bovine papilloma virus (Lacey *et al.*, 1986). It has been shown that skin papillomas develop only in places where the skin has been lacerated and the virion was revealed only in the papilloma cells. Similar results have been obtained in transgenic mice with the *c-H-ras* gene regulated by the suprabasal keratin promoter (Bailleul *et al.*, 1990). Only hyperkeratosis of skin and forestomach, the places of suprabasal keratin expression, was observed in these transgenic mice. However, after biting or scratching the skin between the ears, the transgenic mice developed papillomas. The authors of these studies believe that, because the *H-ras* transgene is expressed in suprabasal cells, the cells which have left the stem cell compartment can be induced to form at least benign tumors *in vivo*.

Tumors were also induced in transgenic mice by weak carcinogens in doses which were not tumorigenic in control animals (Sell *et al.*, 1991). In these experiments, a co-carcinogenic effect was revealed by exposing female transgenic mice to hepatitis B virus secondary to overexpression of the gene for the large envelope polypeptide of hepatitis B or to the hepatocarcinogens, aflatoxin and diethylnitrosamine. In these mice there was more rapid and extensive nodule formation and oval cell proliferation as well as development of adenomas and primary hepatocellular carcinomas than what was seen in the transgenic mice not exposed to carcinogens. The results suggest that the chronic liver damage and repair caused by overexpression of the hepatitis B virus large envelope polypeptide in the hepatocytes of transgenic mice acts synergistically with chemical carcinogens to produce neoplasia in the liver. In our experiments, transgenic mice carrying a gene for a surface antigen regulated by metallothionein promoter showed not only damage and repair but also immunological changes such as hyperplasia of lymphoid tissue (Kapitskaya *et al.*, 1990; Shvemberger *et al.*, 1990). This seems to indicate an influence also of the third, immunological, factor on hepatocarcinogenesis in transgenic mice carrying hepatitis B virus.

A higher probability of tumor development in transgenic mice induced by a change in hormonal status has been reported by many authors. Data from Leder and co-workers (Leder *et al.*, 1986) are particularly convincing. The genetic construction used by these authors consisted of *c-myc* regulated by LTR MMTV. A glucocorticoid-sensitive promoter and enhancer was added directly to the 5' end of the *myc* gene. In the transgenic mice obtained, the expression of MMTV/*c-myc* was detected in some tissues, although no malignant transformation was observed. However, after treatment with dexametasone, dramatic development of tumors was found in these mice. There were tumors of mammary glands, tumors of testicle Sertoli cells, malignant lymphomas of three types, mastocytomas. Neither in the tissues in which the tumors were detected, nor in any other tissues were any disturbances of cell proliferation or normal development found, although the expression of the oncogene did occur and its level was rather high. The results of these experiments have led the authors to conclude that expression of *c-myc* was necessary but not sufficient for tumor development. They believe that an additional action of some dominant oncogenes is needed for the increased role of amplified *c-myc* (Leder *et al.*, 1986).

The influence of hormone status on tumor development in transgenic mice may be traced in the mice that develop tumors of mammary glands (Leder *et al.*, 1986; Andres *et al.*, 1987). Such tumors appear as a rule in females and only after one or several pregnancies. Changes in hormone status must be considered as an additional event in tumorigenesis, but we do not know what the event or rather a chain of events following certain hormonal changes in the organism is.

Of special interest are the results of experiments with interleukin 6 transgenic mice (IL-6) (Suematsu *et al.*, 1992). The transgenic mice C57Bl/6 carrying human cDNA under regulation of H-2L^d (murine major histocompatibility complex class promoter) developed massive polyclonal plasmacytosis but not plasmacytomas. However, the introduction of a BALB/c genetic background into the IL-6 transgenic mice by crossing induced monoclonal transplantable plasmacytomas with the chromosomal translocation (12;15). The data indicate that the developed plasmacytomas have the *c-myc*-induced rearrangements specific for the BALB/c mice plasmacytoma translocations (12;15). Hence the development of the plasmacytomas could be considered the result of a synergistic action of overexpression of the IL-6 gene and certain genetic factors of the BALB/c strain.

Studies on disturbed function of two oncogenes simultaneously have become an independent line of research in carcinogenesis with the use of double transgenic mice. Thus, crosses between two lines of transgenic mice with *c-myc* and *v-H-ras* genes linked to MMTV regulatory sequences yielded hybrids of high oncogenic level (Sinn *et al.*, 1987). They developed mammary adenocarcinomas, malignant lymphomas, hyperplasias and tumors of Harderian glands, salivary gland adenocarcinomas, and tumors of testicular vesicles. The number of tumors in hybrid mice was higher than the total of tumors developed in MMTV/*myc* and MMTV/*v-H-ras* transgenic mice. Since these tumors arose stochastically and apparently were monoclonal in origin, additional somatic events have been suggested to be necessary for their full malignant progression, even in the presence of activated *v-H-ras* and *c-myc* transgenes (Sinn *et al.*, 1987). Multifocal mammary tumors also developed in cotransgenic mice with *H-ras* and *c-myc* genes regulated by the WAP (whey acidic protein) promoter gene (Andres *et al.*, 1988), whereas in lines of mice with only one of these genes (*H-ras* or *c-myc*), even with transgene expression, no tumor was found.

Coexpression of both oncogenes led to a severely obstructed glandular organization, and preneoplastic lesions were observed with high frequency in postlactational glands; however, palpable tumors occurred focally and only after several pregnancies. Andres and his coauthors (1988) believe neoplastic diseases to result from the escape of the tumor cells from this complex growth control.

Double transgenic mouse albumin/*c-myc* and MT/TGFα were examined for tumorigenesis (Murakami *et al.*, 1993). Coexpression of *c-myc* and transforming growth factor α as transgenes in the mouse liver resulted in a tremendous acceleration of neoplastic development in this organ compared with the expression of either of these transgenes alone. A similar effect was observed in other experiments with double transgenic mice. Thus, an increase of liver neoplasia was found in double transgenic mice MMTV/*c-myc* MMTV/*c-ras* (Sandgren *et al.*, 1989). The perturbed development of B lymphocytes in bone marrow was revealed in *c-fos/v-jun* double transgenic mice (Fujita *et al.*, 1993).

The analysis of oncogene cooperation in the tumor development and progression is a promising way to elucidate the role of oncogenes in oncogenesis. Transgenic mice carrying several oncogenes are very useful for effective studies of the problem.

The involvement of *c-myc* and *c-ras* in oncogenesis in transgenic mice harboring certain oncogenes seems to occur quite often (Alexander *et al.*, 1989; van Lohuizen *et al.*, 1989; Haupt *et al.*, 1992). Transgenic mice carrying oncogene *pim*-1, when given upstream immunoglobulin enchancer and downstream LTR of murine leukemia virus (MuLV), developed T lymphomas only in 5–10% of the mice up to 7 months (van Lohuizen *et al.*, 1989). However, after infection of *pim*-1 transgenic mice with MuLV, lymphomas developed more rapidly—in 7–8 weeks. It has been shown that the provirus insertion in T lymphomas activated either *c-myc* or *N-myc*. These data suggest the existence of a strong cooperation between *pim*-1 and *myc* in lymphomagenesis. However, as the authors believe, the coexpression of two genes is also not sufficient for lymphoma development because expression of both genes was found in all lymphocytes, although lymphomas were of monoclonal origin.

Similar results were obtained with retroviral infection in Eμ/*N-ras* transgenic mice (Haupt *et al.*, 1992). Molony murine leukemia virus was used as an insertional mutagen to produce a dramatic increase of T-cell lymphomas in these transgenic mice. It was found that 68% of the tumors bore a proviral insert 5′ to the *c-myc* gene while 13% had an insert within the 3′ untranslated region of the *N-myc* gene. The authors have concluded that activation of the *myc* gene appears to be the dominant pathway to tumorigenesis by insertional mutagenesis in lymphoid cells expressing a mutant *ras* gene. However, since many of the tumors were not transplantable, even the partnership of *myc* and *ras* was not sufficient for the development of full limphoid malignancy (Haupt *et al.*, 1992). Moreover, 32% of the tumors had no insertion 5' to the *c-myc* and 87% of them had no insertion within the 3' region of the *N-myc*. These data suggest that other, still unknown events play an essential role in the malignant transformation of cells.

A virus infection was also used to study cooperating genetic events in the development of mammary carcinomas in Wnt-1 transgenic mice (Shackleford *et al.*, 1993). Mouse mammary tumor virus infection accelerated mammary carcinogenesis. The alteration of several genes has been shown in tumors containing provirus. As a result of provirus insertion, *int*-2 was disturbed in 39% and *hst* in 3% of tumors. Alteration in expression of both genes was found in 3% of the tumors. Both genes are fibroblast growth factors. It is to be expected that further studies of the problem will discover more and more genes involved in the process of malignant transformation and tumor progression. Among these genes there will be proto-oncogenes, supressors, growth factors, and other genes unknown so far and usually not connected with oncogenesis.

IV. Specific Peculiarities of Tumors in Transgenic Mice

Experiments on tumor progression in transgenic mice have demonstrated numerous neoplastic pathological changes in different tissues and organs. It is important to compare the basic characteristics of these hyperplasias and neoplasias with similar changes that are observed in humans and laboratory animals as a result of various carcinogenic effects. Here we discuss both morphological and karyological peculiarities of tumors in transgenic mice and analyze a probable connection between tumor development and embryonal abnormalities.

A. Morphology of Tumors

Most of the researchers who studied oncogenesis in different strains of transgenic mice point out similarities in tumor development and in morphological and biological features of the tumors (Stewart *et al.*, 1984; Palmiter *et al.*, 1985; Messing *et al.*, 1985; Hinrichs *et al.*, 1987; Ermilov and Shvemberger, 1988; Ermilov *et al.*, 1988). If tumors develop in parents, they are observed in their offspring for several generations, although the localization and character of the neoplasia may differ (Hanahan, 1985; Ceci *et al.,* 1991; Shvemberger and Ermilov, 1994). In transgenic mice, tumors develop very early, often in the first months of their life, which is quite different from spontaneous tumors of inbred animals and their hybrids. Most commonly, the tumors grow extremely slowly, which can be seen in tumors of outer localization: carcinomas of skin and of skin derivatives, sarcomas of soft tissues. It seems that during the growth period these tumors do not exert a strong destructive effect on the organism until some critical moment; after this moment, the condition of the mouse shows a dramatic deterioration that is usually associated with the development of multiple tumors or their dissemination.

Tumors in transgenic mice usually develop only as solid nodes, but sometimes multiple tumors of the same origin can be observed. Disseminating tumors also occur, although rather uncommonly and often quite unexpectedly. Thus, metastatic hibernomas (tumors from brown adipose tissue) developed in mice carrying the α-amylase-SV40 T antigen hybrid gene (Fox *et al.,* 1989). Although α-amylase is a liver promoter, the expression of the transgene was revealed not only in the liver, parotid, and pancreas, but also in white and brown adipose tissue. The high frequency of metastases in the liver, lungs, spleen, heart, and adrenals of these mice allows disseminated malignancies to be studied. Stomach carcinomas developed in transgenic mice with the murine pancreatic amylase 2.2 SV40 T antigen fusion gene also had definite features of malignancy (Ceci *et al.,* 1991). Extensive invasion of the mucosa and submucosa was characteristic of these tumors and metastatic foci were observed in the liver, lung, and lymph nodes.

Ocular and cutaneous malignant melanomas arose in new inbred lines of transgenic mice with a transgene composed of the tyrosinase promoter, expressed in pigment cells, and the SV40 early region transforming sequences (Bradl *et al.,* 1991). The eye melanomas developed aggressively, were highly invasive, and metastasized to local and distant sites. Perhaps it was not by chance that all these metastatic tumors occurred in transgenic mice with SV40 sequences in their genome.

Several authors point out similar characteristics of tumor morphology in transgenic mice, carriers of different cellular or viral genes: high cytotypical and histotypical differentiation of the tumor cells, lack of foci of necrosis even at terminal stages, intensive infiltration with lymphocytes and macrophages (Hinrichs *et al.,* 1987; Ermilov *et al.,* 1988). We know that successful tumor transplantation from transgenic mice to hybrids or to animals of parental strains is very difficult. Almost all successful transplantations have been made in nude mice. Successful subcutaneous transplantation of transgenic pituitary tumors to nontransgenic immunocomplement mice may be regarded as an exception to this rule (Helseth *et al.,* 1992). At the same time, the tumor cells developed in transgenic mice can be used to obtain cell lines. Several attempts at establishing cell lines from tumors of rare localizations also were successful. Thus, cell lines were produced from malignant neurofibroblasts (Hinrichs *et al.*, 1987), heart rhabdomyoblasts and osteoblasts (Benringer *et al.*, 1988), brown adipocytes (Ross *et al.*, 1992), Sertoli cells (Paquis-Flucklinger *et al.*, 1993), rhabdomyoblasts (Teitz *et al.*, 1993), and so on.

One of the peculiarities of oncogenesis in transgenic mice is their ability to also develop hyperplasia and dysplasia. In these transgenic mice, such lesions may occur both in the organs in which the tumors arise and in the organs in which the tumors do not arise.

Most reports about the hyperplastic lesions observed deal with transgenic mice harboring the virus sequences in their genome (Tsukamoto *et al.*, 1988; Held *et al.*, 1989; Vogel *et al.*, 1991; Helseth *et al.*, 1992). As a rule, hyperplastic foci were found in the liver, pancreas, mammary, and salivary glands. Less common was localization of hyperplasia in the pituitary gland in transgenic mice carrying a polyoma early region promoter linked to the DNA encoding polyoma large T antigen (Helseth *et al.*, 1992). The development of foci of microadenomas and large adenomas (up to 5 mm) depended on the age of the animals.

A wide spectrum of hyperplasia and dysplasia lesions was seen in transgenic mice carrying transforming growth factor α (Sandgren *et al.*, 1990; Matsui *et al.*, 1990). Some differences were reported in these studies, probably owing to different promoters used in the genetic constructions. MT-directed expression of TGFα (Sandgren *et al.*, 1990) in transgenic mice induced a spectrum of changes in the growth and differentiation of certain adult tissues. In transgenic mice harboring TGFα linked to regulatory sequences of MMTV, a range of morphological abnormalities, including lobular hyperplasia, cystic hyperplasia, adenoma, and adenocarcinoma developed only in the mammary glands of transgenic females. In transgenic mice carrying Δ*gag-myc*, diffuse and focal hyperplasia was found in such organs as the adrenals, salivary glands, and kidneys (Table III; Shvemberger and Ermilov, 1994).

One of the intriguing questions in oncogenesis in transgenic mice is whether the frequency and malignancy of the developed tumors depend on a copy number of transgene. Only a few authors answer this question. The development of malignant melanoma in transgenic mice with an integrated recombinant gene composed of the tyrosinase promoter expressed in pigment cells and the SV40 early region transforming sequences was dependent on the copy number (Bradl *et al.*, 1991). In particular, the earliest formation of melanomas was associated with higher copy numbers of the transgene; mice of different single-copy lines varied greatly in the age of onset and frequency of eye melanomas. The same conclusion has been drawn from a study of oncogenesis in transgenic mice harboring the polyoma virus early region gene (Wang and Bautch, 1991). Both vascular and bone tumors occurred in these animals. The number of transgene expression sites in the mice of a given lineage was believed to correlate with the severity and latency of the tumor phenotype in these animals.

Teitz and his coauthors (1993) studied rhabdomyosarcomas arising in transgenic mice harboring the β-globin locus control region fused with SV40 large T antigen gene. They found the phenotype to depend on the copy number of the transgene, as seen here:

copy number of transgene	phenotype of animals
1–2	normal mice
3–7	rhabdomyosarcoma hyperplasia foci and tumors in pancreas
7	hyperglycemia, growth delay
10	growth delay, death at 2–4 weeks

Moreover, in mice with high copy numbers of the transgene, the rhabdomyosarcomas appeared in more than one site.

All the biological features of oncogenesis mentioned have shown that in transgenic mice with certain genetic peculiarities, this process is different from the process of tumor development in laboratory animals induced by various carcinogenic factors. It is evident that oncogenesis in transgenic mice is similar to the development of hyperplastic and neoplastic formations in man. This allows these mice to be used as models for investigating hereditability, prophylaxis, diagnosis, and treatment of similar tumors in man.

B. Rare Tumors

Tumors of rare localization often occur in transgenic mice quite unexpectedly, without any connection with introduced genes and regulatory sequences. Thus, apart from bilateral osteosarcomas of the petrous part of the temporary bone and rhabdomyosarcomas of the right cardiac atrium, there have been multiply endocrine neoplasias (Theuring et al., 1990; Rindl et al., 1990; Schulz et al., 1992a,b; Shvemberger and Ermilov, 1994) and tumors of the brain and peripheral nervous system (Palmiter et al., 1985; Iwamoto et al., 1993; Shvemberger and Ermilov, 1994).

C. Chromosomal Changes

1. Karyology of Tumors

Only a few papers deal with tumor karyology in transgenic mice. The disturbances of karyotype were accurately described in the pancreatic cancer developed in transgenic mice carrying the El/SV40 ER (Levine *et al.*, 1991). The authors examined karyological changes during the progression from hyperplasia to cancer. This progression was followed by a transformation from diploid to tetraploid and then to aneuploid karyotype associated with biological progression. Studies on plasmacytomas in Eμ/c-*myc* and Eμ/N-*myc* transgenic mice after induction by pristane oil and/or A-MULV did not reveal usual plasmacytoma-associated translocations that juxtapose c-*myc* to heavy- or light-chain immunoglobulin loci in the pristane oil-induced plasmacytomas in nontransgenic mice (Wang *et al.*, 1992).

2. Recombinations in the Transgenic Mouse Genome

Up to now, data about intra-and extratransgene recombinations in transgenic mice genome have been rather scanty. This is probably due, first of all, to this event being quite rare. Second, to reveal such events as recombination, a special arrangement of experiments is required. Therefore it is not suprising that reports about recombination in the transgenic mouse genome have begun to appear only in the past few years. Meanwhile, information about recombination in transgenic mice is very important: it not only characterizes the transgene state in the genome but also allows a study of the possible role of recombination in normal development and oncogenesis (Shvemberger *et al.*, 1995).

Tumor development is known to be accompanied by numerous chromosomal events such as point chromosomal and genomic mutations. So the study of recombination in transgenic mice both characterizes this model and elucidates the role of recombination in different periods of malignant growth, from cell transformation to metastatic malignant tumors.

One of the first studies on recombination in transgenic mice used the recombination signal sequences of the immunoglobulin gene (Kawaichi *et al.*, 1991). The construct pLTR/100 for production of transgenic mice was made in such a manner that human interleukin 2 receptor light (IL-2R L) chain cDNA was placed in the inverted orientation relative to the upstream SV40 promoter and was flanked by the IG gene recombination signal sequences. Gene expression in such a construct would be possible only if the substrate gene IL-2R L in the transgene could invert in normal position, i.e., the recombination took place. Transgenic mice bearing the recombination substrate RNA similar to pLTR/100 expressed the human IL-2L chain in the spleen, thymus, and bone marrow. In other tissues examined, no gene expression was detected.

The special construct containing the bacterial β-galactosidase gene (lactZ) was used to study DNA recombination in the transgenic mouse brain (Matsuoka *et al.*, 1991). Intratransgene recombination is required for expression of the substrate gene. Recombination-positive areas of the adult brain were revealed by color reaction with a substrate of β-galactosidase. Evidence was provided that the staining in the transgenic mouse brain was indeed due to the expression of exogenous lactZ. The staining pattern was persistently inherited from F_0 parent to F_1 and F_2 transgenic mice. The authors predict that rearranging genes in nerve cells would provide a new insight into the role of DNA recombination in the development of the central nervous system. However, Abeliovich and coauthors (1992) believe that it is premature to decide whether developmentally meaningful somatic recombination occurs in the brain. Our data about intratransgene recombination of the neomycin-transferase gene have suggested the possibility of intratransgene recombination in different organs and tissues (Shvemberger *et al.*, 1995).

Cre (the bacteriophage P-1 *Cre* recombinase) was used in transgenic mice to establish whether *Cre* could effectively mediate chromosomal DNA recombination (Orban *et al.*, 1992). Transgenic *Cre*-mediated recombination turned out to be highly efficient. The *Cre* recombinase activity could be tissue-specific and heritable in transgenic mice. The fact that *Cre* could function at different chromosomal sites can be used to study the molecular-biological mechanisms of recombination *in vitro*.

The development of hepatic tumors was detected in transgenic mice carrying a plasmid with a urokinase-type plasminogen activator gene (uPA) (Sandgren *et al.*, 1992). It is interesting that in these experiments the tumors developed only from the hepatocytes that did not express uPA owing to loss of transgene DNA. The uPA is hepatotoxic, and nontransgenic hepatocytes infrequently clonally repopulate the entire liver within 3 postnatal months. All these mice developed hepatic tumors starting at 8 months of age. Analysis of the transgene locus reveals that the tumors arise only from the subclass of transgene-deficient cells in which not only the transgene but, possibly, a significant amount of flanking DNA is a primary carcinogenic determinant in this model of hepatocarcinogenesis. These data are particularly important for the molecular-biological theory of oncogenesis because they indicate an essential role for nonspecific destruction in the chromosomal structure which could be followed by disturbances in intracellular regulatory networks.

D. Tumor Development and Embryonal Abnormalities

Both the occurrence of tumors and the abnormal development of some tissues or organs in transgenic mice represent special phenomena that are rather difficult to interpret. Nevertheless, analysis of these data may be useful in understanding peculiarities of malignant growth not only in transgenic mice but also in spontaneous malignant tumors. Moreover, the results of all these experiments may be helpful for a general theory of oncogenesis.

Abnormal development of the pancreas simultaneously with multifocal well-differentiated hepatocellular carcinomas was observed in transgenic mice harboring the human TGFα gene (Jhappan *et al.*, 1990). In the mammary gland, morphogenetic penetration of epithelial duct cells into the stromal fat pad is defective; the pancreas shows progressive interstitial fibrosis and acinoductular metaplasia. These facts allowed the authors to deduce an important role for TGFα not only in neoplastic transformation but also in organogenesis.

Transgenic mice carrying the homeobox fusion gene E2A-PBX1 under control of the immunoglobulin heavy chain enhancer developed leukemia; they also developed both proliferation and apoptosis in lymphoid cells (Dedera *et al.*, 1993). The authors suggest an association between the nuclear oncogene-induced cell cycle progression and programmed cell death.

Abnormalities of the lens and lens tumors were examined using two oncogenes: one was a virus oncogene SV40 T antigene, the other the proto-oncogene *c-mos* (Westphal, 1988). Both of them were controlled by the regulatory elements of the lens crystalline genes. In the first set of the experiments, lens tumors (phacomas) were produced; in the second one, the expression of the *c-mos* resulted only in specific defects of the lens fiber differentiation. No tumors were observed even in the experiments in which transgenic mice carried *c-mos* regulated by LTR MSV and activated previously in the 3T3 cells. The most prominent phenotype in these mice was cataracts, which resulted from a benign lens defect developed after birth. So it can be concluded that neither tissue-specific nor virus regulatory sequences nor active oncogenes were sufficient for oncogenesis in certain tissues. Moreover, some defective tissues do not necessarily undergo malignant transformation. Thus, the data indicate once more a specific interrelation between the product of the oncogene expression and structural-functional peculiarities of each cell studied for oncogenesis.

Considering what was said earlier, the experiment with transgenic mice bearing a complex gene expressing a fusion protein—retinoic acid receptor-β-galactosidase (RAR-LacZ) (Balkan *et al.*, 1992)—is of special interest. The fusion protein is able to transactivate expression in the absence of retinoic acid. The authors believe that the development of cataracts and microphtalmia is characteristic of retinoid-induced teratogenesis and indicates similiarities of the transgenic phenotype with retinoid teratogenesis owing to expression of the constitutively active RAR-LacZ fusion protein in retinoid-sensitive tissues.

It is important that extensive sexual abnormalities in transgenic mice chronically expressing the human Müllerian inhibiting substance under the control of the mouse metallothionein 1 (MT/MIS) did not lead to tumor development in these mice in spite of pronounced disturbances in their hormonal status (Behringer *et al.*, 1990). Nor was malignant transformation observed in a number of experiments with transgenic mice showing numerous deviations from development (Mahon *et al.*, 1988; Powell and Rogers, 1990; Tepper *et al.*, 1990; Varki *et al.*, 1991).

In our experiments, we performed a phenotypical study of 500 transgenic mice carrying the Δ*gag-myc* gene (unpublished data). As mentioned earlier (Table III), a number of mice developed tumors in different tissues. Moreover, numerous abnormalities of prenatal and postnatal development were found in these transgenic mice. The prenatal changes were megaloencephaly and cranium deformations. In postnatal development, there was gigantism or growth delay as well as hair delay and loci of skin depigmentation. The pronounced disturbances such as, for example, gigantism or cranial deformation in the fetus led to the death of fetuses or neonatal mice. However, such abnormalities as hair or growth delay were observed only in the perinatal period and then all of them disappeared by puberty. It is interesting that these transgenic mice did not develop any tumors.

V. Oncogenesis as a Multistep Process: Facts and Opinions

The results of numerous studies on oncogenesis in transgenic mice stimulated discussion of possible molecular-biological mechanisms of malignant transformation and progression of tumors, and, particularly, of the necessity of two, three, or more events for tumor development.

Based on a detailed analysis of hereditary tumors, such as retinoblastomas, Wilms' tumors, neuroblastomas and others, Knudson (1973, 1991) suggested the necessity of two genetic events for their development. His theory of a two-step carcinogenesis has been rapidly accepted, and the number of its supporters soon multiplied. This was the time of the discovery and intensive study of cellular and viral oncogenes, and the possibility of deciphering the molecular-biological mechanisms of malignant cell transformation and the progression of tumor growth seemed to be quite real. The concept of multistep carcinogenesis is in good agreement with the convincing evidence in favor of a monoclonal origin of tumors both in humans and laboratory animals (Tanooka, 1988; Nowell, 1989; 1991; Wainscoat and Fey, 1990; Stein, 1991, Mok *et al.*, 1992; Fialkow, 1976). However, data on a great number of the known proto-oncogenes involved in oncogenesis made the idea of molecular-biological mechanisms of the malignant cell transformation and tumor progression more complicated and it required a model of multistep carcinogenesis (Freeman *et al.*, 1990; Clark, 1991; Nowell, 1991; Sandberg, 1992; Sugimura, 1992; Wiedemann and Morgan, 1992). According to this concept, the involvement of each proto-oncogene implies a new stage in oncogenesis (Bishop, 1991; Weinberg, 1992). The theory of a multistep carcinogenesis has become much more complicated with the discovery of the suppressor genes, when the number of possible genetic disturbances during a malignant cell transformation has been found to be much higher; and it has become possible to regard each step of malignant growth either as involving another proto-oncogene or as switching off of a suppressor gene (Klein, 1987; Chigira, 1991; Lehman *et al.*, 1991; Wynford-Thomas, 1991; Hollingworth and Lee, 1991; Clemens, 1992; Patterson, 1992; Marx, 1993).

The first results obtained on transgenic mice with viral and cellular oncogenes have clarified the key problems of malignant growth mechanisms: the clonal nature of tumorigenesis and the multistep mechanism of carcinogenesis.

Monoclonality or multiclonality of tumor development had arisen as a problem once again in connection with numerous studies on tumor production in transgenic mice. Whereas in humans the monoclonal origin of tumors can be explained quite sufficiently as the result of a single event in a single cell, in transgenic mice, the premalignant event, such as expression of an oncogene in particular, was observed in a number of cells in different tissues. This fact allowed the consideration of a multiclonal origin for tumors in transgenic mice. Nevertheless, from the very beginning of studies on tumor development in transgenic mice, many researchers accumulated evidence on the monoclonal origin of tumors in transgenic mice (Adams *et al.*, 1985; Brinster and Palmiter, 1986). There were, as a rule, single tumors in a tissue or organ that led to the logical conclusion that the majority of tumors in transgenic mice, as in humans, were monoclonal. Apart from these logical constructions, new data were obtained before long about unique chromosomal translocations in tumors in transgenic mice (Adams *et al.*, 1985; Hanahan, 1988). These data also confirmed the possibility of the monoclonality of tumors even in transgenic mice. The data on the polyclonality of the tumors developed in *c-neu* transgenic mice (Muller *et al.*, 1988) did not refute the previous conclusion about the monoclonality of tumors in transgenic mice in most cases.

Discussions of a number of obligatory states in oncogenesis were based on additional data obtained in studying this process in transgenic mice. Both logical considerations and experimental data were used to reach the conclusion of multistep development of tumors in transgenic mice (Hanahan, 1988; Ando *et al.*, 1992; Guy *et al.*, 1992; Larue *et al.*, 1993). Indeed, in most of the above-cited studies of oncogenesis in transgenic mice, it was shown that the transgene was expressed in many cells and in many tissues, whereas tumors occurred in particular tissues, and originated from single cells. The hypothesis of multistep carcinogenesis was also supported by the results of chromosome analysis carried out on transgenic mice when the secondary involvement of some chromosomes in malignant cell transformation and tumor progression was detected (Lindgren *et al.*, 1989).

Two rather similar studies performed by the same group of researchers (Stewart *et al.*, 1984; Muller *et al.*, 1988) are of great interest with respect to the concept of multistep carcinogenesis. In one of them, transgenic mice carrying four variants of the MMTV/*c-myc* construct developed tumors of mammary glands in females after the second or third pregnancy, the tumors often being detected during pregnancies (Stewart *et al.*, 1984). These data allowed the authors to support the idea of a two-step mechanism for carcinogenesis which, apart from the proto-oncogene expression, includes another event of a mutational or hormonal character.

However, after their study of tumor formation in transgenic mice bearing the proto-oncogene *c-neu* under the control of promotor MMTV, the authors reconsidered the idea of the necessity of several steps in malignant cell transformation (Muller *et al.*, 1988). Quite unexpectedly, all the transgenic mice, both males and females, developed adenocarcinomas in all mammary glands, which contained the tumor tissue only, with no hyperplastic foci. Foci of hyperplasia were detected only in the salivary glands and epididymis, in which no tumors were observed. Based on these results, the authors had to reconsider their conclusions about a multistep carcinogenesis, which were drawn from the earlier experimental data on *c-myc* and *c-ras* transgenic mice, and postulated that it is the combination of oncogene activity and the peculiarities of the tissue that is a major determinant of malignant transformation and progression (Muller *et al.*, 1988).

The abundant oncological data used to put forward the theory of multistep carcinogenesis suggest at the same time in alternative mechanism for carcinogenesis. At present the stochastic character of malignant cell transformation is accepted by many investigators of malignant growth, including some supporters of the theory of multistep carcinogenesis (Shvemberger and Ermilov, 1994). The concept of oncogenesis as a result of one stochastic event has found support in recent studies on oncogenesis in transgenic mice. First of all, it has become evident that the presence of any cellular proto-oncogene or viral oncogene in the genome of the transgenic mouse increases the probability of tumor development. As a rule, tumors develop in tissues expressing a transgene, but even in such tissues this probability remains rather low: tumors usually develop in one or, more rarely, in two tissues, whereas in the rest of organs and tissues, even in the case of transgene expression, either some hyperplastic changes occur or there is no neoplastic pathology at all. At the cellular level, the probability of malignant cell transformation also remains extremely low, if the enormous number of cells in any tissue and the monoclonal development of tumors is taken into account. Thus, it seems quite evident that the significance of disturbances contributed by foreign genes to the fate of the cell is different for different cells of the same tissue as well as for different tissues.

Since the probability of tumor development in transgenic mice varies in different organs and tissues under the same conditions, it is logical to propose a model in which such unique events would appear with a high probability at the cellular level. Indeed, such a model was found by Muller and coauthors (1988). Thus, the results of their work allow the conclusion that a situation in which a normal cell is transformed into a malignant one can be interpreted in such a way that a unique event can be transformed into a regular one. Such an approach views oncogenesis, not as a series of consecutive steps, but as a transformation of the cell as a biological system *in toto* from one qualitative state to another.

VI. Conclusion

The results of numerous studies on oncogenesis performed in transgenic mice have provided important evidence for a hypothesis interpreting mechanisms of malignant cell transformation. In addition, it is the specific features of oncogenesis in transgenic mice that drew attention to the role of multilevel homeostatic mechanisms in this process.

The study of oncogenesis in transgenic mice has shown that tumors develop in those cells in which foreign sequences of oncogenes are expressed. However, even the oncogene expression in the cell does not imply that the cell is on the way to malignant transformation. There is every reason to believe that in most cases the disturbances of regulatory systems of cell metabolism are caused by the involvement of a product of the mutant oncogene; a disturbance of the same kind can occur in cases when a proto-oncogene either overexpresses protein or expresses it at the cell activity stages at which such expression normally is absent. Disturbances of gene activity, nevertheless, can fall within the compensatory possibilities of the cell and the cell, in spite of a disturbed pattern of metabolism, may remain structurally and functionally normal. For this reason, the malignant transformation of the cell still remains a unique event that is not in contradiction with a much higher probability of tumor development at the tissue and organism levels, as has been observed in most of the studies on oncogenesis in transgenic mice.

The situation changes when the transgene expression induces disturbances in cell metabolism which cannot be compensated by the systems responsible for intracellular homeostasis, and the cell loses its normal phenotype. If the changes appear to be incompatible with the cell's life, the cell dies. In other cases, transgene expression in the cell may lead to an increase in cell proliferative activity, formation of diffuse and focal hyperplasia, and benign or malignant tumors. At present, it is well known that products of the proto-oncogene expression perform key functions in cell metabolic networks. Therefore, abnormal oncoproteins may lead to different disturbances in intracellular regulatory systems. Since the resistance of these regulatory networks is different for each cell at each stage of its activity, the consequences of these disturbances also may be different.

Acknowledgments

We are grateful to Dr. L. Z. Pevzner for advice and assistance in preparing the manuscript. Our work was supported in part by the Russian State Program "Frontiers in Genetics" and "National Priorities in Medicine and Health Service: the Tumor Disease."

References

Abeliovich, A., Gerber, D., Katsuki, M., Craybiel, A. M., and Tonegawa, S. (1992). On somatic recombination in the central nervous system of transgenic mice. *Science* **257**, 404–408.

Adams, J. M., Harris, A. W., Pinkert, C. A., Corcoran, L. M., Alexander, W. S., Cory, S., Palmiter, R. D., and Brinster R. L. (1985). The *c-myc* oncogene driven by immunoglobulin enhancers induces lymphoid malygnancy in transgenic mice. *Nature* (*London*) **318**, 533–538.

Adams, T. E., Alpert, S., and Hanahan, D. (1987). Non-tolerance and autoantibodies to a transgenic self antigen espressed in pancreatic β cells. *Nature* (*London*) **325**, 223–228.

Alexander, W. S., Bernard, O., Langdon, W. Y., Harris, A. W., Adams, J. M., and Cory, S. (1989). Oncogene cooperation and B-lymphoid tumorigenesis in Eμ-*myc* transgenic mice. *Hematol. Bluttransfus.* **32**, 423–427.

Al-Shawi, R., Kinnaird, J., Burke, J., and Bishop, J. O. (1990). Expression of a foreign gene in a line of transgenic mice is modulated by a chromosomal position effect. *Mol. Cell. Biol.* **10**, 1192–1198.

Al-Ubaidi, M. R., Font, R. L., Quiambao, A. B., Keener, M. J., Liou, G. I., Overbeek, P. A., and Baehr, W. (1992a). Bilateral retinal and brain tumors in transgenic mice expressing simian virus 40 large T antigen under control of the human interphotoreceptor retinoid-binding protein promoter. *J. Cell Biol.* **119**, 1681–1687.

Al-Ubaidi, M. R., Nollyfield, J. G., Overbeec, P. A., and Baehr, W. (1992). Photoreceptor degeneration induced by the expression of simian virus 40 large tumor antigen in retina of transgenic mice. *Proc. Natl. Acad. Sci. U. S. A.* **89**, 1194–1198.

Ando, K., Saitoh, A., Hino, O., Takahashi, R., Rimura, M., and Katsuki, M. (1992). Chemically induced forestomach papillomas intransgenic mice carry mutant human *c-Ha-ras* transgenes. *Cancer Res.* **52**, 978–982.

Andres, A. C., Schoenenberger, C. A., Groner, B., Henninghausen, L., LeMeur, M., and Gerlinger, P. (1987). *Ha-ras* oncogene expression directed by a milk protein gene promoter tissue specificity, hormonal regulation, and tumor induction in transgenic mice. *Proc. Natl. Acad. Sci. U. S. A.* **84**, 1299–1303.

Andres, A. C., van der Valk, M. A., Schoenenberger, C. A., Flückiger, F., Le Meur, M., Gerlinger, P., amd Groner B. (1988). *Ha-ras* and *c-myc* oncogene expression interferes with morphological and functional differentiation of mammary epithelial cells in single and double transgenic mice. *Gene Dev.* **2**, 1486–1495.

Arbeit, J. M., Munger, K., Howley, P. M., and Hanahan, D. (1993). Neuroepithelial carcinomas in mice transgenic with human papillomavirus type 16E6/E7 ORFs. *Am. J. Pathol.* **142**, 1187–1197.

Babinet, C., Morello, D., and Renard, J. P. (1989). Transgenic mice. *Genome* **31**, 938–949.

Baetscher, M., Schmidt, E., Shimizu, A., Leder, P., and Fishman, M. C. (1991). SV40 T antigen transforms calcitonin cells of the thyroid but not CGRP-containing neurons in transgenic mice. *Oncogene* **6**, 1133–1138.

Bailleul, B., Surani, M. A., White, S., Barton, S. C., Brown, K., Blessing, M., Jorcano, J., and Balmain, A. (1990). Skin hyperkeratosis and papilloma formation in transgenic mice expressing a *ras* oncogene from a suprabasal keratin promoter. *Cell* **62**, 697–708.

Balkan, W., Klintworth, G. K., Bock, C. B., and Linney, E. (1992). Transgenic mice expressing a constitutively active retinoic acid receptor in the lens exhibit ocular defects. *Dev. Biol.* **151**, 622–625.

Bastin, M. (1992). Polyomavirus-mediated transformation: a model of multistep carcinogenesis. *Cancer Invest.* **10**, 85–92.

Bautch, V. L. (1989). Effects of polyoma virus oncogenes in transgenic mice. *Mol. Biol. Med.* **6**, 309–317.

Bautch, V. L., Toda, S., Hassell, J. A., and Hanahan, D. (1987). Endothelial cell tumors develop in transgenic mice carrying polyoma virus middle T oncogene. *Cell* **51**, 529–538.

Beddington, R. (1992). Transgenic mutagenesis in the mouse. *Trends Genet.* **8**, 345–347.

Benriquez, R. R., Peschon, J. J., Messing, A., Dartcide, C. L., Hauschka, S. D., Palmiter, R. D., and Brinster, R. L. (1988). Heart and bone tumors in transgenic mice. *Proc. Natl. Acad. Sci. U. S. A.* **85**, 2648–2652.

Behringer, R. R., Cate, R. L., Froelick, G. J., Palmiter, R. D., and Brinster, R. L. (1990). Abnormal sexual development in transgenic mice chronically expressing Mullerian inhibitory substance. *Nature (London)* **345**, 167–172.

Benvenisty, N., Ornitz, D. M., Bennett, G. L., Sahagan, B. G., Kuo, A., Cardiff, R. D., and Leder, P. (1992). Brain tumours and lymphomas in transgenic mice that carry HTLV-1LTR/ *c-myc* and Ig/*tax*. *Oncogene* **7**, 2399–2405.

Bishop, J. M. (1991). Molecular themes in oncogenesis. *Cell* **64**, 235–248.

Bouchard, L., Lamarre, L., Tremblay, P. J., and Jolicoeur, P. (1989). Stochastic appearance of mammary tumors in transgenic mice carrying the MMTV *c-neu* oncogene. *Cell* **57**, 931–936.

Bowden, G. P., and Krieg, P. (1991). Differential gene expression during multistage carcinogenesis. *Environ. Health Perspect.* **93**, 51–56.

Bradl, M., Klein-Szanto, A., Porter, S., and Mintz, B. (1991). Malignant melanoma in transgenic mice. *Proc. Natl. Acad. Sci. U. S. A.* **88**, 164–168.

Brinster, R. L., and Palmiter, R. D. (1986). Introduction of genes into germ line of animals. *Harvey Lectures,* Series 80, 1–39.

Brinster, R. L., Chen, H. Y., Trumbauer, M., Serear, A. W., Warren, R., and Palmiter R. D. (1981). Somatic expression of herpes thymidine kinase in mice following injection of a fusion gene into eggs. *Cell* **27**, 223–231.

Brinster, R. L., Chen, H. Y., Messing, A., Dyke, T., Levin, A. J., and Palmiter R. D. (1984). Transgenic mice harboring SV40 T-antigen genes develop characteristic brain tumors. *Cell* **37**, 367–379.

Burck, K. B., Liu, E. T., and Larrick, J. W. (1988). "Oncogenes: an introduction to the concept of cancer genes." Springer-Verlag, New York.

Byrne, G. W., Kappen, C., Schughart, K., Utset, M., Bogarad, L., and Ruddle, F. H. (1991). Analysis of regulatory genes using the transgenic mouse system. *Biotechnology* **16**, 135–152.

Cartier, N., Miquerol, L., Tulliez, M., Lepetit, N., Levrat, F., Grimber, G., Briand, P., and Kahn, A. (1992). Diet-dependent carcinogenesis of pancreatic islets and liver in transgenic mice expressing oncogenes under the control of the L-type pyruvate kinase gene promoter. *Oncogene* **7**, 1413–1422.

Ceci, J. D., Kovatch, R. M., Swing, D. A., Jones, J. M., Snow, C. M., Rosenberg, M. P., Jenkins, N. A., Copeland, N. G., and Meisler, M. H. (1991). Transgenic mice carrying a murine amylase 2.2/SV40 T antigen fusion gene develop pancreatic acinar cell and stomach carcinomas. *Oncogene* **6**, 323–332.

Chigira, M. (1991). Freedom from holism in multicellular organisms: a possible role of tumor supressor genes. *Med. Hypotheses* **36**, 146–151.

Chisari, F. V., Klopchin, K., Moryama, T., Pasquinelli, C., Dunsford, H., Sell, S., Pinkert, C. A., Brinster, R. L., and Palmiter, R. D. (1989). Molecular pathogenesis of hepatocellular carcinoma in hepatitis B virus transgenic mice. *Cell* **59**, 1145–1156.

Choi, Y., Lee, I., and Ross, S. R. (1988). Requirement for Simian virus 40 small tumor antigen in tumorigenesis in transgenic mice. *Mol. Cell Biol.* **8**, 3382–3390.

Costantini, F., and Lacy E. (1981). Introduction of a β-globin gene into the mouse germ line. *Nature (London)* **294**, 92–94.

Clark, W. H. (1991). Tumour progression and the nature of cancer. *Br. J. Cancer* **64**, 631–644.

Cleary, M. L. (1991). Oncogenic conversion of transcription factors by chromosomal translocation. *Cell* **66**, 619–622.

Clemens, M. (1992). Tumorigenesis. Suppression with a difference. *Nature (London)* **360**, 210–211.

Dedera, D. A., Waller, E. K., LeBrun, D. P., Sen-Majumdar, A., Steven, M. E., Barsh, G. S., and Cleary, M. L. (1993). Chimeric homeobox gene E2A-PBX1 induces proliferation, apoptosis, and malignant lymphomas in transgenic mice. *Cell* **74**, 833–843.

De Mayo, F. L., Finegold, M. J., Hansen, T. N., Stanley, L. A., Smith, B., and Bullock, D. W. (1991). Expression of SV40 T antigen under control of rabbit uteroglobin promoter in transgenic mice. *Am. J. Physiol.* **261**, L70–L76.

Dildrop, R., Ma, A., Zimmerman, K., Hsu, E., Tesfaye, A., De Pinho, R., and Alt, F. W. (1989). IgH Enchancer-mediated deregulation of *N-myc* gene expression in transgenic mice: generation of lymphoid neoplasias that lack *c-myc* expression. *EMBO J.* **8**, 1121–1128.

Dunsford, H. A., Sell, S., and Chisari, F. V. (1990). Hepatocarcinogenesis due to chronic liver cell injury in hepatitis B virus transgenic mice. *Cancer Res.* **50**, 3400–3407.

Ermilov, A. N., and Shvemberger I. N. (1988). The transgenic mice tumors. *Exp. Oncol.* **10**, 3–8.

Ermilov, A. N., Kapitskaya, M. Z., Shvemberger, I. N., and Tomilin, N. B. (1988). Virus oncogene *myc* in the genome of transgenic mice and their progeny. *Tzitologia* **30**, 589–596.

Faas, S. J., Pan, S., Pinkert, C. A., Brinster, R. L., and Knowles, B. B. (1987). Simian virus 40 (SV-40)-transgenic mice that develop tumors are specifically tolerant to SV40 T antiden. *J. Exp. Med.* **165**, 417–427.

Feriotto, G., Pozzi, L., Piva, R., Deledda, F., Barbieri, R., Nastruzzi, C., Ciucci, A., Natali, P. G., Giacomini, P., and Gambari, R. (1991). Transgenic mice mimic the methylation pattern of the human HLD-DR α gene. *Biochem. Biophys. Res. Commun.* **175**, 459–466.

Fialkow, P. J. (1976). Clonal origin of human tumors. *Biochem. Biophys. Acta* **458**, 283–321.

Fisch, P., Boehm, T., Lavenir, I., Larson, T., Arno, J., Forster, A., and Rabbitts, T. H. (1992). T-cell acute lymphoblastic lymphoma induced in transgenic mice by the RBTN1 LIM-domain genes. *Oncogene* **7**, 2389–2397.

Fox, N., Crooke, R., Hwang, L.-H. S., Schibler, U., Knowles, B. B., and Solter, D. (1989). Metastatic hibernomas in transgenic mice expressing an α-amylase-SV40 T antigen hybrid gene. *Science* **224**, 461–463.

Freeman, C. S., Martin, M. M., and Marks, C. L. (1990). An overview of tumor biology. *Cancer Invest.* **8**, 71–90.

Fujita, K., Miki, N., Mojica, M. P., Takao, S., Phuchareon, J., Nishikawa, S., Sado, T., and Tokuhisa, T. (1993). B cell development is pertubed in bone marrow from c-fos/v-jun doubly transgenic mice. *Int. Immunol.* **5,** 227–230.

Glasner, S., Memoli, V., and Longnecker, D. S. (1992). Characterization of ELSV transgenic mouse model of pancreatic carcinoma. Histologic type of large and small tumors. *Am. J. Pathol.* **140,** 1237–1245.

Gordon, J. W. (1983). Studies of foreign transmitted through the germ lines of transgenic mice. *Exp. Zool.* **228,** 313–324.

Gordon, J. W., and Ruddle, F. H. (1981). Integration and stable germ line transmission of genes injected into mouse pronuclei. *Science* **214,** 1244–1246.

Gordon, J. W., and Ruddle, F. H. (1983). Gene transfer into mouse embryos: production of transgenic mice by pronuclear injection. *Methods Enzymol.* **101,** 411–433.

Gordon, J. W., Scangos, G. A., Plotkin, D. J., Barbosa, J. A., and Ruddle, F. H. (1980). Genetic transformation of mouse embryous by microinjection of purified DNA. *Proc. Natl. Acad. Sci. U. S. A.* **77,** 7380–7384.

Green, J. E., Hinrichs, S. H., Vogel, J., and Jay, G. (1989). Exocrinopathy resembling Sjogren's syndrome in HTLV-I *tax* transgenic mice. *Nature (London)* **341,** 72–74.

Greenhalgh, D. A., Rothnagel, J. A., Wang, X.-J., Quintanilla, M. I., Orengo, C. C., Gagne, T. A., Bundman, D. S., Longley, M. A., Fisher, C., and Roop, D. R. (1993). Hyperplasia, hyperkeratosis and benign tumor production in transgenic mice by a targeted v-fos oncogene suggest a role for *fos* in epidermal differentiation and neoplasia. *Oncogene* **8,** 2145–2157.

Griep, A. E., Herber, R., Jeon, S., Lohse, J. K., Dubielzig, R. R., and Lambert, P. F. (1993). Tumorigenecity of human papillomavirus type 16E6 and E7 in transgenic mice correlates with alterations in epithelial cell growth and differentiation. *J. Virol.* **67,** 1373–1384.

Grigoriadis, A. E., Schellander, K., Wang, Z.-Q., and Wagner, E. F. (1993). Osteoblasts are target cells for transformation in c-fos transgenic mice. *J. Cell Biol.* **122,** 685–701.

Guy, C. T., Webster, M. A., Schaller, M., Parsons, T. J., Cardiff, R. D., and Muller, W. J. (1992). Expression of the *neu* proto-oncogene in the mammary epithelium of transgenic mice induces metastatic disease. *Proc. Natl. Acad. Sci. U. S. A.* **89,** 10578–10582.

Hanahan, D. (1985). Heritable formation of pancreatic β-cell tumours in transgenic mice expressing recombinant insulin/Simian virus 40 oncogenes. *Nature (London)* **315,** 115–122.

Hanahan, D. (1988). Dissecting multistep tumorigenesis in transgenic mice. *Ann. Rev. Genet.* **22,** 479–519.

Hanahan, D. (1989). Transgenic mice as probes into complex systems. *Science* **246,** 1265–1275.

Hariharan, I. K., Harris, A. W., Crawford, M., Abud, H., Webb, E., Cory, S., and Adams, J. M. (1989). A bcr-v-abl oncogene induces lymphomas in transgenic mice. *Mol. Cell. Biol.* **9,** 2798–2805.

Hauft. S. M., Kim, S. H., Schmidt, G. H., Pease, S., Rees, S., Harris, S., Roth, K. A., Hansbrough, J. R., Cohn, S. M., Ahnen, D. J., Wright, N. A., Goodlad, R. A., and Gordon, J. I. (1992). Expression of SV-40 T antigen in the small intestinal epithelium of transgenic mice results in proliferative changes in the crypt and reentry of villus-associated enterocytes into the cell cycle but has no apparent affect on cellular differentiation programs and does not cause neoplastic transformation. *J. Cell Biol.* **117,** 825–839.

Haupt, Y., Alexander, W. S., Barri, G., Klinken, S. P., and Adams, J. M. (1991). Novel zinc finger gene implicated as *myc* collaborator by retrovirally accelerated lymphomagenesis in Eμ-myc transgenic mice. *cell* **65,** 753–762.

Haupt, Y., Harris, A. W., and Adams, J. M. (1992). Retroviral infection accelerates T lymphogenesis in Eμ-N-ras transgenic mice by activating c-myc or N-myc. *Oncogene* **7,** 981–986.

Heisterkamp, N., Jenster, G., ten Hoeve, J., Zovich, D., Pattengale, P. K., and Croffen, J. (1990). Acute leukemia in bcr/abl transgenic mice. *Nature (London)* **344,** 251–253.

Held, W. A., Mulins, J. J., Kuhn, N. J., Gallogher, F., Gu, G. D., and Gross, K. W. (1989). T antigen expression and tumorigenesis in transgenic mice containing a mouse major urinary protein/SV40 T antigen hybrid gene. *EMBO J.* **8,** 183–191.

Helseth, A., Siegal, G. P., Haug. E., and Bautch, V. L. (1992). Transgenic mice that develop pituitary tumors. A model for cushing's disease. *Am. J. Pathol.* **140**, 1071–1080.

Hino, O., Kitagawa, T., Nomura, K., Ohtake, K., Cui, L., Furuta, Y., and Aizawa, S. (1991). Hepatocarcinogenesis in transgenic mice carrying albumin-promoted SV40 T antigen gene. *Jpn. J. Cancer Res.* **82**, 1226–1233.

Hinrichs, S. H., Herenberg, M., Reynolds, G. K., and Jay, G. (1987). A transgenic mouse model for human neurofibromatosis. *Science* **237**, 1340–1343.

Hogan, B., Costantini, F., and Lacy, E. (1986). "Manipulating the mouse embryo." Cold Spring Harbor Laboratory, New York.

Hollingworth, R. E., and Lee, W. H. (1991). Tumor suppressor genes: new prospects for cancer research. *J. Natl. Cancer Inst.* **83**, 91–96.

Howley, P. M. (1991). Tumor progression in transgenic mice containing the bovine papillomavirus genome. *Basic Life Sci.* **57**, 103–107.

Iwamoto, T., Taniguchi, M., Wajjwalku, W., Nakashima, I., and Takahashi, M. (1993). Neuroblastoma in a transgenic mouse carrying a metallothionein/*ret* fusion gene. *Br. J. Cancer* **67**, 504–507.

Jaenisch, R. (1988). Transgenic animals. *Science* **240**, 1468–1474.

Jensen, N. A., Rodriguez, M. L., Garvey, J. S., Miller, C. A., and Hood, L. (1993). Transgenic mouse model for neurocristopathy: Schwannomas and facial bone tumors. *Proc. Natl. Acad. Sci. U. S. A.* **90**, 3192–3196.

Jhappen, C., Stahle, C., Harkins, R. N., Fausto, N., Smith, G. H., and Merlino, G. T. (1990). TGFα overexpression in transgenic mice induces liver neoplasia and abnormal development of the mammary gland and pancreas. *Cell* **61**, 1137–1146.

Kapitskaya, M. Z., Medvedeva, N. D., Ermilov, A. N., Ginkul, L. B., Shvemberger, I. N., and Tomilin, N. V. (1990). The transgenetic particularities of mice containing human virus hepatitis B surface antigen gene. II. The inheritance of transgene and its expression in liver. *Tsitologia* **32**, 720–727.

Kawaichi, M., Oka, C., Keeves, R., Kinoshita, M., and Honjo, T. (1991). Recombination of exogenous interleukin 2 receptor gene flanked by immunoglobulin recombination signal sequences in a pre-B cell line and transgenic mice. *Biol. Chem.* **266**, 18387–18394.

Kelly, F., Kellermann, O., Mechali, F., Gaillard, J., and Babinet, C. (1986). Expression of Simian virus 40 oncogenes in F9 embryonal carcinoma cells, in transgenic mice, and in transgenic embryos. *In* "Cancer cells" (M. Botchan, T. Grodzicker, and P. A. Sharp, eds.), Vol. 4, pp. 363–372. Cold Spring Harbor Symposium, Upton, New York.

Kim, C.-M., Koike, K., Saito, I., Miyamura, T., and Jay, G. (1991). HBx gene of hepatitis B virus induced liver cancer in transgenic mice. *Nature* (*London*) **351**, 371–320.

Kim, S. J., Winokur, T. S., Lee, H.-D., Danielpour, D., Kim, K. Y., Geiser, A. G., Chen, L.-S., Sporn, M. B., Roberts, A. B., and Jay, G. (1991). Overexpression of transforming growth factor-β in transgenic mice carrying the human T-cell lymphotropic virus type 1 *tax* gene *Mol. Cell. Biol.* **11**, 5222–5228.

Klein, G. (1987). The approaching era of the tumor suppressor genes. *Science* **238**, 1539–1545.

Klein-Szanto, A., Bradl, M., Porter, S., and Mintz, B. (1991). Melanosis and associated tumors in transgenic mice. *Proc. Natl. Acad. Sci. U. S. A.* **88**, 169–173.

Knowles, B. B., McCarrick, J., Solter, D., and Damjanov, I. (1990). Osteosarcomas in transgenic mice expressing an α-amylase-SV40 T-antigen hybrid gene. *Am. J. Pathol.* **137**, 259–262.

Knudson, A. G. (1973). Mutation and human cancer. *Adv. Cancer Res.* **17**, 317–352.

Knudson, A. G. (1991). Overview: genes that predispose to cancer. *Mutat. Res.* **247**, 185–190.

Krulewski, Th., Weumann, P. E., and Gordon, J. W. (1989). Insertional mutation in a transgenic mouse. Allelic with Purkinje cell degeneration. *Proc. Natl. Acad. Sci. U. S. A.* **86**, 3709–3712.

Lacey, B., Alpert, S., and Hanahan, D. (1986). Bovine papillomavirus genome elicits skin tumors in transgenic mice. *Cell* **34**, 343–358.

Langdon, W. Y., Harris, A. W., Cory, S., and Adams, J. M. (1986). The *c-myc* oncogene perturbs B lymphocyte development in Eμ-*myc* transgenic mice. *Cell* **47**, 11–18.

Larue, L., Dougherty, N., Bradl, M., and Mintz, B. (1993). Melanocyte culture lines from Tyr-SV40E transgenic mice: models for the molecular genetic evolution of the malignant melanoma. *Oncogene* **8**, 523–531.

Leder, A., Pattengale, P. K., Kuo, A., Stewart, T., and Leder, Ph. (1986). Consequences of widespread deregulation of the *c-myc* gene in transgenic mice: multiple neoplasms and normal development. *Cell* **45**, 485–495.

Lee, G.-H., Li, H., Ohtake, K., Nomura, K., Hino, O., Furuta, Y., Aizawa, S., and Kitagaawa, T. (1990). Detection of activated *c-H-ras* oncogene in hepatocellular carcinomas developing in transgenic mice harboring albumin promotor-regulated Simian virus 40 gene. *Carcinogenesis* **11**, 1145–1148.

Lee, G.-H., Merlino, G., and Fausto, N. (1992). Development of liver tumors in transforming growth factor transgenic mice. *Cancer Res.* **52**, 5162–5170.

Lee, T. H., Finegold, M. J., Shen, R. F., DeMayo, J. L., Woo, S. L., and Butel, J. S. (1990). Hepatitis B virus transactivator X protein is not tumorigenic in transgenic mice. *J. Virol.* **64**, 5939–5947.

Lehman, T. A., Reddel, R., Pfeifer, A. M. A., Spillar, E., Kaighn, M. E., Weston, A., Gervin, B. I., and Harris, C. C. (1991). Oncogenes and tumor-suppressor genes. *Env. Health Perspect.* **93**, 133–144.

Lettmann, C., Schmitz, B., and Doerfler, W. (1991). Persistence or loss of preimposed methylation patterns and *de novo* methylation of foreign DNA integrated in transgenic mice. *Nucleic Acids Res.* **19**, 7131–7137.

Levine, D. S., Sanchez, C. A., Rabinovitch, P. S., and Reid, B. J. (1991). Formation of the tetraploid intermediate is associated with the development of cells with more than four centrioles in the elastase-Simian virus 40 tumor antigen transgenic mouse model of pancreatic cancer. *Proc. Natl. Acad. Sci. U. S. A.* **88**, 6427–6431.

Lindgren, V., Sippola-Thiele, M., Skowronski, J., Weltzel, E., Howley, P. M., Hanahan, D. (1989). Specific chromosomal abnormalities characterize fibrosarcomas of bovine papilloma virus type 1 transgenic mice. *Proc. Natl. Acad. Sci. U. S. A.* **86**, 5025–5029.

van Lohuizen, M., Schejjen, B., Wientjens, E., van der Gulden, H., and Berns, A. (1991). Identification of cooperating oncogenes in Eμ-*myc* transgenic mice by provirus tagging. *Cell* **65**, 737–752.

Mahon, K. A., Overbeek, P. A., and Westphal, H. (1988). Prenatal lethality in a transgenic mouse line is the result of a chromosomal translocation. *Proc. Natl. Acad. Sci. U. S. A.* **85**, 1165–1168.

Mangues, R., Seidman, I., Pellicer, A., and Gordon, J. W. (1990). Tumorigenesis and male sterility in transgenic mice expressing a MMTV/*N-ras* oncogene. *Oncogene* **5**, 1491–1497.

Mangues, R., Saidman, I., Gordon, J. W., and Pellicer, A. (1992). Overexpression of the *N-ras* proto-oncogene, not somatic mutational activation, associated with malignant tumors in transgenic mice. *Oncogene* **7**, 2073–2076.

Marx, J. (1993). Learning how to suppress cancer. Fast and furiously, researchers are identifying new "tumor suppressor genes," which promise a new understanding of cancer - and perhaps a clinical payoff. *Science* **261**, 1385–1387.

Mathews, L. S., Hammer, R. E., Benringer, R. R., D'Ercole, A. J., Bell, G. I., Brinster, R. L., and Palmiter, R. D. (1988). Growth enhancement of transgenic mice expressing human insulin-like growth factor I. *Endocrinology* **123**, 2827–2833.

Matsui, Y., Halter, S. A., Holt, J. T., Hogan, B. L. M., and Coffey, R. J. (1990). Development of mammary hyperplasia and neoplasia in MMTV-TGFα transgenic mice. *Cell* **61**, 1147–1155.

Matsuoka, M., Nagawa, F., Okazaki, K., Kingsbury, L., Yoshida, K., Miller, U., Larue, D. T., Winer, J. A., and Sakao, H. (1991). Detection of somatic DNA recombination in the transgenic mouse brain. *Science* **254**, 81–86.

Mellentin, J. D., Nourse, J., Hunger, S. P., Smith, S. D., and Cleary, M. L. (1990). Molecular analysis of the t(1;19) breakpoint cluster region in pre-B cell ALL. *Genes Chromosome Cancer* **2**, 239–247.

Messing, A., Chen, H. J., Palmiter, R. D., and Brinster, R. L. (1985). Peripheral neuropathies, hepatocellular carcinomas and islet cell adenomas in transgenic mice. *Nature (London)* **316**, 461–463.

Mok, C.-H., Tsao, S.-W., Knapp, R. C., Fishbaugh, P. M., and Lau, C. (1992). Unifocal origin of advanced human epithelial ovarian cancer. [Advances in Brief]. *Cancer Res.* **52**, 5119–5122.

Monica, K., Galili, N., Nourse, J., Saltman, D., and Cleary, M. L. (1991). PBX2 and PBX3, new homeobox genes with extensive homology to the human proto-oncogene. *Mol. Cell. Biol.* **11**, 6149–6157.

Montag, A. G., Oka, T., Baek, K. H., Choi, C. S., Jay, G., and Agarwal, K. (1993). Tumors in hepatobiliary tract and pancreatic islet tissue to transgenic mice haboring gastrin Simian virus 40 large tumor antigen fusion gene. *Proc. Natl. Acad. Sci. U. S. A.* **90**, 6696–6700.

Moriyama, T., Guilhot, S., Klopchin, K., Moss, B., Pinkert, C. A., Palmiter, R. D., Brinster, R. L., Kanagawa, O., and Chisari, F. V. (1990). Immunobiology and pathogenesis of hepatocellular injury in hepatitis B virus transgenic mice. *Science* **248**, 361–364.

Muller, W. J., Sinn, E., Pattengale, P. K., Wallace, R., and Leder, P. (1988). Single-step induction of mammary adenocarcinoma in transgenic mice bearing the activated *c-neu* oncogene. *Cell* **54**, 105–115.

Muller, W. J., Lee, F. S., Dickson, C., Peters, G., Pattengale, P., and Leder, P. (1990). The *int-2* gene product acts as an epithelial growth factor in transgenic mice. *EMBO J.* **9**, 907–913.

Murakami, H., Sanderson, N. D., Nagy, P., Marino, P. A., Merlino, G., and Thorgeirsson, S. S. (1993). Transgenic mouse model for synergistic effects of nuclear oncogenes and growth factor in tumorigenesis: interaction of *c-myc* and transforming growth factor α in hepatic oncogenesis. *Cancer Res.* **53**, 1719–1723.

Nerenberg, M., Hinrichs, S. H., Reynolds, R. K., Khouri, G., and Jay, G. (1987). The *tat* gene of human T-lymphotropic virus type 1 induces mesenchymal tumors in transgenic mice. *Science* **237**, 1324–1329.

Nielsen, L. L., Discafani, C. M., Gurnani, M., and Tyler, R. D. (1991). Histopathology of salivary and mammary gland tumors in transgenic mice expressing a human *Ha-ras* oncogene. *Cancer Res.* **51**, 3762–3767.

Nielsen, L. L., Guranani, M., and Tyler, R. D. (1992). Evaluation of the Wap-*ras* transgenic mouse as a model system for testing anticancer drugs. *Cancer Res.* **52**,N13, 3733–3738.

Nowell, P. C. (1989). The clonal nature of neoplasia. *In* "Cancer cells" (J. Levine, ed.), Vol. 1, pp. 29–30. Cold Spring Harbor Symposium, Upton, New York.

Nowell, P. C. (1991). How many human cancer genes? *J. Natl. Cancer Inst.* **83**, 1061–1064.

Orban, P. C., Chui, D., and Marth, J. D. (1992). Tissue- and site-specific DNA recombination in transgenic mice. *Proc. Natl. Acad. Sci. U. S. A.* **89**, 6861–6865.

Ornitz, D. M., Hammer, R. E., Messing, A., Palmiter, R. D., and Brinster, R. L. (1987). Pancreatic neoplasia induced by SV40 T-antigen expression in acinar cells of transgenic mice. *Science* **238**, 188–193.

Ornitz, D. M., Moreadith, R. W., and Leder, P. (1991). Binary system for regulating transgene expression in mice: targeting *int-2* gene expression with yeast GAL4/UAS control elements. *Proc. Natl. Acad. Sci. U. S. A.* **88**, 698–702.

Palmiter, R. D., and Brinster, R. L. (1985). Transgenic mice. *Cell* **41**, 343–345.

Palmiter, R. D., and Brinster, R. L. (1986). Germ-line transformation of mice. *Ann. Rev. Genet.* **20**, 465–499.

Palmiter, R. D., Chen, H. Y., Messing, A., and Brinster, R. L. (1985). SV-40 enhancer and large T-antigen are instrumental in development of choroid plexus tumours in transgenic mice. *Nature (London)* **316**, 457–460.

Palmiter, R. D., Sandgren, E. P., Avarbock, M. R., and Allen, D. D. (1991). Heterologous introns can enhance expression of transgenes in mice. *Proc. Natl. Acad. Sci. U. S. A.* **88,** 478–482.

Paquis-Flucklinger, V., Michiels, J.-F., Vidal, F., Alquier, C., Pointis, G., Bourdon, V., Cuzin, F., and Rassoulzadegan, M. (1993). Expression in transgenic mice of the large T antigen of polyomavirus induces Sertoli cell tumours and allows the estabilishment of differentiated cell lines. *Oncogene* **8,** 2087–2094.

Parker, P. J. and Katan, M. (1990). "Molecular biology of oncogenes and cell control mechanisms." Ellis Horwood Td., New York.

Patterson, H. (1992). Approaches to proto-oncogene and tumour suppressor gene identification. *Eur. J. Cancer* **28,** 258–263.

Perraud, F., Yoshimura, K., Louis, B., Dalemans, W., Ali-Hadji, D., Schultz, H., Claudepierre, M.-C., Chartier, C., Danel, C., Bellocq, J.-P., Crystal, R. G., Lecocq, J.-P., and Pavirani, A. (1992). The promoter of the human cystic fibrosis transmembrane conductance regulator gene detecting SV 40 T antigen expression induces malignant proliferation of ependymal cells in transgenic mice. *Oncogene* **7,** 993–997.

Pimentel, E. (1989). "Oncogenes." CRC Press, Boca Raton, Florida.

Powell, B. C., and Rogers, G. E. (1990). Cyclic hair-loss and regrowth in transgenic mice overexpressing an intermediate filament gene. *EMBO J.* **9,** 1485–1493.

Quaife, C. G., Pinkert, C. A., Ornitz, D. M., Palmiter, R. D., and Brinster, R. L (1987). Pancreatic neoplasia induced by *ras* expression in acinar cells of transgenic mice. *Cell* **48,** 1023–1034.

Quong, M. W., Massari, M. E., Zwart, R., and Murre, C. (1993). A new transcriptional-activation motif restricted to a class of helix-loop-helix proteins is functionally conserved in both yeast and mammalian cells. *Mol. Cell. Biol.* **13,** 792–800.

Reynolds, R. K., Hoekzema, G. S., Vogel, J., Hinrichs, S. H., and Jay, G. (1988). Multiple endocrine neoplasia induced by the promiscuous expression of a viral oncogene. *Proc. Natl. Acad. Sci. U. S. A.* **85,** 3135–3139.

Rindi, G., Grant, S. G., Yiangou, Y., Ghatei, M. A., Bloom, S. R., Bautch, V. L., Soieia, E., and Polak, J. M. (1990). Development of neuroendocrine tumors in the gastrointestinal tract of transgenic mice. *Am. J. Pathol.* **136,** 1349–1363.

Rosenbaum, H., Welb, E., Adams, J. M., Cory, S., and Harris, A. W. (1989). *N-myc* transgene promotes B lymphoid proliferation, elicits lymphomas and reveals cross regulation with *c-myc*. *EMBO J.* **8,** 749–755.

Ross, S. R., Chey, L., Graves, R. A., Fox, N., Solevjeva, V., Klaus, S., Ricquier, D., and Spiegelman, B. M. (1992). Hibernoma formation in transgenic mice and isolation of a brown adipocyte cell line expressing the uncoupling protein gene. *Proc. Natl. Acad. Sci. U. S. A.* **89,** 7561–7565.

Rusconi, S. (1991). Transgenic regulation in laboratory animals. *Experientia* **47,** 866–877.

Rüther, U., Garber, C., Komitowski, D., Muller, R., and Wagner, E. F. (1987). Deregulated *c-fos* expression interferes with normal bone development in transgenic mice. *Nature (London)* **325,** 412–416.

Rüther, U., Muller, W., Sumida, T., Tokuhisa, T., Rajewsky, K., and Wagner, E. F. (1988). *c-fos* expression interferes with thymus development in transgenic mice. *Cell* **53,** 847–856.

Rüther, U., Komitowski, D., Schubert, F. R., and Wagner, E. F. (1989). *c-fos* expression induces bone tumors in transgenic mice. *Oncogene* **4,** 861–865.

Saitoh, A., Kimura, M., Takahashi, R., Yokoyama, M., Nomura, T., Izawa, M., Sekiya, T., Nishimura, S., and Katsuki, M. (1990). Most tumors in transgenic mice with human *c-Ha-ras* gene contained somatically activated transgenes. *Oncogene* **5,** 1195–1200.

Sandberg, A. A. (1992). The chromosome changes in benign tumors and the multistep process of carcinogenesis. (Abstracts of the Third European Workshop on Cytogenetics and Molecular Genetics of Human Solid Tumors. p. 102–183). *Cancer Genet. Cytogen.* **63,** 104.

Sandgren, E. P., Quaife, C. J., Pinkert, C. A., Palmiter, R. D., and Brinster, R. L. (1989). Oncogene-induced liver neoplasia in transgenic mice. *Oncogene* **4,** 715–724.

Sandgren, E. P., Luetteke, N. C., Palmiter, R. D., Brinster, R. L., and Lee, D. C. (1990). Overexpression of TGFα in transgenic mice: induction of epithelial hyperplasia, pancreatic metaplasia, and carcinoma of the breast. *Cell* **61,** 1121–1135.

Sandgren, E. P., Palmiter, R. D., Heckel, J. L., Brinster, R. L., and Degen, J. L. (1992). DNA rearrangement causes hepatocarcinogenesis in albumin-plasminogen activator transgenic mice. *Proc. Natl. Acad. Sci. U. S. A.* **89,** 11523–11527.

Sapienza, C., Peterson, A. C., Rossant, J., and Balling, R. (1987). Degree of methylation of transgene is dependent on gamete of origin. *Nature (London)* **328,** 251–254.

Schmidt, E. V., Pattengale, P. K., Weir, L., and Leder, Ph. (1988). Transgenic mice bearing the human *c-myc* gene activated by an immunoglobulin enhancer: A pre-B-cell lymphoma model. *Proc. Natl. Acad. Sci. U. S. A.* **85,** 6047–6051.

Schoenenberger, C.-A., Andres, A.-C., Groner, B., van der Valk, M., LeMeur, M., and Gerlinger, P. (1988). Targeted *c-myc* gene expression in mammary glands of transgenic mice induces mammary tumors with constitutive milk protein gene transcription. *EMBO J.* **7,** 169–175.

Schuh, A. C., Keating, S. J., Monteclaro, F. S., Vogt, P. K., and Breitman, M. L. (1990). Obligatory wounding requirement for tumorigenesis in *v-jun* transgenic mice. *Nature (London)* **346,** 756–760.

Schuh, A. C., Keating, S. J., Yeung, M. C., and Breitman, M. L. (1992). Skeletal muscle arises as a late event during development of wound sarcomas in *v-jun* transgenic mice. *Oncogene* **7,** 667–676.

Schulz, N., Propst, F., Rosenberg, M. P., Linnoila, R. I., Paules, R. S., Kovatch, R., Ogiso, Y., and Woude, G. V. (1992a). Pheochromocytomas and C-cell thyroid neoplasms in transgenic mice: a model for the human multiple endocrine neoplasia type 2 syndrome. *Cancer Res.* **52,** 450–455.

Schulz, N., Propst, F., Rosenberg, M. M., Linnoila, R. I., Paules, R. S., Schulte, D., and Vande Woude, G. F. (1992b). Patterns of neoplasia in *c-mos* transgenic mice and their relevance to multiple endocrine neoplasia. *Henry Ford Hosp. Med. J.* **40,** 307–311.

Sell, S., Hunt, J. M., Dunsford, H. A., and Chisari, F. V. (1991). Synergy between hepatitis B virus expression and chemical hepatocarcinogens in transgenic mice. *Cancer Res.* **51,** 1278–1285.

Shackleford, G. M., MacArthur, C. A., Kwan, H. C., and Varmus, H. E. (1993). Mouse mammary tumor virus infection accelerates mammary carcinogenesis in Wnt-1 transgenic mice by insertional activation of int-2/Fgf-3 and hst/Fgf-4. *Proc. Natl. Acad. Sci. U. S. A.* **90,** 740–744.

Shvemberger, I. N., and Ermilov, A. N. (1994). The peculiarities of neoplastic growth in transgenic mice. *Tzitologia* **36,** 131–146.

Shvemberger, I. N., Ermilov, A. N., Medvedeva, N. D., Kapitskaya, M. Z., and Tomilin, N. V. (1990). Production and characteristics of transgenic mice expressing a gene encoding surface antigen of human hepatitis B virus. *Tsitologia* **32,** 252–255.

Shvemberger, I. N., Ermilov, A. N., Ginkul, L. B., Smagina, L., Romanov, S., and Glebov, O. K. (1995). Recombinations of neomycintransferase gene in transgenic mice. *Tzitologia* (in press).

Sidman, C. L., Denial, T. M., Marshall, J. D., and Roths, J. B. (1993). Multiple mechanisms of tumorigenesis in Eμ-myc transgenic mice. *Cancer Res.* **53,** 1665–1669.

Sinn, E., Muller, W., Pattengale, P., Tepler, I., Walloce, R., and Leder, Ph. (1987). Coexpression of MMTV/*Ha ras* and MMTV/*c-myc* genes in transgenic mice: synergistic action of oncogenes *in vivo. Cell* **49,** 465–475.

Small, J. A., Blair, D. G., Showalter, S. D., and Seangos, G. A. (1985). Analysis of a transgenic mouse containing Simian virus 40 and *v-myc* sequences. *Mol. Cell. Biol.* **5,** 642–648.

Stewart, T. A., Pattengale, P. K., and Leder, Ph. (1984). Spontaneous mammary adenocarcinomas in transgenic mice that carry and express MTV/*myc* fusion genes. *Cell* **83**, 624–637.

Stein, W. D. (1991). Analysis of cancer incidence data on the basis of multistage and clonal growth models. *Adv. Cancer Res.* **56**, 161–213.

Stöcklin, E., Botteri, F., and Groner, B. (1993). An activated allele of the *c-erbB*-2 oncogene impairs kidney and lung function and causes early death of transgenic mice. *J. Cell Biol.* **122**, 199–208.

Suematsu, S., Matsusaka, T., Matsuda, T., Ohno, S., Miyazaki, J., Yamamura, K., Hirano, T., and Kishimoto, T. (1992). Generation of plasmacytomas with the chromosomal translocation t(12;15) in interleukin 6 transgenic mice. *Proc. Natl. Acad. Sci. U. S. A.* **89**, 232–235.

Suda, Y., Aizawa, S., Hirai, S., Inoue, T., Furuta, Y., Suzuki, M., Hirohashi, S. and Ikawa, Y. (1987). Driven by the same Ig enhancer and SV40 T promoter *ras* induced lung adenomatous tumors, *myc* induced pre-B cell lymphomas and SV40 large T gene a variety of tumors in transgenic mice. *EMBO J.* **6**, 4055–4065.

Sugimura, T. (1992). Multistep carcinogenesis: a 1992 perspective. *Science* **258**, 603–607.

Sutrave, P., Kelly, A. M., and Hughes, S. H. (1990). *ski* can cause selective growth of skeletal muscle in transgenic mice. *Genes Dev.* **4**, 1462–1472.

Takagi, H., Sharp R., Hammermeister, C., Goodrow, T., Bradlay, M. O., Fausto, N., and Merlino, G. (1992). Molecular and genetic analysis of liver oncogenesis in transforming growth factor α transgenic mice. *Cancer Res.* **52**, 5171–5177.

Tanooka, H. (1988). Monoclonal growth of cancer cells: experimental evidence. *Jpn. J. Cancer Res.* **79**, 657–665.

Teitz, T., Chang, J. C., Kitamura, M., Yen, B., and Kan, J. W. (1993). Rhabdomyosarcoma arising in transgenic mice harboring the β-globin locus control region fused with Simian virus 40 large T antigen gene. *Proc. Natl. Acad. Sci. U. S. A.* **90**, 2910–2914.

Tepper, R. I., Levinson, D. A., Stanger, B. Z., Campos-Torres, J., Abbas, A. K., and Leder, P. (1990). IL-4 induces allergic-like inflammatory disease and alters T cell development in transgenic mice. *Cell* **62**, 457–467.

Theuring, F., Götz, W., Balling, R., Korf, H.-W., Schulze, F., Herken, R., and Gruss, P. (1990). Tumorigenesis and eye abnormalities in transgenic mice expressing MSV-SV40 large T-antigen. *Oncogene* **5**, 225–232.

Tremblay, P. J., Pothier, F., Hoang, T., Tremblay, G., Brownstein, S., Liszauer, A., and Jolicouer, P. (1989). Transgenic mice carrying the mouse mammary tumor virus *ras* fusion gene: distinct effects in various tissues. *Mol. Cell. Biol.* **9**, 854–859.

Tsukamoto, A. S., Grosschedl, R., Guzman, R. C., Parslow, T., and Varmus, H. E. (1988). Expression of the *int*-1 gene in transgenic mice is associated with mammary gland hyperplasia and adenocarcinomas in male and female mice. *Cell* **55**, 619–625.

van Lohuizen, M., Verbeek, S., Krimpenfort, P., Domen, J., Sarie, C., Radaszkiewicz, T., and Berns, A. (1989). Predisposition to lymphomagenesis in *pim*-1 transgenic mice: cooperation with *c-myc* and *N-myc* in murine leukemia virus-induced tumors. *Cell* **56**, 673–682.

Varki, A., Hooshmand, F., Diaz, S., Varki, N. M., and Hedrick, S. M. (1991). Developmental abnormalities in transgenic mice expressing a sialic acid-specific 9-*o*-acetylesterase. *Cell* **65**, 65–74.

Verbeek, S., van Lohuizen, M., van der Valk, M., Domen, J., Kraal, G., and Berns, A. (1991). Mice bearing the Eμ-*myc* and Eμ-*pim*-1 transgenes develop pre-B-cell leukemia prenatally. *Mol. Cell. Biol.* **11**, 1176–1179.

Vogel, J., Hinrichs, S. H., Napolitano, L. A., Ngo, L., and Jay, G. (1991). Liver cancer in transgenic mice carrying the human immunodeficiency virus *tat* gene. *Cancer Res.* **51**, 6686–6690.

Voncken, J. W., Morris, C., Pattengale, P., Dennert, G., Kikly, C., Groffen, J., and Heisterkamp, N. (1992). Clonal development and karyotype evolution during leukemogenesis of BCR/ABL transgenic mice. *Blood* **79**, 1029–1036.

Wainscoat, J. S., and Fey, M. F. (1990). Assessment of clonality in human tumors: a review. *Cancer Res.* **50,** 1355–1360.

Wang, R., and Bautch, V. L. (1991). The polyomavirus early region gene in transgenic mice causes vascular and bone tumors. *J. Virol.* **65,** 5174–5183.

Wang, Y., Sugiyama, H., Axelson, H., Panda, C. K., Bobonits, M., Ma, A., Steinberg, J. M., Alt, F. W., Klein, G., and Wiener, F. (1992). Functional homology between *N-myc* and *c-myc* in murine plasmacytomagenesis: plasmacytoma development in *N-myc* transgenic mice. *Oncogene* **7,** 1241–1247.

Weinberg, R. A. (1992). The integration of molecular genetics into cancer management. *Cancer* **70** Suppl., 1653–1658.

Went, D. F., and Stranzinger, G. (1991). Transgenic vertebrates. Conclusions and outlook. *Experientia* **47,** 934–936.

Westphal, H. (1987). Transgenic mice. *Bioessays* **6,** 73–76.

Westphal, H. (1988). Perturbations of lens development in the transgenic mouse. *Cell Diff. Dev.* **25** Suppl., 33–37.

Wiedemann, L. M., and Morgan, G. J. (1992). How are cancer-associated genes activated or inactivated? *Eur. J. Cancer* **28,** 248–251.

Wilkie, Th. M., Brinster, R. L., and Palmiter, R. D. (1986). Germline and somatic mosaicism in transgenic mice. *Devel. Biol.* **118,** 9–18.

Wynford-Thomas, D. (1991). Oncogenes and anti-oncogenes; the molecular basis of tumour behaviour. *J. Pathol.* **165,** 187–201.

Integration of Intermediate Filaments into Cellular Organelles

Spyros D. Georgatos*† and Christèle Maison*
*Program of Cell Biology, European Molecular Biology Laboratory, 69012 Heidelberg, Germany and †Department of Basic Sciences, Faculty of Medicine, The University of Crete, 711 10 Heraklion, Greece

The intermediate filaments represent core components of the cytoskeleton and are known to interact with several membranous organelles. Classic examples of this are the attachment of keratin filaments to the desmosomes and the association of the lamin filament meshwork with the inner nuclear membrane. At this point, the molecular mechanisms by which the filaments link to membranes are not clearly understood. However, since a substantial body of information has been amassed, the time is now ripe for comparing notes and formulating working hypotheses. With this objective in mind, we review here pioneering studies on this subject, together with work that has appeared more recently in the literature.

KEY WORDS: Intermediate filaments, Lamins, Nuclear envelope, Plasma membrane, Centrosome, Membrane skeleton, Cytoskeleton.

I. Introduction

Numerous essays and reviews on intermediate filaments (IFs) have been published in the past 5 years. This literature covers the structure of IFs and their presumed role in the pathogenesis of disease. Nonetheless, another interesting aspect of IF biology, i.e., the association of IFs with intracellular organelles, has not been discussed sufficiently in recent texts. In fact, the last comprehensive review on this topic was published 15 years ago (Lazarides, 1980).

Since that time, ideas have evolved. As the field has advanced, a deeper understanding has replaced our initial naivete, which was based almost exclusively on descriptive studies. For instance, the chemical heterogeneity of IF proteins is now comprehended in molecular and evolutionary terms; several IF-associated proteins have been characterized; IFs have been shown to be dynamic entities and are no longer scorned as "static" structures; finally, thanks to genetic approaches, there are now specific clues concerning the role of IFs during development and differentiation. The new information offers exciting prospects for future research on IFs and prepares the grounds for far-reaching medical applications.

In this review, we have made an effort to critically discuss previous and more recent ideas concerning the cellular interactions of IFs. As the reader would realize, despite the profusion of data, many questions remain open and several mechanistic problems can only be addressed on a speculative or hypothetical basis.

II. Molecular Properties and Presumed Functions of IF Proteins

IFs, together with microtubules and actin microfilaments, make up the cytoskeleton of most eukaryotic cells. They consist of heterogeneous protein subunits which share the ability to polymerize into 10-nm rope-like structures. Biochemical studies and sequence comparisons have demonstrated that IF proteins consist of three structural domains: a central, largely α-helical segment ("rod") and two nonhelical end regions ("head" and "tail" at the NH_2- and COOH-terminals respectively). The helical domain is subdivided into four parts (termed coil 1a, 1b, 2a and 2b), and has a defined length of either 310 or 352 amino acid residues. Whereas the rod contains conserved motifs and sequence principles, the end domains of different IF proteins vary markedly both in size and in primary structure.

Unlike other cytoskeletal elements, IFs do not depolymerize at low temperature and survive treatment with chaotropic agents, high salt, and low concentrations of urea. In physical terms, IFs are more deformable than actin microfilaments and can withstand more stress than microtubules without being ruptured (Janmey et al., 1991). There is good reason to believe that these unique physicochemical features are due to the special type of hydrophobic and ionic forces that hold IF polymers together.

IF proteins immediately after biosynthesis form dimeric coiled-coils, i.e., left-handed superhelices consisting of two intertwined α-helices. The formation of coiled-coils is entropically driven and can be explained by invoking hydrophobic interactions. The rod domains of IF proteins contain multiple heptad repeats, i.e., blocks of seven amino acids in which the first and the fourth residues are apolar (Fuchs and Weber, 1994). In a helical configuration, the apolar residues align along one side of the helix and create a long hydrophobic stripe. This hydrophobic surface exhibits a strong tendency to pair with the corresponding hydrophobic stripe of another IF helix, so that apolar residues in the two molecules are buried away from the aqueous solvent. In addition to this, IF oligomers higher than dimers are probably stabilized by a multiplicity of ionic interactions. Basic and acidic residues in the rod domain of IF proteins are periodically repeated at intervals of ~9.5 amino acids (Parry *et al.*, 1977; Conway and Parry, 1988, 1990; Fraser *et al.*, 1985;). Since IF coiled-coils are thought to form staggered and antiparallel tetramers before fully polymerizing into filaments (Fuchs and Weber, 1994), the periodic arrangement of charged amino acids may allow pairing of acidic and basic residues occurring in partially overlapping rods (interhelical bonding). Finally, interactions between ionic pairs may also stabilize individual α-helices of IF proteins by way of intramolecular bonding (intrahelical ion pairing) (Letai and Fuchs, 1995).

It should be stressed that the multiple noncovalent connections among IF subunits do not render the polymer immune to treatments which shift the assembly equilibrium. Thus, IFs would disassemble upon dilution and repolymerize to a various extent depending on subunit concentration (Zackroff and Goldman, 1979; Angelides *et al.*, 1989). Furthermore, as could be expected from equilibrium polymers, preformed filaments can be disassembled *in vitro* by an excess of short synthetic peptides which correspond to different parts of the α-helical domains of IF proteins (Hatzfeld and Weber, 1991; Kouklis *et al.*, 1992; Chipev *et al.*, 1992). Presumably, these peptides exchange with normal subunits, but after incorporation into the polymer cannot sustain filament structure.

Based on intracellular location, IFs can be divided into two general classes: the cytoplasmic IFs and the nuclear lamin filaments. The current understanding is that the nuclear lamins, which represent the ancestors of cytoplasmic IF proteins, are involved in universal cellular functions (e.g., the reversible disassembly of the nuclear envelope during mitosis), whereas cytoplasmic IFs play more subtle, tissue- or cell-specific roles (Fuchs and Weber, 1994; Klymkowsky, 1995). Cytoplasmic IFs, contrary to the nuclear lamins, do not depolymerize in a systemic fashion during mitosis. In most cases, the filaments thicken and sometimes form a cagelike structure around the chromosomes (Maison *et al.*, 1993).

Sixty different genes coding for IF proteins have so far been identified in the human genome and the list continues to grow. IF genes are differentially expressed (alone, or in combination) in the various cells and tissues of vertebrates and invertebrates (for cataloging, see Steinert and Roop, 1988; Fuchs and Weber, 1994). Early embryos usually express keratins, whereas later during embryonic life most nonepithelial cells express transiently either vimentin or other IF proteins (e.g., nestin). Finally, when animals reach adulthood, each cell or group of cells express differentiation-specific

IF proteins. For instance, epithelial cells express keratins (over 30 types); neuronal cells express the neurofilament triplet proteins (NFL, NFM, and NFH), α-internexin, and peripherin; glial cells contain the glial fibrillary acidic protein (GFAP); mesenchymal cells (blood cells, fibroblasts, endothelial cells) express vimentin; muscle cells express desmin; lens fiber cells express (in addition to vimentin) filensin and phakinin; finally, all cells express B-type lamins (B1, B2, or B3), whereas only differentiated cells (independent of origin) express A-type lamins (lamin A and lamin C).

Why the subunits of IFs are heterogeneous is not clear. One explanation may be that the various types of IFs, despite their morphological similarities, have unique physicochemical properties and bind to different ligands, depending on cellular context. The cell- and tissue-specific functions of cytoplasmic IFs become apparent in disease. For example, mutations in skin keratins which compromise filament assembly *in vivo* cause increased fragility of epidermal cells. In these cases, rubbing or scratching of the skin causes lysis of keratinocytes and separation of epithelial sheets in the epidermis, precipitating blistering (Fuchs, 1994). Sometimes, inappropriate expression of cytoplasmic IF proteins can also create pathology: For instance, overaccumulation of neurofilaments in the somata of neurons causes lesions reminiscent of amyotrophic lateral sclerosis, while overaccumulation of vimentin IFs in the cells of the eye lens arrests differentiation and causes cataracts (Xu *et al.*, 1993; Cote *et al.*, 1993; Capetanaki *et al.*, 1989). Interestingly, in certain systems the lack of tissue-specific IF proteins during development is apparently sensed by "quality control" mechanisms which abort *in vitro* differentiation. Thus, glial cells treated with antisense oligonucleotides which block the expression of glial fibrillary acidic protein fail to develop the characteristic cytoplasmic processes of astrocytes when stimulated by neurons (Weinstein *et al.*, 1991). Similarly, blockade of desmin expression in myogenic cells inhibits myoblast fusion and formation of myotubes (Li *et al.*, 1994).

It is not understood how the various types of IFs act to accomplish their specific functions. One school of thought believes that IFs are "skeletal" elements which keep in place transmembrane devices involved in cell–cell interactions. However, others believe that IFs maybe adaptable components of various "cellular machines" which carry out cell-specific functions. Although these ideas are not mutually exclusive, it is fair to say that cell-biological assays measuring specific IF functions are not yet available.

III. Interactions of Cytoplasmic IFs with the Cell Nucleus

Cytoplasmic IFs are known to form dense filament networks radiating from the cell nucleus and extending to the plasma membrane. Whether the filaments are physically integrated into the nuclear envelope has been debated in the literature for quite some time. To discuss this problem intelligently, we ought to adopt certain criteria: thus, if cytoplasmic IFs are indeed coupled to the cell nucleus, there must be morphologically apparent connections between those two structures; furthermore, in such a case the subunits of cytoplasmic IFs should exhibit a measurable affinity for specific components of the nuclear envelope; finally, a dynamic connection between cytoplasmic IFs and the nuclear envelope would mean that obliteration of cytoplasmic IFs must somehow affect nuclear architecture.

A. Spatial Associations of Cytoplasmic IFs with the Nuclear Envelope

Examination of higher eukaryotic cells by conventional thin section/transmission electron microscopy reveals that cytoplasmic IFs approach the nuclear surface very closely and terminate "blindly" at the level of the outer nuclear membrane. Looking at previously published micrographs, one has the impression that sometimes arrays of IFs anchor "end-on" at the nuclear pores, whereas at other times the filaments seem to make tangential contacts with the nuclear envelope (Harris and Brown, 1971; Franke, 1971; Lehto et al., 1978; Woodcock, 1980; Jones et al., 1982; Goldman et al., 1985). However, it is difficult to decide from such images whether the filaments really insert into the nuclear membrane and the nuclear pores, or just spiral around the nucleus. To our knowledge, a three-dimensional reconstruction of the outer boundary of the nuclear envelope has not been published yet.

Whole-mount electron microscopy on detergent/high salt-extracted cells and analysis of resinless sections have been employed to obtain a more distinctive picture of the spatial associations between cytoplasmic IFs and the cell nucleus. When such methods are used (Fig. 1), it often appears that cytoplasmic IFs are continuous with the nuclear lamina meshwork, or that they connect to the nuclear pores via short, ~5-nm-thick fibers (Capco et al., 1982; Fey et al., 1984; Katsuma et al., 1987; Carmo-Fonseca et al., 1987; Fujitani et al., 1989; French et al., 1989; Wang et al., 1989; Carmo-Fonseca and David-Ferreira, 1990). Judging from the morphology of these short fibers, it would be reasonable to suggest that they correspond to the recently identified fibrils which project from the outer (cytoplasmic) annulus of the nuclear pore (Ris, 1991). It has been claimed that among the components of the pore-associated fibrils is Tpr, a large coiled-coil protein (Byrd et al., 1994). It is conceivable that Tpr may interact with the coiled-coil regions of cytoplasmic IF proteins. Alternatively, the "core" of the pore-associated fibrils may consist of IF proteins, whereas the outer

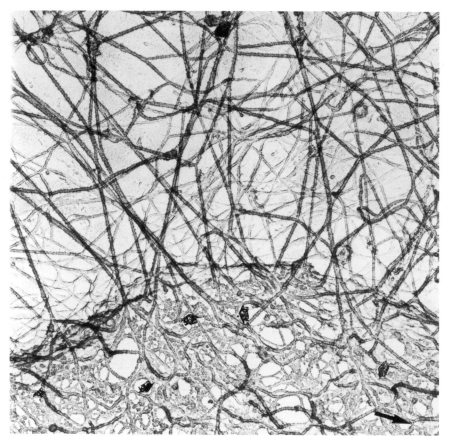

FIG. 1   IF-nuclear matrix scaffold from prostatic fibroblast as visualized after critical-point drying and metal shadowing. Long, individual, approximately 15-nm filaments surround the nucleus but are never seen to fuse with the nuclear matrix. Short, 7-nm filaments extend laterally from IFs and abut upon the nuclear lamina (arrowheads) and pore complexes (arrows). ×56,000. (Courtesy of Dr. M. Carmo-Fonseca, reproduced from Carmo-Fonseca *et al.* 1987, by permission of the *European Journal of Cell Biology.*)

shell of these structures may contain specific nucleoporins which act as IF-binding proteins.

The possibility that the pore fibrils contain cytoplasmic IF proteins is supported by immunoelectron microscopy data. At least in some cell types, pore-associated fibers are decorated by specific antibodies recognizing the nonhelical tail domain of vimentin (Fig. 2 and our unpublished observations).

Concerns have been expressed about critical point drying and detergent/high salt extraction, i.e., the methods that have been used in many of the ultrastructural studies described earlier: there is a general feeling that such manipulations may alter cellular architecture and induce fortuitous associations between neighboring organelles. In addition to this, since filaments crossing the nuclear membranes or the channels of the nuclear pores have not been observed in conventionally processed specimens, the idea that cytoplasmic IFs gain access to the nuclear interior and link directly to the nuclear lamina is currently viewed with skepticism. Nevertheless, there are several rational ways around this topological problem. First, taking into account the fibrillar substructure of IFs, it seems likely that the filaments unravel into thinner subfibers (e.g., protofibrils) as they approach the nuclear surface. These subfibers could easily break or collapse during sample preparation. A precedent for this exists in the case of the pore complex-associated fibrils. These structures were originally identified as "particles" or "granules" because they had apparently collapsed or recoiled during specimen preparation by conventional methods (cf. Unwin and Milligan, 1982; Jarnik and Aebi, 1991). Second, it is conceivable that cytoplasmic IFs establish connections with intranuclear elements during mitosis, when the nuclear envelope is broken down and the physical barrier between the cytoplasm and the nucleoplasm disappears. Indeed, recent studies on cultured cells show that during mitosis fragments of the nuclear envelope which contain B-type lamins are anchored to vimentin filaments (Maison et al., 1993). Along the same lines, it has been observed that the nuclear matrix protein NMP 125, which is segregated away from cytoplasmic IFs during interphase, docks on vimentin filaments during cell division (Marugg, 1992). Finally, an intriguing possibility is that IF oligomers are imported into the nucleus during interphase by "piggy-backing" on karyophilic proteins. This can be deduced from recent experiments showing that keratin and vimentin subunits, which normally localize in the cytoplasm, are readily transported into the nucleus and accumulate in the nucleoplasm after truncation of their COOH-terminal domains (Gill et al., 1990; Bader et al., 1991; Eckelt et al., 1992). That wild-type IF proteins do not accumulate in the nucleoplasm to the same extent as truncated subunits can be explained by a competing process by which the COOH-terminal part of these molecules anchors the filaments to cytoplasmic structures.

B. In Vitro Binding of Cytoplasmic IF Proteins to Nuclear Components

Those who have fractionated nucleated cells know first-hand that cytoplasmic IFs cannot be easily dissociated from the nuclear envelope. This is often attributed to nonspecific binding, but this explanation is rather naive: during subcellular fractionation, the cytoskeletal structures

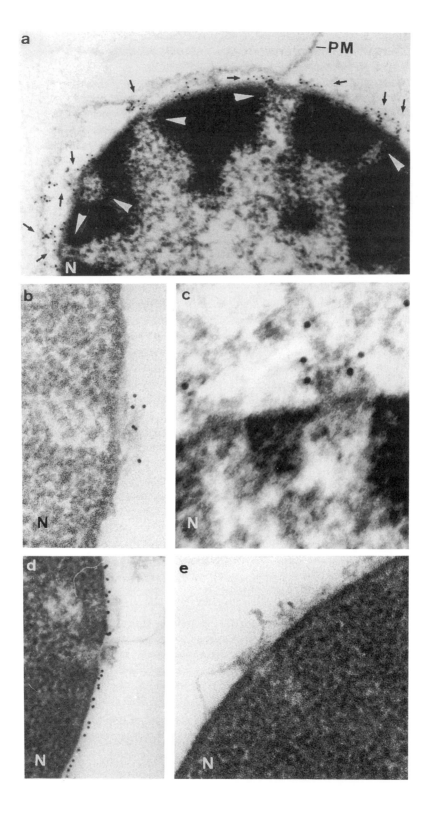

are subjected to enormous shearing stress and diluted abysmally with detergent-containing media. Such treatments are expected to cause dissociation of fortuitous complexes held together by mere "adhesion." Therefore, that cytoplasmic IFs remain connected to a "nuclear ghost" structure, while most of the other cytoskeletal elements are removed, implies a very strong link of the IF-based cytoskeleton and the lamin-based karyoskeleton.

Consistent with the idea of such a linkage, vimentin, desmin, and peripherin, three different cytoplasmic IF proteins expressed in distinct cell types, have been shown to bind to B-type lamins *in vitro* (Georgatos and Blobel, 1987; Georgatos *et al.*, 1987; Djabali *et al.*, 1991; Papamarcaki *et al.*, 1991). This binding can be detected by a variety of experimental means ranging from affinity chromatography, coimmunoprecipitation, and ligand blotting assays. The interactions of desmin, vimentin, and peripherin with nuclear lamin B do not reflect "annealing" of cytoplasmic and nuclear IF proteins through their homologous, coiled-coil regions. Instead, binding involves the nonhelical COOH-terminal tail domains of cytoplasmic IF proteins and some (yet unidentified) site at the COOH-terminal domain of B-type lamins (Djabali *et al.*, 1991).

Desmin and vimentin also associate with two plasma membrane proteins cross-reactive with antilamin B antibodies, but clearly distinct from the nuclear lamins (Cartaud *et al.*, 1989, 1990, 1995). Interestingly, one of the two proteins is a component of the desmosomal plaque and localizes exactly at the area of IF contact (see next section). On the basis of these observations, it has been proposed that a conserved structural motif, termed "the lamin B fold," may occur in a variety of cellular proteins and confer on them an ability to interact with cytoplasmic IFs (Papamarcaki *et al.*, 1991). Obviously, to confirm or refute this idea, we need to determine the amino acid sequence of the lamin B-like polypeptides and the three-dimensional

---

FIG. 2 Association of vimentin IFs with the nuclear pores as revealed by immunoelectron microscopy. All panels show turkey erythrocyte ghosts permeabilized by Triton X-100, decorated with various antibodies and protein A-gold, thin-sectioned and processed for electron microscopy. Panels (a), (b), and (c) show immunolabeling with antibodies recognizing the non-helical tail domain of vimentin. Panel (d) depicts immunostaining with antibodies specific for the lamina and lamin B-receptor associated polypeptide p18. Panel (e) shows staining with nonimmune sera. Arrowheads in (a) indicate nuclear pores, while small arrows point to vimentin material. Note the intimate association of antibody-decorated IFs with the nuclear pores in (b) and (c) and the lack of filament or pore decoration in (d) and (e). Unit membranes do not appear here due to detergent extraction of the lipid. PM indicates the plasma membrane-associated spectrin/actin skeleton and N denotes the cell nucleus. Magnifications are: (a) = 39,000; (b) = 88,000; (c) = 171,000; (d) = 121,000; (e) = 110,000.

structure of the region of the B-type lamins which is responsible for binding cytoplasmic IF proteins.

Other biochemical studies have established that nonepithelial IF proteins bind to single-stranded, double-stranded, and supercoiled DNA (Traub *et al.*, 1983, 1985a, 1992; Traub and Nelson, 1983; Vorgias and Traub, 1986; Kühn *et al.*, 1987; Shoeman and Traub, 1990). The avid and highly reproducible binding to single-stranded DNA has been exploited by many investigators for affinity-purifying IF proteins on a preparative scale. Favoring the idea that this interaction is physiologically meaningful is the finding that cytoplasmic IF proteins can be cross-linked to DNA *in situ* (Wedrychowski *et al.*, 1986a,b; Cress and Kurath, 1988). It is possible that the binding to DNA reflects a "primordial" property of the nuclear lamins also conserved among cytoplasmic IF proteins (see Section VIII).

Finally, binding of cytoplasmic IF proteins to the core histones H3 and H4 has been reported to occur under *in vitro* conditions (Traub *et al.*, 1986a). Considering the high positive charge of the histones and the negative charge of most cytoplasmic IF proteins, one may question the physiological relevance of these data. However, a fact suggesting specificity is that the highly basic linker histones (e.g., H1) do not bind to cytoplasmic IF proteins when histones H3 and H4 are present. Thus, the IF–histone interactions do not appear to be solely electrostatic.

C. Modulation of Nuclear Architecture by Cytoplasmic IFs

An illustrative example which reveals the potential for transmembrane interactions between cytoplasmic IFs and intranuclear structures is provided by recent studies on HL-60 human promyelocytic leukemia cells (Collard *et al.*, 1992). Undifferentiated HL-60 cells exhibit two peculiar properties: (1) they do not synthesize vimentin, the IF protein normally expressed in all nucleated blood cells, and (2) they possess an unusual "cap" of lamin A asymmetrically disposed along the inner side of the inner nuclear membrane. Upon treatment of such cells with phorbol esters, vimentin synthesis is induced and IFs start assembling in the cytoplasm. Interestingly, the newly formed filaments seem to originate from a cap structure adjacent to the nuclear envelope and neighboring to the lamin A cap. With time, both caps, one on the nucleoplasmic side and one on the cytoplasmic side, start growing in a coordinated fashion. In the end, vimentin filaments spread in the cytoplasm, whereas lamin A forms a uniform layer underlying the inner nuclear membrane. The coordinated migration of the intranuclear and cytoplasmic caps clearly suggests a functional coupling between the perinuclear system of IFs and the karyoskeleton.

Several other studies indicate that cytoplasmic IFs may modulate the architecture and the physical properties of the nuclear envelope. For instance, it has been observed that cells deficient in cytoplasmic IFs often possess nuclei that are either very fragile, or deeply invaginated (lobules) (Wang and Traub, 1991; Sarria *et al.*, 1994). When wild-type vimentin is transfected in such IF-null cells, the nuclear surface becomes smooth; yet, this does not occur when the transfected vimentin has been truncated and therefore is unable to form normal IF networks (Sarria *et al.*, 1994). Lobulation of the cell nucleus is also observed when dividing fibroblasts are microinjected with antivimentin antibodies (Kouklis *et al.*, 1993) and when transgenic mice are forced to express mutated keratins in the basal layer of the epidermis (Fuchs *et al.*, 1992). The simplest interpretation of these results is that cytoplasmic IFs actively "stretch" the nuclear envelope.

Another indication that cytoplasmic IFs have some influence on nuclear shape comes from *in situ* work on heart and skeletal muscle. It is widely known that the nuclei of muscle fibers possess regularly spaced indentations. These indentations seem to be in register with the Z-bands of neighboring myofibrils and constitute anchorage sites where arrays of Z-disk-associated desmin filaments attach (Tokuyasu *et al.*, 1985b). Since the shape and surface morphology of muscle nuclei change during muscle contraction and relaxation (Bloom and Cincilla, 1969), it is tempting to speculate that desmin filaments pinching on the nucleus "pull" the nuclear envelope.

Franke and Schinko (1969) have studied the morphology of isolated skeletal muscle nuclei that have been separated from myofibrils and cytoskeletal elements. Surprisingly, they found that nuclei from which the nuclear envelope had been stripped retained their indentations. One interpretation of this finding could be that the nuclear indentations reflect effects of the ionic milieu on the packing of chromatin and not stretching of the nuclear membrane by IFs. Nonetheless, it might be that the deformation of the nuclear envelope by attaching IFs changes the arrangement of the underlying heterochromatin and that this alteration is then fixed under certain ionic conditions (i.e., a form of "rigor state"). Consistent with the idea that IFs affect, directly or indirectly, the global properties of chromatin is also the fact that nuclear rotation is enhanced when neuronal cells are treated with acrylamide (Hay and De Boni, 1991). This agent is known to depolymerize the neurofilaments, but pleiotropic effects cannot be excluded.

Challenging those interpretations, recent experiments in which a single mouse vimentin gene has been knocked out indicate that genetically manipulated animals develop and reproduce without an obvious pathological phenotype (Colucci-Guyon *et al.*, 1994). Thus, the absence of cytoplasmic IFs in vimentin-expressing cells does not seem to cause major problems in nuclear metabolism. Of course, at this stage one cannot rigorously exclude the possibility that the lack of vimentin in these animals is compensated for by expression of a new, yet-uncharacterized, IF protein, nor can one rule out changes in nuclear architecture which are phenotypically "silent" in early stages of life (see Section VIII).

IV. Anchorage of Cytoplasmic IFs to the Plasma Membrane

A. Association of IFs with the Desmosomes

The best example of plasma membrane–IF attachment is the association of IFs with the cytoplasmic face of the desmosome (Farquhar and Palade, 1963; Kelly, 1966; Staehelin, 1974; Skerrow and Matolsy, 1974; Drochmans *et al.,* 1978). Desmosomes are disk-shaped junctions involved in cell–cell adhesion. They are found primarily in epithelial cells, but also occur in meningeal cells, myocardial and Purkinje cell fibers of the heart, dentritic reticulum cells of lymph nodes and spleen, and certain glial cells of lower vertebrates (Schwarz *et al.,* 1990).

The desmosomes are 0.1–2.0 μm in size and are symmetrically disposed across two adjacent cells. In between the two plasma membranes there is a space of about 20–30 nm, the desmoglea, which harbors the extracellular domains of desmosomal transmembrane glycoproteins. The cytoplasmic portion of the desmosome consists of an electron-dense plaque. This plaque has a variable thickness, depending on tissue type (Rayns *et al.,* 1969; Hull and Staehelin, 1979), and comprises three different regions (Jones *et al.,* 1986): a highly electron-dense zone subjacent to the plasma membrane, an intermediate area, and an innermost fibrillar layer with bundles of IFs attached. The filaments which associate with desmosomes are of different types. Keratin IFs interact with epithelial cell desmosomes and desmosomes present in astrocytes of amphibians (Rungger-Brandle *et al.,* 1989); desmin IFs associate with the desmosomes of cardiac muscle cells (Kartenbeck *et al.,* 1983); vimentin IFs associate with desmosomes of meningeal and dentritic reticulum cells (Kartenbeck *et al.,* 1984; Schmelz and Franke, 1993). IFs do not terminate at the plaque in an "end-on" fashion; instead, they loop through the fribrillar material of the desmosomal plaque (Kelly, 1966).

The desmosomal membrane contains two transmembrane glycoproteins, desmoglein and desmocollin, which represent members of the cadherin family (Koch *et al.,* 1990; Nilles *et al.,* 1991; Mechanic *et al.,* 1991). The desmocollins display a higher degree of homology with the classical cadherins, whereas the desmogleins possess a unique COOH-terminal domain containing 20 amino acid repeats which are not present in the other cadherins (Koch *et al.,* 1992; Schafer *et al.,* 1994). The desmosomal cadherins belong to multigene families and are expressed in a cell type-specific fashion (Schmidt *et al.,* 1994).

The desmosomal plaque contains the cytoplasmic domains of desmoglein and desmocollin as well as the peripheral membrane proteins plakoglobin and desmoplakin I (Cowin and Garrod, 1983; Cowin *et al.,* 1986; Franke *et al.,* 1982; Muller and Franke, 1983). Strictly speaking, plakoglobin is not a desmosome-specific component and is also found in diverse forms of intercellular junctions (Cowin *et la.,* 1986; Kapprell *et al.,* 1990; Franke *et al.,* 1987b). It forms a family with β-catenin and the *Drosophila* segment polarity gene product *Armadillo* (Peifer *et al.,* 1992; Butz *et al.,* 1992), and is known to interact with cadherins (Knudsen and Wheelock, 1992) and desmoglein (Mathur *et al.,* 1994).

Apart from plakoglobin and desmoplakin I, which are constitutive components of the desmosomal plaque, a variety of other proteins appear to be facultative components of the desmosome. For example, desmoplakin II, a splicing variant of desmoplakin I (Cowin *et al.,* 1985; Green *et al.,* 1988, 1990), has been detected primarily in stratified, complex, and transitional epithelia (Cowin *et al.,* 1985), but not in the desmosome-rich myocardium (Angst *et al.,* 1990; Cowin *et al.,* 1985). Another protein, plakophilin 1 (also known as "band 6 protein"), has been detected in desmosomal plaques of stratified, complex, and transitional epithelia, and has been shown to be a member of the plakoglobin family (Kapprell *et al.,* 1988; Schmidt *et al.,* 1994). Cartaud *et al.* (1990) have identified a 140-kDa protein of the desmosomal plaque in bovine muzzle epithelium that shares epitopes with nuclear lamin B (see Section III,B), while another plaque protein of 140 kDa, called "08L antigen," has been recently identified by Ouyang and Sugrue (1992). It is at present unclear whether these proteins occur in nonepithelial cells. A giant polypeptide termed desmoyokin ($M_r = 680$ kDa) has been identified as a plaque component in desmosomes of several stratified epithelia and cell lines, and has been proposed to play a role in desmosomal stability (Heida *et al.,* 1989). In addition, the calmodulin-binding proteins, desmocalmin (Tsukita and Tsukita, 1885) and keratocalmin (Fairley *et al.,* 1991), have been characterized as plaque components.

The cytoplasmic domains of desmogleins have been shown by antibody microinjection experiments to traverse the desmosomal plaque in such a way that the characteristic region that contains the repeats is exposed on the cytoplasmic plaque surface (Koch *et al.,* 1990; Schmelz *et al.,* 1986). To find out whether desmosomal cadherins are involved in plaque assembly and IF anchorage, Troyanovsky *et al.* (1993) transfected cultured epithelial cells with cDNA constructs encoding chimeric proteins. The chimeras contained the transmembrane region of the gap junction protein connexin 32 (for targeting to the plasma membrane) and the cytoplasmic domains of either desmocollins, or desmogleins. This elegant experiment showed that chimeric proteins comprising the cytoplasmic domain of the larger desmocollin variant (Dsc1a) form extensive gap junction-like structures in the plasma membrane. These structures associate with plakoglobin/desmoplakin-rich plaques and appear to anchor bundles of cytoplasmic IFs. In contrast to this, constructs encoding the cytoplasmic portion of desmoglein did not induce plaque assembly and, in fact, disrupted all endogenous desmosomes of the transfected cells. Thus, the cytoplasmic domain of desmocollin Dsc1a is sufficient to recruit components required for the formation of a desmosomal plaque and this plaque can serve as a docking site for IFs. Interestingly, connexin 32-Dsc1a constructs lacking the COOH-terminal 38 amino acids neither bound plakoglobin nor anchored IFs. Leaving aside the possibility of deletion-induced misfolding, these data indicate that the short segment of the 38 amino acids contains specific information for plaque assembly and anchoring of IFs.

Desmoplakin I is considered a prime candidate for physically linking IFs to the desmosome. Analysis of the predicted amino acid sequence of desmoplakin I (Green *et al.*, 1990) reveals that its COOH-terminal tail contains 4.6 copies of a 38-residue repeat. The periodicity of the acidic and basic residues in these repeats is the same as that found in the coil 1b subdomain of IF proteins (see Section II,C), providing a basis for potential ionic interactions between these two categories of molecules. The same type of 38-residue repeat has also been found in the COOH-terminal domain of plectin, a well-characterized IF-associated protein (Wiche *et al.*, 1991), and the 230-kDa bullous pemphigoid antigen (Green *et al.*, 1990; Tanaka *et al.*, 1991), which is located in the cytoplasmic plaque of the hemidesmosome (see Section IV,B). This said, it should be pointed out that *in vitro* assays with intact desmoplakins have so far failed to demonstrate a direct interaction with IFs (Kapprell *et al.*, 1988; O'Keefe *et al.*, 1989).

However, recent experiments with COOH-terminal fragments of desmoplakin I have documented a specific and direct binding to the NH$_2$-terminal domain of type II keratins (Kouklis *et al.*, 1994). This is in line with results reported by Stappenbeck and Green (1992), who have transfected cDNAs encoding specific domains of desmoplakin I into COS-7 and NIH-3T3 cells. They have elegantly demonstrated that truncated proteins containing the rod and COOH-terminal domain of desmoplakin I coalign with and disrupt both keratin and vimentin IF networks. It should, nevertheless, be noted that vimentin does not bind to the COOH-terminal domain of desmoplakin I *in vitro* (Kouklis *et al.*, 1994) and that transfected derivatives of desmoplakin bind very weakly to vimentin IFs *in vivo* (Stappenbeck *et al.*, 1993). Deletion experiments have shown that a segment of 68 amino acids located at the very COOH-terminal of desmoplakin I is necessary for coalignment with and disruption of keratin (but not of vimentin) filaments

(Stappenbeck *et al.*, 1993). This region shares sequence similarity with both plectin and the 230-kDa bullous pemphigoid antigen. A very interesting observation pertinent to the IF–desmosome interaction is that the keratin II binding site for desmoplakin contains a conserved *GSRS* motif, five copies of which are also found in the desmoplakin tail (Kouklis *et al.*, 1994). SR motifs have been identified in various splicing factors and are thought to provide protein–protein interaction sites usually regulated by phosphorylation with SR-specific kinases (Gui *et al.*, 1994). Taking all of these observations at face value, one should consider the possibility that the COOH-terminal portion of desmoplakin I contains a cryptic keratin-binding site which is "censored" by some other part of the desmoplakin molecule. The opening and closing of this site may be regulated by phosphorylation, or by interactions of desmoplakin I with other plaque proteins. It is clear that structural analysis of the desmoplakin tail by X-ray diffraction or multidimensional NMR will help in clarifying this question.

The currently available information does not specify whether desmoplakin I represents the sole attachment site for IFs at the desmosomal plaque, or whether it provides a facultative anchorage point utilized in a subset of desmosomes. Intuitively, the latter possibility seems more likely than the former, given that vimentin does not bind to desmoplakin I *in vitro*, whereas in some cell types vimentin filaments clearly connect to the desmosomes (see earlier discussion). In fact, other plaque proteins have been found to interact directly with IFs. These proteins include desmocalmin (Tsukita and Tsukita, 1985), keratocalmin (Fairley *et al.*, 1991), and plakophilin 1 (Kapprell *et al.*, 1988), which have been found to bind keratin proteins *in vitro*. Another potential IF–plaque linker is the lamin B-like 140-kDa protein which binds to vimentin (Cartaud *et al.*, 1990) and a 140-kDa glycoprotein recognized by bullous pemphigoid autoantibodies (Jones, 1988). Finally, a particularly interesting case is the polypeptide IFAP-300, a high-molecular-weight protein which coisolates with IFs. Using both morphological and biochemical techniques, Skalli *et al.* (1994) have demonstrated that IFAP-300 is present in the plaques of desmosomes and hemidesmosomes and that it binds polymerized keratins *in vitro*.

It is quite possible that IFs connect to the desmosomal plaque through multiple adaptor proteins. To decide which of those proteins are the "bolts" and which the "Scotch tape," one may have to use genetic approaches. For example, desmoplakin, plakophilin, and plakoglobin knockouts may reveal the relative contribution of these proteins in the anchorage of IFs and in the organization of the desmosomal plaque.

B. Association of IFs with the Hemidesmosomes

Hemidesmosomes are punctate junctions that mediate cell adhesion to the basal membrane in stratified and transitional epithelia (Farquhar and

Palade, 1963; Staehelin, 1974). Despite the morphological similarities between desmosomes and hemidesmosomes, the latter are not "half-desmosomes" as their name implies. The hemidesmosome possesses a cytoplasmic plaque which is linked to IFs (Jones et al., 1986) and a subbasal dense plate anchoring to the basal lamina or collagen fibrils. In contrast to what is seen in desmosomes, it is generally thought that only keratin IFs interact with hemidesmosomes. In mammalian hemidesmosomes, it has been suggested that IFs may terminate at the hemidesmosomal plaque in an "end-on fashion" (Jones et al., 1986).

The characterization of the structural components of the hemidesmosome has lagged behind, in part because of difficulty in isolating these structures. The principal adhesion molecules in hemidesmosomal membranes are the α_6/β_4 integrins (Stepp et al., 1990; Jones et al., 1991). The α_6 subunit is a transmembrane protein with a small cytoplasmic domain and shows homology to other integrin α-chains. Unlike other β integrins, the β_4-subunit has an enormous cytoplasmic domain and occurs in three alternatively spliced forms (Tamura et al., 1990; Suzuki and Naitoh, 1990). It has been suggested that this domain represents a molecular specialization necessary for plaque formation (Sonnenberg et al., 1991), but the caveat in this hypothesis is that β_4 also occurs in epithelia which do not possess hemidesmosomes.

Using affinity-purified autoantibodies from patients with bullous pemphigoid, Klatte et al. (1989) have shown that the hemidesmosomal plaque contains a 230-kDa (called BPA) and an 180-kDa polypeptide. The 180-kDa protein is a type II transmembrane component (Hopkinson et al., 1992) possessing an extracellular region which is homologous to collagen (Guidice et al., 1991). Using another monoclonal antibody, Kurpakus and Jones (1991) have characterized a 200-kDa polypeptide and localized it in the innermost region of the hemidesmosome plaque. Finally, Hieda et al. (1992) have characterized a 500-kDa protein (HD1) as a hemidesmosomal component. Immunoelectron microscopy of corneal epithelium indicates that at least a portion of the HD1 molecule is located in the innermost region of the hemidesmosomal plaque. However, HD1 is also found in endothelial and glial cells which lack hemidesmosomes.

A putative IF-plaque linker in hemidesmosomes is the $\beta 4$ integrin. Spinardi et al. (1993) have shown that an internal deletion mutant of its large cytoplasmic domain perturbs the association of hemidesmosomes with IFs in transiently transfected keratinocytes. Another candidate protein is the 230-kDa bullous pemphigoid antigen (BPA). As described earlier, BPA belongs to a family of proteins which includes desmoplakins and plectin. Hence, BPA may occupy the corresponding position and perform the same IF-linking role in hemidesmosomes as desmoplakin I in the desmosomes. Striking evidence favoring this idea has been provided recently by Guo et al. (1995). These workers have found that the hemidesmosomes of BPA-null mice have no IFs attached. In retrospect, this finding reinforces the idea that BPA, plectin, and desmoplakin contain common structural motifs which mediate the connections of cytoplasmic IFs with other cellular structures.

C. Association of IFs with the Erythrocyte Membrane

The interaction of cytoplasmic IFs with the plasma membrane is not limited to the desmosomes and hemidesmosomes because IFs are often associated with the plasma membrane of cells that do not possess such specialized junctions. Owing to the complexity of the subplasmalemmal region, two simple models have been used for studying the association of IFs with the plasma membrane: the mature erythrocyte (this section) and the lens fiber cell (Section IV,D).

Granger *et al.* (1982) were the first to unveil an extensive IF system associated with the plasma membrane of avian erythrocytes. This became possible by attaching erythrocytes to cationized glass, opening the cells by sonication, and viewing the specimens *en face*. Indirect immunofluorescence and rotary shadowing of membrane "mats" showed that IFs were multiply connected to the plasma membrane. A similar association between cytoplasmic IFs and the plasma membrane has been seen in amphibian erythrocytes (Centonze *et al.*, 1986).

It has long been thought that IFs may anchor to the membrane skeleton, a proteinaceous meshwork lining the inner face of the plasma membrane in most eukaryotic cells. The major constituent of the erythrocyte membrane skeleton is the protein spectrin. Spectrin participates in multiple associations, including binding to actin, protein 4.1, and ankyrin, its principal membrane attachment site (Bennett and Gilligan, 1993). It exists as a heterotetramer and consists of two subunits (α and β spectrin). Spectrin is not unique to erythrocytes and is expressed in almost all cells of vertebrates. An analog of erythrocyte spectrin called fodrin has been initially characterized in brain (Levine and Willard, 1981), while another spectrin-like protein, termed TW 260/240, has been identified in the intestinal terminal web (Glenney *et al.*, 1982). Similar proteins have been characterized in invertebrates, including *Drosophila* (Byers *et al.*, 1992; Dubreuil *et al.*, 1989), echinoderms (Wessels and Chen, 1993), *Dictyostelium* (Bennett and Condeelis, 1988), and *Caenorhabditis elegans* (Loer, 1992). Recent studies show that mammals express two related but distinct α spectrins, a tissue-invariant form (α_G) and an erythrocyte-specific form (α_{RBC}). In contrast, birds and other vertebrates express a single form of α spectrin which is found in erythrocytes and in other tissues. β Spectrins are diverse and comprise at least five different isoforms (β_R, β_G, β_{NM}, β_{TW}, and β_H) (Bennett and Gilligan, 1993).

Microinjection of antispectrin antibodies into Madin-Darby Canine Kidney (MDCK) cells and fibroblasts causes intracellular precipitation of spectrin and results in aggregation and altered distribution of vimentin filaments (Mangeat and Burridge, 1984). Along the same lines, Hirokawa *et al.* (1982, 1983) have observed structural links between actin microfilaments and IFs in mouse intestinal brush border and have suggested that these filament linkers may belong to the spectrin family. Finally, in differentiating PC12 cells, fodrin is localized at the perinuclear region closely associated with peripherin and neurofilaments (Takemura *et al.*, 1993). Taken together, these observations constitute strong circumstantial evidence suggesting that IFs interact with spectrin-associated proteins.

Spectrin-dependent binding of vimentin and desmin filaments to human erythrocyte inside-out vesicles (IOVs) has been detected using *in vitro* binding assays (Langley and Cohen, 1986, 1987). It has also been found that spectrins of different origins bind preferentially types of IFs (Langley and Cohen, 1987). Thus, vimentin filaments bind more avidly to erythrocyte spectrin, while neurofilaments bind preferentially to brain spectrin (fodrin). Despite this, the interpretation of the available binding data is not straightforward. On one hand, the binding of vimentin and desmin filaments to spectrin-reconstituted vesicles is highly substoichiometric (molar ratio of spectrin : desmin = 1 : 230), consistent with a focal, or "end-on" anchorage of the filaments to the membranes. However, this does not fit morphological observations showing that vimentin/desmin IFs associate with the surfaces of spectrin-reconstituted vesicles through multiple lateral contacts (Langley and Cohen, 1986).

Other experiments have shown that, in contrast to fully polymerized vimentin, oligomeric vimentin (which contains more "ends" than an equivalent amount of filamentous vimentin) binds in a saturable manner to IOVs depleted of spectrin and actin. Extraction of protein 4.1 or reconstitution with purified spectrin does not seem to affect the association of vimentin with the erythrocyte membrane, but removal of ankyrin significantly lowers the binding. Finally, vimentin has been found to bind under *in vitro* conditions to human and avian erythrocyte ankyrin (Georgatos *et al.*, 1985, 1987). On the basis of this information, it has been proposed that ankyrin rather than spectrin provides the attachment sites for IF proteins in the erythrocytic membrane skeleton. Ankyrin-like polypeptides have been detected in many nonerythroid cells (Bennett and Gilligan, 1993).

D. Association of IFs with the Plasma Membrane of Lens Fiber Cells

The eye lens consists of two principal cell populations, the anterior epithelial cells and the lens fiber cells (LFCs), which extend from the anterior to the posterior surface of the lens. The LFCs originate from the lens epithelium. As they terminally differentiate, LFCs acquire a hexahedral shape, develop an extensive system of intracellular junctions, accumulate crystallins, and lose their nuclei (Rafferty, 1985; Maisel *et al.,* 1981).

The membrane skeleton of the LFCs and of the mammalian erythrocyte can be considered analogous. Most of the typical components of the erythrocyte membrane skeleton (spectrin, actin, protein 4.1, ankyrin, etc.) also occur in the lens (Allen *et al.,* 1987; Aster *et al.,* 1984a,b, 1986). The cytoskeleton of the LFCs includes a few microtubules and numerous vimentin IFs (Benedetti *et al.,* 1981; Ramaekers and Bloemendal, 1981). A tight link between vimentin filaments and the LFC membrane has been reported on the basis of *in situ* and biochemical observations (Bradley *et al.,* 1979; Ramaekers *et al.,* 1982). Electron microscopy has shown that vimentin filaments connect in an "end-on" fashion to the plasma membrane. Moreover, nascent vimentin synthesized *in vitro* from total lens mRNA becomes associated with lens membranes added immediately after translation (Ramaekers *et al.,* 1982).

To characterize lens-specific proteins involved in the anchorage of IFs to the plasma membrane, Merdes *et al.* (1991) have studied the properties of a 100–110-kDa protein, termed filensin, which they identified as a component of the LFC membrane skeleton by biochemical methods and immunoelectron microscopy. Purified filensin can reassociate with urea-stripped lens cell membranes, suggesting an interaction with intrinsic elements of the membrane (Brunkener and Georgatos, 1992). Radiolabeled filensin binds specifically to lens vimentin and a 47-kDa peripheral membrane protein called phakinin. Phakinin exactly colocalizes with filensin *in situ* (Merdes *et al.,* 1993).

Avian and mammalian filensin, as well as phakinin, have been recently cloned and sequenced (Remington, 1993; Gounari *et al.,* 1993; Merdes *et al.,* 1993; Hess *et al.,* 1993; Masaki and Watanabe, 1992). Their primary structure reveals that both are members of the IF superfamily of proteins. Moreover, filensin and phakinin have been shown to be the components of lens-specific cytoskeletal structures, known as the beaded-chain filaments (BFs) (Georgatos *et al.,* 1994a). These unique filamentous structures line the inner aspect of the plasma membrane of the LFCs and probably connect to vimentin filaments at focal points. The submembranous filensin/phakinin filament lattice may link transcytoplasmic IFs to the plasma membrane and stabilize the LFC membrane against mechanical stress.

V. Interactions of Cytoplasmic IFs with Cytoplasmic Organelles

A. Association of IFs with the Dense Bodies of Smooth Muscle and the Z-disk of Skeletal Muscle Fibers

IFs are present in smooth, skeletal, and cardiac muscle, but of the three muscle types IFs are most abundant in the former. In adult smooth muscle,

IFs made principally of desmin (Lazarides and Hubbard, 1976; Small and Sobieszek, 1977) form a network that links the dense bodies with membrane-bound dense plaques (Uehara et al., 1971; Cooke and Chase, 1971; Cooke, 1976). The dense bodies are elongated structures sometimes appearing as chains up to 1.5 μm in length; they are often continuous across the cell for 200 to 300 nm and are interconnected by filamentous material (Bond and Somlyo, 1982).

The structural arrangement of IFs in smooth muscle cells is still controversial. Small and Sobieszek (1980) have proposed that smooth muscle IFs and associated dense bodies form an intracellular domain separate from the domain formed by the actin and myosin filaments and that the two domains could be observed in cross sections. However, actin filaments are also found attached to dense bodies (Bond and Somlyo, 1982; Somlyo and Franzini-Armstrong, 1985). Using high-resolution immunocytochemistry on fresh avian gizzard smooth muscle, Stromer and Bendayan (1988) have observed that IFs are organized into an axial bundle associated with both the nucleus and the mitochondria. The bundle, which probably extends along the entire length of the cell, may support and define the position of various organelles. In addition, IFs are concentrated around and between cytoplasmic dense bodies in arrays that are predominantly parallel to the myofilament axis, and also between cytoplasmic dense bodies and membrane-associated plaques. Fay et al. (1983) have indicated that the dense bodies are linked to each other by filaments that are oriented at an oblique angle with respect to the cell axis. Freezing and deep-etching of smooth muscle (Somlyo and Franzini-Armstrong, 1985) show that cytoplasmic dense bodies are linked with IFs by "struts" to their sides.

There are no reports to indicate the existence of an axial IF bundle in embryonic or neonatal smooth muscle. In studying the mechanism of assembly of the contractile and cytoskeletal elements in developing smooth muscle, Chou et al. (1992) could first identify the axial IF bundle in 1-day postnatal cells. In fact, myofilaments (actin and thick filaments) and filament attachment sites (cytoplasmic and membrane-associated dense bodies) are assembled before the axial IF bundle. These data imply that the axial IF bundle may have a minor role in smooth muscle cell development.

It has been suggested that dense bodies may be analogous to striated muscle Z-lines (Lazarides, 1980). Using immunocytochemical techniques on longitudinal sections of myofibrils, desmin has been localized, together with α-actinin and actin, at the level of the Z-disk (Lazarides and Hubbard, 1976; Lazarides and Granger, 1978; Bennett et al., 1978; Campbell et al., 1979). In adult chicken skeletal muscle fibers, desmin is found at the peripheral domain of each Z-disk (Granger and Lazarides, 1978; Richardson et al., 1981). Moreover, by high-resolution immunoelectron microscopy, it has been observed that transverse desmin filaments surround myofibrils at the level of Z-disks, while longitudinal desmin filaments connect the peripheries of successive Z-disks and extend to the sarcolemma (Tokuyasu *et al.*, 1983; Wang and Ramirez-Mitchell, 1983). It has been proposed that in skeletal muscle the desmin filaments serve to integrate various components of the contractile apparatus by linking individual myofibrils laterally at the level of their Z-disks as well as connecting Z-disks with the plasma membrane (Lazarides, 1980; Tokuyasu *et al.*, 1983). In cardiac muscle, IFs are much more abundant than in skeletal muscle and localize in the intrafibrillar spaces, in addition to the transverse IFs which surround myofibrils at the level of the Z-disk (Ferrans and Roberts, 1973; Behrendt, 1977; Tokuyasu, 1983).

IFs seem to be more abundant in embryonic than in adult skeletal muscle. In studying cultures of myogenic cells and their *in vitro* progression to myotubes, Bennett *et al.* (1979) have observed by immunofluorescence microscopy that both desmin and vimentin are present in immature myotubes in the form of longitudinal filaments, but vimentin gradually disappears as the myotubes mature, whereas the distribution of desmin changes to a cross-striated pattern after the emergence of striated myofibrils. In contrast, Granger and Lazarides (1979) have claimed that the two proteins coexist at the periphery of the Z-disk in isolated myofibrils of adult chicken skeletal muscle. Moreover, desmin and vimentin have been reported to codistribute in developing myotubes in culture and may play a role in the lateral alignment and organization of myofibrils (Gard and Lazarides, 1980). Further studies have generally supported the thesis of Bennett and coworkers showing that vimentin is undetectable in mature myotubes (Tokuyasu *et al.*, 1984, 1985a). However, realizing that the amount of vimentin in adult muscle is greatly reduced, the discrepancy here may be explained by differences in experimental technique and the use of the antibodies.

Using immunoelectron microscopy on ultrathin frozen sections of muscle of different ages, Tokuyasu *et al.* (1985a) have demonstrated that the transverse orientation of IFs in mature myotubes results from a continuous shifting of the positions and directions of the filaments. Unfortunately, little is known about the signals that stimulate the translocation of longitudinally deployed IFs to their final transverse association at the level of the Z-band (Holtzer *et al.*, 1985).

To examine whether changes in IF subunit composition influence filament organization, Cary and Klymkowsky (1994a) have used the dorsal myotome of the *Xenopus laevis* embryo as a model system. In this type of muscle, desmin filaments assemble in the absence of a preexisting vimentin filament system (Cary and Klymkowsky, 1994b). It turns out that desmin organization in the absence of vimentin is quite different from that seen in vimentin/desmin-containing cells. For example, longitudinal IF arrays are not formed; rather, desmin is initially concentrated at the intersomite junction and the lateral

sarcolemma. As myogenesis proceeds, a reticular network of desmin filaments appears, followed by the association of desmin with Z-lines. Using the
same system, Cary and Klymkowsky (1994a) have examined the behavior of
epitope-tagged vimentin and desmin in embryonic cells. Myocytes expressing
exogenous vimentin were found to contain longitudinal IF arrays, whereas
cells expressing desmin contained a reticular IF meshwork and nonfilamen-
tous aggregates. The head domains of desmin and vimentin appear to be
responsible for their different behavior in muscle cells. Finally, in dorsal myotome cells, overexpression of rod/tail truncated desmin and vimentin causes
major defects in the lateral membranes and sometimes provokes detachment
from the intersomite junction (Cary and Klymkowsky, 1995).

To investigate the putative functions of desmin filaments in the differentiation of muscle cells, Schultheiss *et al.* (1991) have used an expression
vector containing a cDNA coding for a truncated version of desmin. When
expressed in muscle and nonmuscle cells, the truncated protein blocks the
assembly of desmin and vimentin filaments and also causes the disassembly
of preexisting desmin and vimentin filaments. Despite the complete lack
of intact IFs, myoblasts and myotubes expressing truncated desmin assemble and laterally align normal striated myofibrils that contract spontaneously
in a manner indistinguishable from that of control myogenic cells. Because
myoblasts or myotubes lacking IFs are still able to express their normal
differentiation program, the authors argue that desmin filaments may not
play a role in myogenesis.

In contrast to this, a recent study has revealed that silencing desmin
expression in cultured myoblasts arrests fusion and formation of myotubes
(Li *et al.,* 1994). The catch in this system is that treatment of the cells
with desmin antisense oligonucleotides dramatically lowers expression of
muscle-specific transcriptional factors such as MyoD and myogenin. However, the effect seems to be specific because expression of desmin by transfection of the arrested cells restores their ability to fuse into myotubes.
Disruption of the desmin gene in ES cells and examination of *in vitro*
myogenesis in embryoid bodies (EBs) yielded results confirming and extending the previous observations. Skeletal and smooth muscle formation
in desmin-null EBs was severely affected. However, as paradoxical as it
might seem, reduced expression of desmin in heterozygous EBs caused
arrhythmic beating of cardiomyocytes, whereas the complete lack of desmin
in desmin-null EBs did not lead to any significant malfunction in comparison
to the wild type (Y. Capetanaki, personal communication).

B. Potential Interactions of Cytoplasmic IFs with the Microtubules and the Centrosome

Various observations have indicated that the extended configuration of
cytoplasmic IF networks depends on microtubules. Specifically, it has been

found that perturbance of the microtubules with drugs or antitubulin antibodies causes retraction of IFs from the cell periphery and collapse into a perinuclear "cap" structure (Ishikawa *et al.*, 1968; Goldman, 1971; Geiger and Singer, 1980; Goldman *et al.*, 1980; Maro and Bornens, 1982; Maro *et al.*, 1982; Geuens *et al.*, 1983; Blose *et al.*, 1984). The same type of IF "collapse" has been seen when antibodies against the microtubular motor kinesin were microinjected into living cells (Gyoeva and Gelfand, 1991). Interestingly, *in vitro* binding of neurofilaments to MAPs has been reported (Heinmann *et al.*, 1983), while other studies have implicated a 210-kDa IF-associated protein in IF–microtubule interactions. Together, these findings suggest that IFs, or IF-associated proteins, may interact with microtubules or microtubule motors (for elaborations on this, see Skalli and Goldman, 1991; Draberova and Draber, 1993).

Opposite to this idea is the fact that the collapse of cytoplasmic IFs induced by certain agents (e.g., acrylamide) or anti-IF antibodies does not affect the distribution of microtubules in living cells (Eckert, 1986). In addition, although microtubules and IFs partially coalign at the level of indirect immunofluorescence microscopy, the two systems seem to be dissociated when cells are examined by electron microscopy. Of course, this does not rule out focal contacts of IFs with the microtubules, but the data could be better explained if we assume that IFs and microtubules compete for a common set of accessory protein which regulate their attachments to membranous organelles. For example, it is interesting to note here that the membrane-skeletal protein ankyrin binds to both tubulin and IF proteins (Davis and Bennett, 1984; Georgatos and Marchesi, 1985). Massive depolymerization of the microtubules may create a vast excess of tubulin subunits which would compete for plasma membrane binding sites also utilized by IFs. This situation may trigger the detachment of IFs from the plasma membrane and collapse upon the nucleus. In contrast to this, collapse of IFs antibody cross-linking should not be expected to influence the distribution of microtubules, because the pool of soluble IF subunits will not change and, therefore, the microtubules will not be competed for by IF proteins.

The idea that IFs are somehow associated with the centrioles or the pericentriolar material (PCM) has also been mentioned several times in the literature. In many instances, IFs seem to emanate from a juxtanuclear "organizing center" located in the area of the centrosome (Celis *et al.*, 1984; Eckert *et al.*, 1984). However, a stable interaction between the centrosome and the IFs was not observed when systematic studies addressed this problem in detail. Specifically, Maro *et al.* (1984), using HeLa cells as a model system, have shown that vimentin filaments do not codistribute with the centrioles when the centrosome is dispersed by inducing microtubule polymerization with taxol or the microtubules are depolymerized with nocodazole. A more recent investigation in SW13 cells yielded basically the same results (Trevor *et al.*, 1995). However, the new twist here was that a partial colocalization of the two structures could be observed when low amounts of vimentin were expressed in vimentin-free SW13 cells. Under such circumstances, newly made vimentin accumulated at a juxtanuclear focus which contained γ-tubulin and "infiltrated" the pericentriolar region. These results are not necessarily contradictory to the previous observations and may suggest a "facultative" interaction of IFs with the centrosome. An alternative interpretation of the above mentioned results may be that at low levels of vimentin expression, vimentin network organization is abnormal and the little amount of vimentin present is collapsed by minus-end directed movement along the microtubules.

C. Association of IFs with Mitochondria

Close spatial associations between IFs and mitochondria have been frequently observed in the course of electron microscopic surveys (Buckley and Porter, 1967; Ishikawa et al., 1968; Nickerson et al., 1970; Bernstein and Wollman, 1975; Mose-Larsen et al., 1982; Lazarides and Granger, 1983). An exceptionally lucid example of IF–mitochondria connections has been provided by Hirokawa et al. (1982), who have demonstrated the existence of short fibrillar cross-bridges between mitochondria and neurofilaments in unfixed frog axons. In addition, Harris and Brown (1971) have described fibers connecting the mitochondria with the plasma membrane of avian erythrocytes, while interconnections between IFs and mitochondria have been seen in undifferentiated leukemia cells (Felix and Strauli, 1976), epithelial cells (Zerban and Franke, 1977), Leydig cells (David-Ferreira and David-Ferreira, 1980; Pelliniemi et al., 1982), adrenal cells (Almahbobi et al., 1992a), smooth muscle cells (Stromer and Bendayan, 1990), and in fibroblasts (Toh et al., 1980). In the lactating cow mammary secretory cells, the mitochondria appear to associate with the keratin-desmosome complex (Lee et al., 1979). However, there are some exceptions to this rule: for instance, in cultured human skin fibroblasts, the connections of IFs with mitochondria are less obvious than in other systems (Mose-Larsen et al., 1982).

When cells are treated with organic solvents such as methanol and acetone or with Triton X-100, the mitochondria (which are normally long, filamentous structures) collapse into beadlike structures. Examination of methanol/acetone-treated cells by indirect immunofluorescence reveals that collapsed mitochondria coalign with IFs (Mose-Larsen et al., 1982). Furthermore, treatment of cultured cells with antimicrotubule drugs collapses the IF network around the nucleus and induces the disruption of mitochondria (Heggeness et al., 1978; Toh et al., 1980; Summerhayes et al., 1983). In some of these cases the mitochondria redistribute and localize along the perinuclear ring of IFs. For example, in embryonic fibroblasts, the distribution of mitochondria and vimentin filaments changes in a similar manner when cells are treated with colchicine (Toh *et al.,* 1980). In gerbil fibroma cells, the location of mitochondria and vimentin filaments is strongly correlated in both untreated cells and during the perinuclear retraction of IFs upon colcemid treatment (Summerhayes *et al.,* 1983). However, there are, again, a few exceptions: kangaroo epithelial cells (PtK2) treated with colchicine show no apparent change in either the mitochondria or in the distribution of keratin filaments (Summerhayes *et al.,* 1983).

The effects of antimicrotubule drugs on mitochondria do not appear to be a direct consequence of microtubule depolymerization. Thus, cycloheximide-treated CV-1 cells, in which the IF system has collapsed but microtubules have not been affected (Sharpe *et al.,* 1980), reorganize IFs and mitochondria in the same way (Chen *et al.,* 1981).

Recently, Shyy *et al.* (1989) have used heat-shocked mouse mammary epithelial cells to show that both keratin IFs and mitochondria retract simultaneously around the same perinuclear zone. These results have been confirmed by Collier *et al.* (1993), who demonstrated that severe heat shock of chicken embryo fibroblasts induces changes in both mitochondria and IF networks. They have also observed that the redistribution of IFs and mitochondria occurs even in the presence of taxol, a microtubule-stabilizing drug.

Together, these observations imply that, at least in some systems, IFs are tightly associated with mitochondria. Clearly, this interaction is regulated differently, depending on cell type, and does not appear to be a major determinant of mitochondrial mobility *in vivo* (Chen, 1988). Unfortunately, information about the molecules that mediate the IF–mitochondria connections is not available.

D. Association of IFs with Lipid-Containing Structures

Several reports indicate that vimentin IFs may interact with lipids and lipid-containing organelles. For example, the presence of an IF meshwork around lipid droplets in adipocytes and a close spatial association of IFs with lipid droplets in cholesterol-loaded macrophages have been described (Luckenbill and Cohen, 1966; Novikoff *et al.,* 1980; McGookey and Anderson, 1983). Furthermore, during adipose conversion, vimentin filaments rearrange to assemble a cage-like structure of extraordinary regularity around nascent lipid globules (Franke *et al.,* 1987a). In steroid-producing cells, the lipid droplets in which cholesterol is stored have been found to be attached to IFs (Almahbobi and Hall, 1990; Almahbobi *et al.,* 1992b). Finally, Gillard and coworkers (1991, 1992) have found that glycosphingolipids colocalize with IFs in a variety of cell types, e.g., fibroblasts, smooth muscle cells, hepatoma cells, and keratinocytes. It has been suggested that glycolipids may associate with IFs either as a components of membrane vesicles, or directly.

New observations suggest a role for vimentin IFs in cholesterol metabolism. This conclusion has been reached in studies involving clones of SW13 human adrenal carcinoma cells. Subclones of these cells missing or expressing vimentin have been isolated and synthesis of neutral lipids has been carefully studied (Sarria *et al.*, 1992). It turns out that cells lacking vimentin filaments synthesize cholesterol at a higher rate compared with vimentin-expressing cells or vimentin-minus cells transfected with constructs encoding mouse vimentin. Furthermore, the vimentin-minus cells exhibit decreased level of cholesterol esterification. The molecular basis of this phenomenon is not known.

Consistent with the fact that IFs are closely associated with lipid-containing organelles *in vivo* is the observation that IF proteins bind lipids *in vitro*. Neutral lipids and phospholipids remain associated with keratin and vimentin IFs after extraction of the cells with detergent (Traub *et al.*, 1985b; Asch *et al.*, 1990). Moreover, Traub *et al.* (1985b, 1986b, 1987) have reported direct interactions between purified vimentin and various lipid preparations, including cholesterol. However, the *in vitro* binding of IF proteins to pure lipids should be interpreted with caution. First, it is rather unlikey that IF subunits ever "see" naked lipids *in vivo*. Instead, it is very likely that IFs (i.e., core filaments and associated proteins) touch on membranes containing proteolipid. Second, it is rather improbable that assembled IF proteins expose hydrophobic domains to the cytoplasmic environment. In fact, the only (relatively) hydrophobic region of IF proteins is the rod domain which, as explained in the introduction, forms coiled-coils and buries its apolar side inside the inner core of the polymer.

VI. Interactions of the Nuclear Lamins with the Inner Nuclear Membrane

Nuclear lamins represent a major branch of the IF protein family and form the nuclear lamina, a polymeric structure that lines the nucleoplasmic surface of the inner nuclear membrane. They can be divided into type-A and type-B subunits and, similarly to other IF proteins, possess two nonhelical NH_2- and COOH-terminal regions (head and tail) and a coiled-coil middle domain (rod). The lamins have unique characteristics, i.e., a larger rod (see the introduction), distinct cdc2 kinase phosphorylation sites, a nuclear localization signal (NLS) in the tail domain, and a CAAX box at the COOH-terminal for post-translational isoprenylation and carboxymethylation. The CAAX box occurs in all B-type lamins and in a short-lived lamin A precursor, but does not exist in mature A-type lamins (Georgatos et al., 1994b; Nigg, 1992).

During interphase, the lamins are thought to interact with chromatin (Glass and Gerace, 1990; Burke, 1990; Yuan et al., 1991; Glass et al., 1993; Luderus et al., 1992) and with the inner nuclear membrane (Worman et al., 1988; Foisner and Gerace, 1993). Initially, the association of the nuclear lamina with the inner nuclear membrane was attributed to direct interactions between the isoprenoid tail of the B-type lamins and the lipid bilayer (Krohne et al., 1989; Holtz et al., 1989; Kitten and Nigg, 1991). However, more recently, Firmbach-Kraft and Stick (1993) have shown that isoprenylation, although necessary for membrane targeting, is not sufficient for the stable association of lamins with membranes and that additional factors are involved in lamin–membrane interactions.

Biochemical studies have identified several integral membrane proteins that may link the lamins to the nuclear envelope. One such protein is the lamin B receptor (LBR or p58), a polypeptide originally detected in bird erythrocyte nuclear envelopes (Worman et al., 1988; Bailer et al., 1991) and now known to be present in a variety of organisms and cell types (Georgatos et al., 1994b). In vertebrates, LBR possesses a long NH_2-terminal domain located in the nucleoplasm, eight predicted membrane-spanning segments, and a hydrophilic COOH-terminal domain (Worman et al., 1990; Ye and Worman, 1994). The NH_2-terminal domain of LBR contains DNA-binding motifs and cdc2 kinase phosphorylation sites (Worman et al., 1990; Courvalin et al., 1992).

LBR binds directly to lamin B in vitro (Worman et al., 1988; Ye and Worman, 1994) and in vivo (Simos and Georgatos, 1992; Meier and Georgatos, 1994; Smith and Blobel, 1993). In nucleated erythrocytes, LBR forms a multimeric complex which includes the lamins, a specific nuclear envelope kinase, and and three other polypeptides with molecular masses of 150 kDa (p150), 34 kDa (p34), and 18 kDa (p18) (Simos and Georgatos, 1992). The identity of the p150 protein remains unknown. However, determination of the NH_2-terminal amino acid sequence of p34 by Simos and Georgatos (1994) has shown that this polypeptide is homologous to p32, a splicing factor 2 (SF2)-associated protein, originally identified in HeLa cells (Krainer et al., 1991). Biochemical experiments and microsequencing of p18 have revealed that this protein is an intrinsic membrane component related to a 17-kDa isoquinoline-binding protein which is part of the peripheral-type benzodiazepine receptors (PBRs). p18 has been localized at the nuclear envelope by immunoelectron microscopy and shown to bind to LBR and type-B lamins in vitro (G. Simos, C. Maison and S. D. Georgatos, unpublished observations). The LBR kinase, which is distinct from cdc2, kinase A, and casein kinase, has been recently characterized and found to phosphorylate serine-arginine (SR) motifs present in the NH_2-terminal domain of the LBR and in splicing factors. Interestingly, phosphorylation of SR sites in LBR abolishes binding of LBR to p34 (E. Nikolakaki, G. Simos, S. D. Georgatos, and T. Giannakouros, unpublished observations).

Apart from LBR, four lamina-associated polypeptides (LAPs) of the inner nuclear membrane have been identified in rat liver nuclei (Senior and Gerace, 1988; Foisner and Gerace, 1993). The LAPs consist of LAP 1A (75 kDa), 1B (68 kDa) and 1C (55 kDa), and LAP 2 (53 kDa). Recently, LAP2 and LAP1 C have been cloned and sequenced (Furukawa et al., 1995; Martin et al., 1995). Both polypeptides seem to be type II integral membrane proteins. The isotypes of LAP 1 (which may arise from alternative splicing) show different binding affinities for the lamins. LAPs 1A and 1B interact with both type-A and type-B lamins, while LAP 1C does not bind to lamins under in vitro conditions. LAP 2 binds to lamin B and to mitotic chromosomes. LAP 1 and LAP 2 are phosphorylated during interphase and mitosis. Mitotic phosphorylation of LAP 2 abolishes in vitro binding to lamin B and chromosomes.

During mitosis, the association of the lamins with the inner membrane is highly modulated. Whereas the bulk of the A-type lamins dissociate from the nuclear envelope and disperse in the cytoplasm, B-type lamins remain bound to membraneous structures. Previous morphological studies using mitotic chicken erythroid cells have suggested an interaction of B-type lamins with extended membrane cisternae that the morphologically resembled elements of endoplasmic reticulum (ER) (Stick et al., 1988). In addition, employing mitotic HeLa cells, Chaudhary and Courvalin (1993) have claimed that B-type lamins dissociate from nuclear envelope vesicles and reassemble independently from LBR at the end of mitosis. However, since these studies did not involve simultaneous monitoring of LBR and B-type lamins in the same mitotic cell, it is hard to tell whether the interpretation of the data is correct.

Recent observations in mitotic CHO cells indicate that the mitotic vesicles which carry lamin B are depleted of ER, Golgi, and endosomal membrane proteins (Maison et al., 1993). This suggests that B-type lamins do not dissociate from their inner nuclear membrane receptors during mitosis and is consistent with the findings of Meier and Georgatos (1994), who have shown that LBR colocalizes and coimmunoprecipitates with type-B lamins during mitosis. The vimentin-associated vesicles have been isolated using magnetic immunobeads and shown to bind specifically to chromosomes in a lamin B, cytosol, and phosphorylation-dependent manner (Maison et al., 1995). From these data it can be inferred that vimentin IFs serve as a transient docking site where nuclear membrane vesicles are sequestered during mitosis.

VII. *In Vivo* Assembly of IFs

In general, there are two modes of IF assembly *in vivo:* assembly of new filaments into a system of preexisting IFs and *de novo* assembly in the absence of a preexisting IF scaffold. Forced expression of IF proteins in cultured cells has revealed that newly made subunits can integrate into preexisting polymers (Albers and Fuchs, 1987, 1989; Chin and Liem, 1989; Sarria *et al.,* 1990; Ngai *et al.,* 1990; Raats *et al.,* 1992). This mechanism maybe relevant to events occurring during embryonic development when differentiation-specific IFs gradually substitute for the "primordial" IFs. Nevertheless, incorporation of exogenous IF proteins into the endogenous IF system is not always possible and requires subunit "compatibility" (for example, NF-L or desmin readily incorporate into vimentin filaments, but vimentin does not incorporate into keratin filaments).

De novo assembly of IFs seems to occur from a variety of sites. In most cases, when IF subunits unable to incorporate into the endogenous IFs are expressed in cultured cells, the nascent IFs appear to form at random cytoplasmic sites (Kreis *et al.,* 1983). Similarly, *de novo* assembly of vimentin in SW13 cells appears to occur from random cytoplasmic sites (Sarria *et al.,* 1990). However, when keratins are transfected into cells that have their IFs obliterated by previous transfection with dominant-negative, assembly-disruptive mutants, the new system of IFs seems to assemble from the nuclear envelope toward the plasma membrane (Albers and Fuchs, 1987). The same type of vectorial assembly has been noticed when new peripherin IFs start assembling in hippocampal neurons that have been cocultured with muscle cells (Djabali *et al.,* 1993; see Fig. 3).

Finally, assembly of nascent IFs from preferred sites has been reported to occur in embryonic cells following fertilization of mammalian (Plancha *et al.,* 1989) or frog oocytes (Dent *et al.,* 1992). In those cases, new IFs emerge first around the nuclear envelope or the plasma membrane. Recent transfection studies with the lens-specific proteins, filensin and phakinin (see earlier discussion), reveal the same theme: when these proteins are cotransfected into nonlenticular epithelial cells, filaments emerge from aster-like structures closely associated with the plasma membrane and the nuclear envelope (G. Goulielmos, F. Gaunari, S. Remington, S. Müller, M. Häner, U. Aebi, and S. D. Georgatos, unpublished observations).

The differences seen in the various cell types could be explained in a logical fashion if we assume that vectorial assembly and stochastic assembly from many cytoplasmic sites are both possible, depending on the specific intracellular circumstances. For instance, in fertilized eggs and developing neurons, filament nucleation sites on membranous organelles may become available at a fine point when the endogenous system of IFs has been

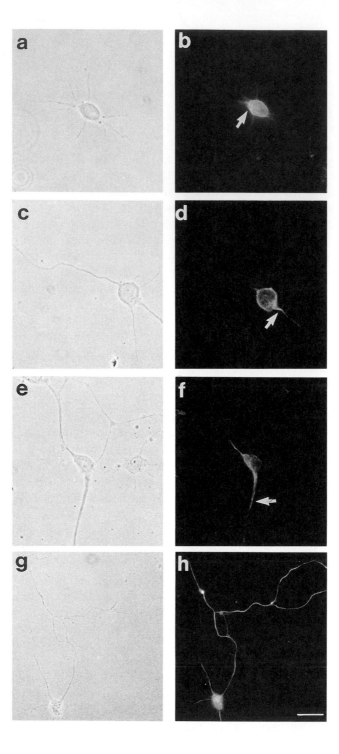

dismantled and before the new IF subunits accumulate in enough quantities. Conversely, assembly from multiple cytoplasmic sites may occur when new filament subunits are made in cells possessing another intact IF system which presumably occupies all IF-binding sites at the plasma membrane and the nuclear envelope.

VIII. Conclusions and Prospects

A concept which seems to emerge as new data accumulate is that IFs are physically coupled to various membranous organelles. The filament–membrane interactions, typified by the attachment of the nuclear lamins to the nuclear envelope and the anchorage of cytoplasmic organelles to the desmosomal plaque, seem to be dynamic and highly regulated. Therefore, it would be reasonable to suspect that IFs or IF proteins, by modulating the properties of membranes, may play pivotal roles during development and differentiation.

Looking critically at the published literature, one realizes that the associations and spatial interelations between filaments and membranes are sometimes difficult to trace by just looking at stationary cells. However, the tight connections between IFs and membranous organelles become apparent when the cells arc challenged. A graphic example of this has been provided by early studies on chicken melanocytes. Mayerson and Brumbaugh (1981) subjected these cells to centrifugation at $300 \times g$, a force that is enough to cause organelle displacement but less than that required for cell lysis. Serial sectioning and electron microscopy on centrifuged cells showed that IFs "stream" behind moving melanosomes, mitochondria, and nuclei, and attach firmly to the membranous envelopes of these organelles.

By now, there is plenty of evidence suggesting that IFs play an important role in tissue organization, perhaps by modulating intercellular junctions and cell surface specializations; however, IFs do not appear to function as a structure-reinforcing "endoskeleton" at the level of individual cells. Thus, although mutations in epidermal keratins affect the structure of epithelial

---

FIG. 3 Peripherin assembly in hippocampal neurons cocultured with muscle cells. Hippocampal neurons growing in the presence of muscle cells for 4 days were fixed and processed for immunofluorescence microscopy. Even though all cells have processes, as evidenced by phase contrast (a, c, e, g), the pattern of peripherin labeling is different. (b, d) Labeling restricted to the cell body and the initial segments of the neurites (arrows). (f) Labeling in the proximal segment of the processes (arrow). (h) Labeling of the entire cell. Bar, 10 μm. (Courtesy of Dr. K. Djabali.)

sheets (Fuchs, 1994), collapse or disassembly of cytoplasmic IFs in cultured cells does not lead to cytolysis (Klymkowsky, 1981). Based on this, one may speculate that IF *proteins,* irrespectively of their degree of polymerization, may have intracellular functions distinct from those of IF *networks* (see later discussion).

A large body of evidence supports the notion that IFs are involved in various specialized functions, but there seems to be a paucity of ideas as to how processes as diverse as cell–cell fusion, sorting of mitotic vesicles, and organization of epithelial sheets in the epidermis could engage the same type of biopolymers. Even more pressing is the question of why genetic ablation of IF genes often yields no pathological phenotype. The answers to these questions are neither black-and-white, nor trivial. For instance, knockout of the keratin 8 gene leads to embryonic lethality in some strains of mice (Baribault *et al.,* 1993), but this lethal phenotype is reversed when the keratin-null mice are of a different genetic background. We would like to suggest here two potential solutions to these problems: first, it is conceivable that one type of IF protein can partially substitute for the lack of another IF protein; second, it is equally probable that even *non-IF proteins* may functionally overlap with IF proteins.

The first point is based on an educated guess: that the members of the IF family may be many more than the so-far characterized IF proteins. If this is true, it would be reasonable to expect that when an IF protein gene is knocked out, other IF genes whose identity is not yet known take over. Compensatory overexpression of "look-alike" molecules is commonplace when animals are afflicted by genetic diseases. For example, in thalassemia major, the lack of a functional β-globin gene is partially compensated for by overexpression of the embryonic γ-globin gene. Obviously, substituting one IF type for another might be suboptimal for cell physiology. Thus, genetically manipulated animals that express "look-alike" molecules instead of the differentiation-specific IFs should be compromised in some aspect of their biology. This may not become apparent until the animals age.

The second idea is based on the realization that many cellular components have a molecular organization similar to that of IF proteins (i.e., a tripartite domain substructure and an α-helical, coiled-coil domain). For instance, several transcription factors possess a so-called "leucine zipper" domain which, in reality, is a specialized form of a coiled-coil. Similar to IF proteins, these factors form stable dimers, are regulated by phosphorylation/dephosphorylation, and bind DNA. Furthermore, a multiplicity of cytoskeletal proteins (myosin, tropomyosin, microtubule motors, etc.) possess long coiled-coil domains, often associate with membranous organelles, and are as abundant as IF proteins. Whether such molecules can functionally substitute for IFs when the latter are absent would be a problem worth investigating.

Closing this discussion, it might be useful to briefly discuss some aspects of the evolution of IF proteins. As noted by previous authors, the cytoplasmic IFs probably originated from lamin-like anscestors which lost the exons coding for the nucleophilic signal region and the CAAX box (Dodemont *et al.*, 1990; Riemer *et al.*, 1992). However, it is conceivable that cytoplasmic IF proteins have maintained lamin-like features such as DNA-binding motifs and membrane association sites. These features are not immediately obvious when one looks at the compilation of amino acid sequences, but become apparent when the binding properties of lamins and cytoplasmic IFs are tested using *in vitro* assays. Since the three-dimensional structure of IF subunits has not been determined yet, it may be useful to subject each newly identified IF protein to a "ligand screening" using substrates known to associate with IF proteins and enzymes known to motify IF subunits. In this way, we might be able to classify emerging members of the IF family according to their ligand-binding properties and not only according to their primary structures.

Acknowledgments

We thank M. Carmo-Fonseca (Faculty of Medicine, University of Lisbon, Portugal) for providing the micrograph shown in Fig. 1 and K. Djabali (College de France, Paris, France) for providing the pictures shown in Fig. 3. We acknowledge Y. Capetanaki (Baylor College, Houston Texas, USA) for communicating unpublished results and M. W. Klymkowsky (University of Colorado at Boulder, Colorado, USA) for many critical comments. Finally, we thank A. Pyrpasopoulou (EMBL) for help in electron microscopy. C. M. was supported by the Biotechnology Program of the European Union. This article is dedicated to Adamandia and Stavros Politis.

References

Albers, K., and Fuchs, E. (1987). The expression of mutant epidermal keratin cDNA transfected in simple epithelial and squamous cell carcinoma lines. *J. Cell Biol.* **105,** 791–806.

Albers, K., and Fuchs, E. (1989). Expression of mutant keratin cDNAs in epithelial cells reveals possible mechanisms for initiation and assembly of intermediate filaments. *J. Cell Biol.* **108,** 1477–1493.

Allen, D. P., Low, P. S., Dola, A., and Maisel, H. (1987). Band 3 and ankyrin homologues are present in the eye lens: Evidence for all major erythrocyte membrane components in the same non-erythroid cell. *Biochem. Biophys. Res. Commun.* **149,** 266–275.

Almahbobi, G., and Hall, P. F. (1990). The role of intermediate filaments in adrenal steroidogenesis. *J. Cell Sci.* **97,** 679–687.

Almahbobi, G., Williams, L. J., and Hall, P. F. (1992a). Attachment of mitochondria to intermediate filaments in adrenal cells: Relevance to the regulation of steroid synthesis. *Exp. Cell Res.* **200,** 361–369.

Almahbobi, G., Williams, L. J., and Hall, P. F. (1992b). Attachment of steroidogenic lipid droplets to intermediate filaments in adrenal cells. *J. Cell Sci.* **101**, 383–393.

Angelides, K. J., Smith, K. E., and Takeda, M. (1989). Assembly and exchange of intermediate filament proteins of neurons: Neurofilaments are dynamic structures. *J. Cell Biol.* **198**, 1495–1506.

Angst, B. D., Nilles, L. A., and Green, K. J. (1990). Desmoplakin II expression is not restricted to stratified epithelia. *J. Cell Sci.* **97**, 247–257.

Asch, H. L., Mayhew, E., Lazo, R. O., and Asch, B. B. (1990). Lipids noncovalently associated with keratins and other cytoskeletal proteins of mouse mammary epithelial cells in primary culture. *Biochim. Biophys. Acta* **1034**, 303–308.

Aster, J. C., Brewer, G. J., Hanash, S. M., and Maisel H. (1984a). Band 4.1-like proteins of the bovine lens. *Biochem. J.* **224**, 609–616.

Aster, J. C., Welsh, M. J., Brewer, G. J., and Maisel H. (1984b). Identification of spectrin and protein 4.1-like proteins in mammalian lens. *Biochem. Biophys. Res. Commun.* **119**, 726–734.

Aster, J. C., Brewer, G. J., and Maisel H. (1986). The 4.1-like proteins of the bovine lens: Spectrin-binding proteins closely related in structure to the RBC protein 4.1. *J. Cell Biol.* **103**, 115–122.

Bader, B. L., Magnin, T. M., Freudenmann, M., Stumpp, S., and Franke, W. W. (1991). Intermediate filaments formed de novo from tail-less cytokeratins in the cytoplasm and in the nucleus. *J. Cell Biol.* **115**, 1293–1307.

Bailer, S. M., Eppenberger, H. M., Griffiths, G., and Nigg, E. A. (1991). Characterization a 54-kD protein of the inner nuclear membrane: Evidence for cell cycle-dependent interaction with the nuclear lamina. *J. Cell Biol.* **114**, 389–400.

Baribault, H., Price, J., Miyai, K., and Oshima, R. G. (1993). Mid-gestational lethality in mice lacking keratin 8. *Genes Dev.* **7**, 1191–1201.

Behrendt, H. (1977). Effects of anabolic steroids on rat heart muscle cells. I. Intermediate filaments. *Cell Tissue Res.* **180**, 303–315.

Benedetti, E. L., Dunia, L., Ramaekers, F. C. S., and Kibbelaar, M. A. (1981). Lenticular plasma membranes and cytoskeleton. *In* "Molecular and Cellular Biology of the Eye Lens" (H. Bloemendal, ed.), pp. 137–188. Wiley, New York.

Bennett, G. S., Fellini, J. A., and Holtzer, H. (1978). Immunofluorescent visualization of 100 A filaments in different cultured chick embryo cell types. *Differentiation (Berlin)* **12**, 71–81.

Bennett, G. S., Fellini, S. A., Toyama, Y., and Holtzer, H. (1979). Redistribution of intermediate filament subunits during skeletal myogenesis and maturation in vitro. *J. Cell Biol.* **82**, 577–584.

Bennett, H., and Condeelis, J. (1988). Isolation of an immunoreactive analogue of brain fodrin that is associated with the cell cortex of Dictyostelium amoebae. *Cell Motil. Cytoskel.* **11**, 303–317.

Bernstein, L. H., and Wollman, S. H. (1975). Association of mitochondria with desmosomes in the rate thyroid gland. *J. Ultrastruct. Res.* **53**, 87–92.

Bloom, S., and Cincilla, P. A. (1969). Conformational changes in myocardial nuclei of rats. *Circ. Res.* **24**, 189–196.

Blose, S. H., Meltzer, D. I., and Feramisco, J. R. (1984). 10-nm filaments are induced to collapse in living cells microinjected with monoclonal and polyclonal antibodies against tubulin. *J. Cell Biol.* **98**, 847–858.

Bond, M., and Somlyo, A. V. (1982). Dense bodies and actin polarity in vertebrate smooth muscle. *J. Cell Biol.* **95**, 403–413.

Bradley, R. H., Ireland, M., and Maisel, H. (1979). The cytoskeleton of chick lens cells. *Exp. Eye Res.* **28**, 441–453.

Brunkener, M., and Georgatos, S. D. (1992). Membrane-binding properties of filensin, cytoskeletal protein of the lens fiber cells. *J. Cell Sci.* **103**, 709–718.

Buckley, I. K., and Porter, K. R. (1967). Cytoplasmic fibrils in living cultured cells. A light and electron microscope study. *Protoplasma* **64**, 349–380.

Burke, B. (1990). On the cell-free association of lamins A and C with metaphase chromosomes. *Exp. Cell Res.* **186**, 169–176.

Butz, S., Stappert, J., Weissig, H., and Kemler, R. (1992). Plakoglobin and β-catenin: Distinct but closely related. *Science* **257**, 1442–1444.

Byers, T. J., Brandin, E., Lue, R., Winograd, E., and Branton, D. (1992). The complete sequence of Drosophila beta spectrin reveals supramotifs comprising eight 106-residue repeats. *Proc. Natl. Acad. Sci. U.S.A.* **89**, 6187–6191.

Byrd, D. A., Sweet, D. J., Pante, N., Konstantinov, K. N., Guan, T., Saphire, A. C., Mitchell, P. J., Cooper, C. S., Aebi, U., and Gerace, L. (1994). Trp, a larger coiled coil protein whose amino terminus involved in activation oncogenic kinases, is localized to the cytoplasmic surface of the nuclear pore complex. *J. Cell Biol.* **127**, 1515–1526.

Campbell, G. R., Campbell, J. C., Stewart, U. G., Small, J. V., and Andersen, P. (1979). Antibody staining of 10 nm (100 A) filaments in cultured smooth cardiac and skeletal muscle cells. *J. Cell Sci.* **37**, 303–322.

Capco, D. G., Wan, K. M., and Penman, S. (1982). The nuclear matrix: Three-dimensional architecture and protein composition. *Cell (Cambridge, Mass.)* **29**, 847–858.

Capetanaki, Y., Smith, S., and Health, J. P. (1989). Overexpression of the vimentin gene in transgenic mice inhibits normal lens cell differentiation. *J. Cell Biol.* **109**, 1653–1664.

Carmo-Fonseca, M., and David-Ferreira, J. F. (1990). Interactions of intermediate filaments with cell structures. *Electron Microsc. Rev.* **3**, 115–141.

Carmo-Fonseca, M., Cidadao, A. J., and David-Ferreira, D. F. (1987). Filamentous crossbridges link intermediate filaments to the nuclear pore complexes. *Eur. J. Cell Biol.* **45**, 282–290.

Cartaud, A., Courvalin, J. C., Ludosky, M. A., and Cartaud, J. (1989). Presence of a protein immunologically related to lamin B in the postsynaptic membrane of *Torpedo marmorata* electrocyte. *J. Cell Biol.* **109**, 1745–1752.

Cartaud, A., Ludosky, M. A., Courvalin, J. C., and Cartaud, J. (1990). A protein antigenically related to nuclear lamin B mediates the association of intermediate filaments with desmosomes. *J. Cell Biol.* **111**, 581–588.

Cartaud, A., Jasmin, B. J., Changeux, J.-P., and Cartaud, J. (1995). Direct involvement of a lamin-B related (54KDa) protein in the association of intermediate filaments with the postsynaptic membrane of the *Torpedo marmorata* electrocyte. *J. Cell Sci.* **108**, 153–160.

Cary, R. B., and Klymkowsky, M. W. (1994a). Differential organization of desmin and vimentin in muscle is due to differences in their head domains. *J. Cell Biol.* **126**, 445–456.

Cary, R. B., and Klymkowsky, M. W. (1994b). Desmin organization during the differentiation of the dorsal myotome in *Xenopus laevis*. *Differentiation (Berlin)* **56**, 31–38.

Cary, R. B., and Klymkowsky, M. W. (1995). Disruption of intermediate filament organization leads to structural defects at the intersomite junction in Xenopus myotomal muscle. *Development* **121**, 1041–1052.

Celis, J. E., Small, J. V., Larsen, P. M., Fey, S. J., De Mey, J., and Celis, A. (1984). Intermediate filaments in monkey kidney TC7 cells: Focal centers and interrelationship with other cytoskeletal systems. *Proc. Natl. Acad. Sci. U.S.A.* **81**, 1117–1121.

Centonze, V. E., Ruben, G. C., and Sloboda, R. D. (1986). Structure and composition of the cytoskeleton of nucleated erythrocytes: III. Organization of the cytoskeleton of Bufo maranis erythrocytes as revealed by freeze-dried platinum carbon replicas and immunofluorescence microscopy. *Cell Motil. Cytoskel.* **6**, 376–388.

Chaudhary, N., and Courvalin, J.-N. (1993). Stepwise reassembly of the nuclear envelope at the end of mitosis. *J. Cell Biol.* **122**, 295–306.

Chen, L. B. (1988). Mitochondrial membrane potential in living cells. *Annu. Rev. Cell Biol.* **4**, 155–181.

Chen, L. B., Summerhayes, I. C., Johnson, L. V., Walsh, M. L., Bernal, S. D., and Lampidis, T. J. (1981). Probing mitochondria in living cells with rhodamine 123. *Cold Spring Harbor Symp. Quant. Biol.* **46,** 141–155.

Chin, S. S. M., and Liem, R. K. H. (1989). Expression of rat neurofilament proteins NF-L and NF-M in transfected non-neuronal cells. *Eur. J. Cell Biol.* **50,** 475–490.

Chipev, C. C., Korge, B. P., Markova, N., Bale, S. J., DiGiovanna, J. J., Compton, J. G., and Steinert, P. M. (1992). A leucine-proline mutation in the H1 subdomain of keratin 1 causes epidermolytic hyperkeratosis. *Cell (Cambridge, Mass.)* **70,** 821–828.

Chou, R. R., Stromer, M. H., Robson, R. M., and Huiatt, T. W. (1992). Assembly of contractile and cytoskeletal elements in developing smooth muscle cells. *Dev. Biol.* **149,** 339–348.

Collard, J.-F., Senecal, J. L., and Raymond, Y. (1992). Redistribution of nuclear lamin A is an early event associated with differentiation of human promyelocytic leukemia HL-60 cells. *J. Cell Sci.* **101,** 657–670.

Collier, N. C., Sheetz, M. P., and Schlesinger, M. J. (1993). Concomitant changes in mitochondria and intermediate filaments during heat shock and recovery of chicken embryo fibroblasts. *J. Cell. Biochem.* **52,** 297–307.

Colucci-Guyon, E., Portier, M.-M., Dunia, I., Paulin, D., Pourin, S., and Babinet, C. (1994). Mice lacking vimentin develop and reproduce without an obvious phenotype. *Cell (Cambridge, Mass.)* **79,** 679–694.

Conway, J. F., and Parry, D. A. D. (1988). Intermediate filament structure: 3. Analysis of sequence homologies. *Int. J. Biol. Macromol.* **10,** 79–98.

Conway, J. F., and Parry, D. A. D. (1990). Structural features in the heptad substructure and longer range repeats of two-stranded α-fibrous proteins. *Int. J. Biol. Macromol.* **12,** 328–334.

Cooke, P. (1976). A filamentous cytoskeleton in vertebrate smooth muscle fibers. *J. Cell Biol.* **68,** 539–556.

Cooke, P. H., and Chase, R. H. (1971). Potassium chloride-insoluble myofilaments in vertebrate smooth muscle cells. *Exp. Cell Res.* **66,** 417–425.

Cote, F., Collard, J.-F., and Julien, J.-P. (1993). Progressive neuronopathy in transgenic mice expressing the human neurofilament heavy gene: A mouse model of amyotrophic lateral sclerosis. *Cell (Cambridge, Mass.)* **73,** 35–46.

Courvalin, J.-C., Segil, N., Blobel, G., and Worman, H. J. (1992). The lamin B receptor of the inner nuclear membrane undergoes mitosis-specific phosphorylation and is a substrate for p34cdc2-type protein kinase. *J. Biol. Chem.* **267,** 19035–19038.

Cowin, P., and Garrod, D. (1983). Antibodies to epithelial desmosomes show wide tissue and species cross-reactivity. *Nature (London)* **302,** 148–150.

Cowin, P., Kapprell, H.-P., and Franke, W. W. (1985). The complement of desmosomal plaque proteins in different cell types. *J. Cell Biol.* **101,** 1442–1454.

Cowin, P., Kapprell, H.-P., Franke, W. W., Tamkun, J., and Hynes, R. O. (1986). Plakoglobin: A protein common to different kinds of intercellular adhering junctions. *Cell (Cambridge, Mass.)* **46,** 1063–1073.

Cress, A. E., and Kurath, K. M. (1988). Identification of attachment proteins for DNA in Chinese hamster ovary cells. *J. Biol. Chem.* **263,** 19678–19683.

David-Ferreira, K. L., and David-Ferreira, J. F. (1980). Association between intermediate-sized filaments and mitochondria in rat Leydig cells. *Cell Biol. Int. Rep.* **4,** 655–662.

Davis, J. Q., and Bennett, V. (1984). Brain ankyrin—a membrane-associated protein with binding sites for spectrin, tubulin and the cytoplasmic domain of the erythrocyte anion channel. *J. Biol. Chem.* **259,** 13550–133559.

Dent, J. A., Cary, R. B., Bachant, J. B., Domingo, A., and Klymkowsky, M. W. (1992). Host cell factors controlling vimentin organization in the *Xenopus* oocyte. *J. Cell Biol.* **119,** 855–866.

Djabali, K., Portier, M. M., Gros, F., Blobel, G., and Georgatos, S. D. (1991). Network antibodies identify nuclear lamin B as physiological attachment size for peripherin intermediate filaments. *Cell (Cambridge, Mass.)* **64,** 109–121.

Djabali, K., Zissopoulou, A., de Hoop, M. J., Georgatos, S. D., and Dotti, C. G. (1993). Peripherin expression in hippocampal neurons induced by muscle soluble factor(s). *J. Cell Biol.* **123,** 1197–1206.

Dodemont, H., Riemer, D., and Weber, K. (1990). Structure of an invertebrate gene encoding cytoplasmic intermediate filament (IF) proteins: Implications for the origin and the diversification of IF proteins. *EMBO J.* **9,** 4083–4094.

Draberova, E., and Draber, P. (1993). A microtubule-interacting protein involved in coalignment of vimentin intermediate filaments with microtubules. *J. Cell Sci.* **106,** 1263–1273.

Drochmans, P., Freudenstein, C., Wanson, J.-C., Laurent, L., Keenan, T. W., Stadler, J., Leloup, R., and Franke, W. W. (1978). Structure and biochemical composition of desmosomes and tonofilaments isolated from calf muzzle epidermis. *J. Cell Biol.* **79,** 427–443.

Dubreuil, R. R., Byers, T. J., Sillman, A. L., Bar-Zvi, D., Goldstein, L. S. B., and Branton, D. (1989). The complete sequence of Drosophila alpha-spectrin: Conservation of structural domains between alpha-spectrins and alpha-actinin. *J. Cell Biol.* **109,** 2197–2205.

Eckelt, A., Herrmann, H., and Franke, W. W. (1992). Assembly of a tail-less mutant of the intermediate filament protein, vimentin, in vitro and in vivo. *Eur. J. Cell Biol.* **58,** 319–330.

Eckert, B. S. (1986). Alteration of the distribution of intermediate filaments in PtK1 cells by acrylamide. II: Effect on the organization of cytoplasmic organelles. *Cell Motil. Cytoskel.* **6,** 15–24.

Eckert, B. S., Caputi, S. E., and Brinkley, B. R. (1984). Co-localization of the centriole and keratin intermediate filaments in PtK1 cells. *Cell Motil.* **4,** 241–248.

Fairley, J. A., Scott, J. A., Jenson, G. A., Goldsmith, K. D., and Diaz, L. A. (1991). Characterization of keratocalmin, a calmodulin binding protein from human epidermis. *J. Clin. Invest.* **88,** 315–322.

Farquhar, M. G., and Palade, G. E. (1963). Junctional complexes in various epithelia. *J. Cell Biol.* **17,** 375–412.

Fay, F. S., Fujiwara, K., Rees, D. D., and Fogarty, K. E. (1983). Distribution of alpha-actinin in single isolated smooth muscle cells. *J. Cell Biol.* **96,** 783–795.

Felix, H., and Strauli, P. (1976). Different distribution pattern of 100 A filaments in resting and locomotive leukaemia cells. *Nature (London)* **261,** 604–606.

Ferrans, V. J., and Roberts, W. C. (1973). Intermyofibrillar and nuclear-myofibrillar connections in human and canine myocardium. An ultrastructural study. *J. Mol. Cell Cardiol.* **5,** 247–257.

Fey, E. G., Wan, K. M., and Penman, S. (1984). Epithelial cytoskeletal framework and nuclear matrix-intermediate filament scaffold: Three-dimensional organization and protein composition. *J. Cell Biol.* **98,** 1973–1984.

Firmbach-Kraft, I., and Stick, R. (1993). The role of CaaX-dependent modifications in membrane association of Xenopus lamin B3 during meiosis and fate of B3 in transfected mitotic cells. *J. Cell Biol.* **123,** 1661–1670.

Foisner, R., and Gerace, L. (1993). Integral membrane proteins of the nuclear envelope interact with lamins and chromosomes, and binding is modulated by mitotic phosphorylation. *Cell (Cambridge, Mass.)* **73,** 1267–1279.

Franke, W. W. (1971). Relationship of nuclear membranes with filaments and microtubules. *Protoplasma* **73,** 263–292.

Franke, W. W., and Schinko, W. (1969). Nuclear shape in muscle cells. *J. Cell Biol.* **42,** 326–331.

Franke, W. W., Moll, R., Schiller, D. L., Schmid, E., Kartenbeck, J., and Mueller, H. (1982). Desmoplakins of epithelial and myocardial desmosomes are immunologically and biochemically related. *Differentiation (Berlin)* **25,** 115–127.

Franke, W. W., Hergt, M., and Grund, C. (1987a). Rearrangement of the vimentin cytoskeleton during adipose conversion: Formation of an intermediate filament cage around lipid globules. *Cell (Cambridge, Mass.)* **49,** 131–141.

Franke, W. W., Kapprell, H.-P., and Cowin, P. (1987b). Plakoglobin is a component of the filamentous subplasmalemmal coat of lens cells. *Eur. J. Cell Biol.* **43,** 301–315.

Fraser, R. D. B., MacRae, T. P., Suzuki, E., and Parry, D. A. D. (1985). Intermediate filament structure: 2. Molecular interactions in the filament. *Int. J. Biol. Macromol.* **7,** 258–274.

French, S. W., Kawahara, H., Katsuma, Y., Ohta, M., and Swierenga, S. H. H. (1989). Interaction of intermediate filaments with nuclear lamina and cell periphery. *Electron Microsc. Rev.* **2,** 17–51.

Fuchs, E. (1994). Intermediate filaments and disease: Mutations that cripple cell strength. *J. Cell Biol.* **125,** 511–516.

Fuchs, E., and Weber, K. (1994). Intermediate filaments: Structure, dynamics, function, and disease. *Annu. Rev. Biochem.* **63,** 345–382.

Fuchs, E., Esteves, R. A., and Coulombe, P. A. (1992). Transgenic mice expressing a mutant keratin 10 gene reveal the likely genetic basis for epidermolytic hyperkeratosis. *Proc. Natl. Acad. Sci. U.S.A.* **89,** 6909–6910.

Fujitani, Y., Higaki, S., Sawada, H., and Hirosawa, K. (1989). Quick-freeze, deep-etch visualization of the nuclear pore complex. *J. Electron Microsc.* **38,** 34–40.

Furukawa, K., Pante, N., Aebi, U., and Gerace, L. (1995). Cloning of a cDNA for lamina-associated polypeptide 2 (LAP2) and identification of regions that specify targeting to the nuclear envelope. *EMBO J.* **14,** 1626–1636.

Gard, D. L., and Lazarides, E. (1980). The synthesis and distribution of desmin and vimentin during myogenesis in vitro. *Cell (Cambridge, Mass.)* **19,** 263–275.

Geiger, B., and Singer, S. J. (1980). Association of microtubules and intermediate filaments in chicken gizzard cells as detected by double immunofluorescence. *Proc. Natl. Acad. Sci. U.S.A.* **77,** 4769–4773.

Georgatos, S. D., and Blobel, G. (1987). Lamin B constitutes an intermediate filament attachment site at the nuclear envelope. *J. Cell Biol.* **105,** 117–125.

Georgatos, S. D., and Marchesi, V. T. (1985). The binding of vimentin to human erythrocyte membranes: A model system for the study of intermediate filament membrane interactions. *J. Cell Biol.* **100,** 1955–1961.

Georgatos, S. D., Weaver, D. C., and Marchesi, V. T. (1985). Site-specificity in vimentin-membrane interactions: Intermediate filament subunits associate with the plasma membrane via their head domains. *J. Cell Biol.* **100,** 1962–1967.

Georgatos, S. D., Weber, K., Geisler, N., and Blobel, G. (1987). Binding of two desmin derivatives to the plasma membrane and the nuclear envelope of avian erythrocytes: Evidence for a conserved site-specificity in intermediate filament-membrane interactions. *Proc. Natl. Acad. Sci. U.S.A.* **84,** 6780–6784.

Georgatos S. D., Gounari, F., and Remington S. (1994a). The beaded intermediate filaments and their potential functions in eye lens. *BioEssays* **16,** 413–418.

Georgatos, S. D., Meier, J., and Simos, G. (1994b). Lamins and lamin-associated proteins. *Curr. Opin. Cell Biol.* **6,** 347–353.

Geuens, G., DeBrabander, M., Nuydens, R., and DeMey, J. (1983). The interaction between microtubules and intermediate filaments in cultured cells treated with taxol and nocodazole. *Cell Biol. Int. Rep.* **7,** 35–47.

Gill, S. R., Wong, P. C., Monteiro, M. J., and Cleveland, D. W. (1990). Assembly properties of dominant and recessive mutations in the small mouse neurofilament (NF-L) subunit. *J. Cell Biol.* **111,** 2005–2019.

Gillard, B. K., Heath, J. P., Thurmon, L. T., and Marcus, D. M. (1991). Association of glycosphingolipids with intermediate filaments of human umbilical vein endothelial cells. *Exp. Cell Res.* **192,** 433–444.

Gillard, B. K., Thurmon, L. T., and Marcus, D. M. (1992). Association of glycosphingolipids with intermediate filaments of mesenchymal, epithelial, glial, and muscle cells. *Cell Motil. Cytoskel.* **21,** 255–271.

Giudice, G. J., Squiquera, H. L., Elias, P. M., and Diaz, L. (1991). Identification of two collagen domains within bullous pemphigoid autoantigen, BP180, *J. Clin. Invest.* **87,** 734–738.

Glass, C. A., Glass, J. R., Taniura, H., Hasel, K. W., Blevitt, J. M., and Gerace, L. (1993). The α-helical rod domain of human lamins A and C contains a chromatin binding site. *EMBO J.* **12,** 4413–4424.

Glass, J. R., and Gerace, L. (1990). Lamins A and C bind and assemble at the surface of mitotic chromsomes. *J. Cell Biol.* **111,** 1047–1057.

Glenney, J. R., Osborn, M., and Weber, K. (1982). The intracellular localization of the microvillus 110 K protein, a component considered to be involved in side-on membrane attachment of F-actin. *Exp. Cell Res.* **138,** 199–205.

Goldman, R. D. (1971). The role of three cytoplasmic fibers in BHK-21 cell motility. *J. Cell Biol.* **51,** 752–769.

Goldman, R. D., Hill, B. F., Steinert, P., Whitman, M. A., and Zackroff, R. V. (1980). Intermediate filament-microtubule interactions: Evidence in support of a common organization center. *In* "Microtubules and Microtubule Inhibitors" (M. de Brabander and J. De Mey, eds.), pp. 91–102. Elsevier/North-Holland, Amsterdam.

Goldman, R. D., Goldman, A., Green, K., Jones, J., Lieska, N., and Yang, H.-Y. (1985). Intermediate filaments: Possible functions as cytoskeletal connecting links between the nucleus and the cell surface. *Ann. N. Y. Acad. Sci.* **455,** 1–17.

Gounari, F., Merdes, A., Quinlan, R., Hess, J. Fitz-Gerald, P. G., Quzounis, C. A., and Georgatos, S. D. (1993). Bovine filensin possesses primary and secondary structure similarity to intermediate filaments proteins. *J. Cell Biol.* **121,** 847–853.

Granger, B. L., and Lazarides, E. (1978). The existence of an insoluble Z disc scaffold in chicken skeletal muscle. *Cell (Cambridge, Mass.)* **15,** 1253–1268.

Granger, B. L., and Lazarides, E. (1979). Desmin and vimentin coexist at the periphery of the myofibril Z disc. *Cell (Cambridge, Mass.)* **18,** 1053–1063

Granger, B. L., Repasky, E. A., and Lazarides, E. (1982). Synemin and vimentin are components of intermediate filaments in avian erythrocytes. *J. Cell Biol.* **92,** 299–312.

Green, K. J., Goldman, R. D., and Chisholm, R. L. (1988). Isolation of cDNAs encoding desmosomal plaque proteins: Evidence that bovine desmoplakins I and II are derived from two mRNAs and a single gene. *Proc. Natl. Acad. Sci. U.S.A.* **85,** 2613–2617.

Green, K. J., Parry, D. A. D., Steinert, P. M., Virata, M. L. A., Wagner, R. M., Angst, B. D., and Nilles, L. A. (1990). Structure of the human desmoplakins. Implications for function in the desmosomal plaque. *J. Biol. Chem.* **265,** 2603–2612.

Gui, J. F., Lane, W. S., and Fu, X. D. (1994). A serine kinase regulates intracellular localization of splicing factors in the cell cycle. *Nature (London)* **23,** 678–682.

Guo, L., Degenstein, L., Dowling, J., Yu, Q. C., Wollmann, R., Perman, B., and Fuchs, E. (1995). Gene targeting of BPAG1: Abnormalities in mechanical strength and cell migration in stratified epithelia and neurologic degeneration. *Cell (Cambridge, Mass.)* **81,** 233–243.

Gyoeva, F. K., and Gelfand, V. I. (1991). Coalignment of vimentin intermediate filaments with microtubules depends on kinesin. *Nature (London)* **353,** 445–448.

Harris, J. R., and Brown, J. N. (1971). Fractionation of the avian erythrocyte: An ultrastructural study. *J. Ultrastruct. Res.* **36,** 8–23.

Hatzfeld, M., and Weber, K. (1991). Modulation of keratin intermediate filament assembly by single amino acid exchanges in the consensus sequence at the C-terminal end of the rod domain. *J. Cell Sci.* **99,** 351–362.

Hay, M., and De Boni, U. (1991). Chromatin motion in neuronal interphase nuclei: Changes induced by disruption of intermediate filaments. *Cell Motil. Cytoskel.* **18,** 63–75.

Heggeness, M. H., Simon, M., and Singer, S. J. (1978). Association of mitochondria with microtubules in cultured cells. *Proc. Natl. Acad. Sci. U.S.A.* **75,** 3863–3866.

Heida, Y., Tsukita, S., and Tsukita, S. (1989). A new high molecular mass protein showing unique localization in desmosomal plaque. *J. Cell Biol.* **109,** 1511–1518.

Heinmann, R., Shelanski, M.L., and Liem, R. K. H. (1983). Specific binding of MAPs to the 68,000 dalton neurofilament protein. *J. Cell Biol.* **97**(5, Pt.2), 286a.

Hess, J. F., Casselman, J. T., and FitzGerald, P. G. (1993). cDNA analysis of the 49 kDa lens fiber cell cytoskeletal protein: A new lens specific member of the intermediate filament family? *Curr. Eye Res.* **12**, 77–88.

Hieda, Y., Nishizawa, Y., Uematsu, J., and Owaribe, K. (1992). Identification of a new major hemidesmosomal protein, HD1: A major, high molecular mass component of isolated hemidesmosomes. *J. Cell Biol.* **116**, 1497–1506.

Hirokawa, N., Tilney, L. B., Fujiwara, K., and Heuser, J. E. (1982). The organization of actin, myosin, and intermediate filaments in the brush border of intestinal epithelial cells. *J. Cell Biol.* **94**, 425–443.

Hirokawa, N., Cheney, R. E., and Willard, M. (1983). Location of a protein of the fodrin-spectrin-TW260/240 family in the mouse intestinal brush border. *Cell (Cambridge, Mass.)* **32**, 953–965.

Holtz, D., Tanaka, R. A., Hartwig, J., and McKeon, F. (1989). The CaaX motif of lamin A functions in conjunction with the nuclear localization signal to target assembly to the nuclear envelope. *Cell (Cambridge, Mass.).* **59**, 969–977.

Holtzer, H., Forry-Schaudies, S., Dlugosz, A., Antin, P., and Dubyak, G. (1985). Interactions between IFs, microtubules, and myofibrils in fibrogenic and myogenic cells. *Ann. N.Y. Acad. Sci.* **455**, 106–125.

Hopkinson, S. B., Ridelle, K. S., and Jones, J. C. R. (1992). Cytoplasmic domain of the 180-kD bullous pemphigoid antigen, a hemidesmosomal component: Molecular and cell biologic characterization. *J. Invest. Dermatol.* **99**, 264–270.

Hull, B. E., and Staehelin, L. A. (1979). The terminal web: A reevaluation of its structure and function. *J. Cell Biol.* **81**, 67–82.

Ishikawa, H., Bischoff, R., and Holtzer, H. (1968). Mitosis and intermediate-sized filaments in developing skeletal muscle. *J. Cell Biol.* **38**, 538–555.

Janmey, P. A., Euteneuer, U., Traub, P., and Schliwa, M. (1991). Viscoelastic properties of vimentin compared with other filamentous biopolymer networks. *J. Cell Biol.* **113**, 155–160.

Jarnik, M., and Aebi, U. (1991). Toward a more complete 3-D structure of the nuclear pore complex. *J. Struct. Biol.* **107**, 291–308.

Jones, J. C. R. (1988). Characterization of a 125K glycoprotein associated with bovine epithelial desmosomes. *J. Cell Sci.* **89**, 207–216.

Jones, J. C. R., Goldman, E., Steinert, P. M., Yuspa, S., and Goldman, R. D. (1982). The dynamic aspects of the supramolecular organization of the intermediate filament networks in cultured epidermal cells. *Cell Motil. Cytoskel.* **2**, 197–213.

Jones, J. C. R., Yokoo, K. M., and Goldman, R. D. (1986). Is the hemidesmosome a half desmosome? An immunologocal comparison of mammalian desmosomes and hemidesmosomes. *Cell Motil. Cytoskel.* **6**, 560–569.

Jones, J. C. R., Kurpakus, M. A., Cooper, H. M., and Quaranta, V. (1991). A function for the integrin $\alpha_6\beta_4$. *Cell Regul.* **2**, 427–438.

Kapprell, H.-P., Owaribe, K., and Franke, W. W. (1988). Identification of a basic protein of M_r 75,000 as an accessory desmosomal plaque protein in stratified and complex epithelia. *J. Cell Biol.* **106**, 1679–1691.

Kapprell, H.-P., Duden, R., Owaribe, K., Schmelz, M., and Franke, W. W. (1990). Subplasmalemmal plaques of intercellular junctions: Common and distinguishing proteins. *In* "Morphoregulatory Molecules" (G. M. Edelman, B. A. Cuningham, and J. P. Thiery, eds.), pp. 285–314. Wiley, New York.

Kartenbeck, J., Franke, W. W., Moser, J. G., and Stoffels, U. (1983). Specific attachment of desmin filaments to desmosomal plaques in cardiac myocytes. *EMBO J.* **2**, 735–742.

Kartenbeck, J., Schwechheimer, K., Moll, R., and Franke, W. W. (1984). Attachment of vimentin filaments to desmosomal plaques in human meningioma cells and arachnoidal tissue. *J. Cell Biol.* **98**, 1072–1081.

Katsuma, Y., Swierenga, S. H. H., Marceau, M., and French, S. W. (1987). Connections of intermediate filaments with the nuclear lamina and the cell periphery. *Biol. Cell.* **59,** 193–204.

Kelly, D. E. (1966). Fine structure of desmosomes, hemidesmosomes, and an adepidermal globular layer in developing newt epidermis. *J. Cell Biol.* **28,** 51–72.

Kitten, G. T., and Nigg, E. A. (1991). The CaaX motif is required for isoprenylation, carboxyl methylation and nuclear membrane association of lamin B2. *J. Cell Biol.* **113,** 13–23.

Klatte, D. H., Kurpakus, M. A., Grelling, K. A., and Jones, J. C. R. (1989). Immunochemical characterization of three components of the hemidesmosome and their expression in cultured epithelial cells. *J Cell Biol.* **109,** 3377–3390.

Klymkowsky, M. W. (1981). Intermediate filaments in 3T3 cells collapses after intracellular injection of a monoclonal anti-intermediate filament antibody. *Nature (London)* **291,** 249–251.

Klymkowsky, M. W. (1995). Intermediate filaments: New proteins, some answers, more questions. *Curr. Opin. Cell Biol.* **7,** 46–54.

Knudsen, K. A., and Wheelock, M. J. (1992). Plakoglobin, or an 83-kD homologue distinct from b-catenin, interacts with E-cadherin and N-cadherin. *J. Cell Biol.* **118,** 671–679.

Koch, P. J., Walsh, M. J., Schmelz, M., Goldschmidt, M. D., Zimbelmann, R., and Franke, W. W. (1990). Identification of desmoglein, a constitutive desmosomal glycoprotein, as a member of the cadherin family of cell adhesion molecules. *Eur. J. Cell Biol.* **53,** 1–12.

Koch, P. J., Goldschmidt, M. D., Zimbelmann, R., Troyanovsky, R., and Franke, W. W. (1992). Complexity and expression patterns of the desmosomal cadherins. *Proc. Natl. Acad. Sci. U.S.A.* **89,** 353–357.

Kouklis, P. D., Traub, P., and Georgatos, S. D. (1992). Involvement of the consensus sequence motif at coil 2b in the assembly and stability of vimentin filaments. *J. Cell Sci.* **102,** 31–41.

Kouklis, P. D., Merdes, A., Papamarcaki, T., and Georgatos, S. D. (1993). Transient arrest of 3T3 cells in mitosis and inhibition of nuclear lamin reassembly around chromatin induced by anti-vimentin antibodies. *Eur. J. Cell Biol.* **62,** 224–236.

Kouklis, P. D., Hutton, E., and Fuchs, E. (1994). Making a connection: Direct binding between keratin intermediate filaments and desmosomal proteins. *J. Cell Biol.* **127,** 1049–1060.

Krainer, A. R., Mayeda, A., Kozak, D., and Binns, G. (1991). Functional expression of cloned human splicing factor SF2: Homology to RNA-binding proteins, U1 70K, and Drosophila splicing regulators. *Cell (Cambridge, Mass.)* **66,** 383–394.

Kreis, T. E., Geiger, B., Schmid, E., Jorcano, J. L., and Franke, W. W. (1983). De novo synthesis and specific assembly of keratin filaments, in nonepithelial cells after microinjection of mRNA for epidermal keratin. *Cell (Cambridge, Mass.)* **32,** 1125–1137.

Krohne, G., Waizenegger, I., and Höger, T. H. (1989). The conserved carboxy-terminal cysteine of nuclear lamins is essential for lamin association with the nuclear envelope. *J. Cell Biol.* **109,** 2003–2013.

Kühn, S., Vorgias, C. E., and Traub, P. (1987). Interaction in vitro of non-epithelial intermediate filament proteins with supercoiled plasmid DNA. *J. Cell Sci.* **87,** 543–554.

Kurpakus, M. A., and Jones, J. C. R. (1991). A novel hemidesmosomal plaque component: Tissue distribution and incorporation into assembling hemidesmosomes in an in vitro model. *Exp. Cell Res.* **194,** 139–146.

Langley, R. C., Jr., and Cohen, C. M. (1986). Associations of spectrin with desmin intermediate filaments. *J. Cell. Biochem.* **30,** 101–109.

Langley, R. C., Jr., and Cohen, C. M. (1987). Cell type-specific association between two types of spectrin and two types of intermediate filaments. *Cell Motil. Cytoskel.* **8,** 165–173.

Lazarides, E. (1980). Intermediate filaments as mechanical integrators of cellular space. *Nature (London)* **283,** 249–256.

Lazarides, E., and Granger, B. L. (1978). Fluorescent localization of membrane sites in glycerinated chicken skeletal muscle fibers and the relationship of these sites to the protein composition of the Z-discs. *Proc. Natl. Acad. Sci. U.S.A.* **75,** 3683–3687.

Lazarides, E., and Granger, B. L. (1983). Transcytoplasmic integration in avian erythrocytes and striated muscle; the role of intermediate filaments. *Mod. Cell Biol.* **2**, 143–162.

Lazarides, E., and Hubbard, B. D. (1976). Immunological characterization of the subunit of the 100 A filaments from muscle cells. *Proc. Natl. Acad. Sci. U.S.A.* **73**, 4344–4348.

Lee, C. S., Morgan, G., and Wooding, F. B. P. (1979). Mitochondria and mitochondria-tonofilament-desmosomal association in the mammary gland secretory epithelium of lactating cows. *J. Cell Sci.* **38**, 125–135.

Lehto, V. P., Virtanen, I., and Kurki, P. (1978). Intermediate filaments anchor the nuclei in nuclear monolayers of cultured human fibroblasts. *Nature (London)* **272**, 175–177.

Letai, A., and Fuchs, E. (1995). The importance of intramolecular ion pairing in intermediate filaments. *Proc. Natl. Acad. Sci. U.S.A.* **92**, 92–96.

Levine, J., and Willard, M. (1981). Fodrin: Axonally transported polypeptides associated with the internal periphery of many cells. *J. Cell Biol.* **90**, 631–643.

Li, H., Choudhary, S. K., Milner, D. J., Munir, M. I., Kuisk, I. R., and Capetanaki, Y. (1994). Inhibition of desmin expression blocks myoblast fusion and interferes with myogenic regulators myoD and myogenin. *J. Cell Biol.* **124**, 827–841.

Loer, C. (1992). C. elegans spectrin/fodrin-like clone free to a good home. *Worm Breeder's Gaz.* **12**, 54.

Luckenbill, L. M., and Cohen, A. S. (1966). The association of lipid droplets with cytoplasmic filaments in avian subsynovial adipose cells. *J. Cell Biol.* **31**, 195–199.

Luderus, M. E. E., De Graaf, A., Mattia, E., Den Blaauwen, J. L., Grande, M. A., De Jong, L., and Van Driel, R. (1992). Binding of matrix attachment regions to lamin B1. *Cell (Cambridge, Mass.)* **70**, 949–959.

Maisel, H., Harding, C. V., Alcala, J. R., Kuszak, J., and Bradley, R. (1981). The morphology of the lens. *In* "Molecular and Cellular Biology of the Eye Lens" (H. Bloemendal, ed.), pp. 49–84. Wiley, New-York.

Maison, C., Horstmann, H., and Georgatos, S. D. (1993). Regulated docking of nuclear membrane vesicles to vimentin filaments during mitosis. *J. Cell Biol.* **123**, 1491–1505.

Maison, C., Pyrpasopoulou, A., and Georgatos, S. D. (1995). Vimentin-associated mitotic vesicles interact with chromosomes in a lamin B and phosphorylation-dependent manner. *EMBO J.* **14**, 3311–3324.

Mangeat, P. H., and Burridge, K. (1984). Immunoprecipitation of nonerythrocyte spectrin within live cells following microinjection of specific antibodies: Relation to cytoskeletal structures. *J. Cell Biol.* **98**, 1363–1377.

Maro, B., and Bornens, M. (1982). Reorganization of HeLa cell cytoskeleton induced by an uncoupler of oxidative phosphorylation. *Nature (London)* **295**, 334–336.

Maro, B., Sauron, M.-E., Paulin, D., and Bornens, M. (1982). Taxol induced redistribution of vimentin filaments in HeLa cells. *Biol. Cell.* **45**, 204.

Maro, B., Paintrand, M., Sauron, M.-E., Paulin, D., and Bornens, M. (1984). Vimentin filaments and centrosomes. Are they associated? *Exp. Cell Res.* **150**, 452–458.

Martin, L., Crimaudo, C., and Gerace, L. (1995). cDNA cloning and characterization of lamina-associated polypeptide 1C (LAP1C), and integral protein of the inner nuclear-membrane. *J. Biol. Chem.* **270**, 8822–8828.

Marugg, R. A. (1992). Transient storage of a nuclear matrix protein along intermediate-type filaments during mitosis: A novel function of cytoplasmic intermediate filaments. *J. Struct. Biol.* **108**, 129–139.

Masaki, S., and Watanabe, T. (1992). cDNA sequence analysis of CP94: Rat lens fiber cell beaded-filament structural protein shows homology to cytokeratins. *Biochem. Biophys. Res. Commun.* **186**, 190–198.

Mathur, M., Goodwin, L., and Cowin, P. (1994). Interactions of the cytoplasmic domain of the desmosomal cadherin Dsg1 with plakoglobin. *J. Biol. Chem.* **269**, 14075–14080.

Mayerson, P. L., and Brumbaugh, J. A. (1981). Lavender, chick melanocyte mutant with defective melanosome translocation: A possible role for 10 nm filaments and microfilaments but not microtubules. *J. Cell Sci.* **51,** 25–51.

McGookey, D. J., and Anderson, R. G. W. (1983). Morphological characterization of the cholesteryl ester cycle in cultured mouse macrophage foam cells. *J. Cell Biol.* **97,** 1156–1168.

Mechanic, S., Raynor, K., Hill, J. E., and Cowin, P. (1991). Desmocollins form a distinct subset of the cadherin family of cell adhesion molecules. *Proc. Natl. Acad. Sci. U.S.A.* **88,** 4476–4480.

Meier, J., and Georgatos, S. D. (1994). Type B lamins remain associated with the integral nuclear envelope protein p58 during mitosis: Implications for nuclear reassembly. *EMBO J.* **13,** 1888–1898.

Merdes, A., Brunkener, M., Horstmann, H., and Georgatos, S. D. (1991). Filensin: A new vimentin-binding, polymerization-competent, and membrane-associated protein of the lens fiber cell. *J. Cell Biol.* **115,** 397–410.

Merdes, A., Gounari, F., and Georgatos, S. D. (1993). The 47-kD lens-specific protein phakinin is a tailless intermediate filament protein and an assembly partner of filensin. *J. Cell Biol.* **23,** 1507–1516.

Mose-Larsen, P., Bravo, R., Fey, S. J., Small, J. V., and Celis, J. E. (1982). Putative association of mitochondria with a subpopulation of intermediate-sized filaments in cultured human skin fibroblasts. *Cell (Cambridge, Mass.)* **31,** 681–692.

Muller, H., and Franke, W. W. (1983). Biochemical and immunological characterization of desmoplakins I and II, the major polypeptides of the desmosomal plaque. *J. Mol. Biol.* **163,** 647–671.

Ngai, J., Coleman, T. R., and Lazarides, E. (1990). Localization of newly synthesized vimentin subunits reveals a novel mechanism of intermediate filament assembly. *Cell (Cambridge, Mass.)* **60,** 415–427.

Nickerson, P. A., Skelton, F. R., and Molteni, A. (1970). Observation of filaments in the adrenal of androgen-treated rats. *J. Cell Biol.* **47,** 277–280.

Nigg, E. A. (1992). Assembly and disassembly of the nuclear lamina. *Curr. Opin. Cell Biol.* **4,** 105–109.

Nilles, L. A., Parry, D. A. D., Powers, E. E., Angst, B. D., Wagner, R. M., and Green, K. J. (1991). Structural analysis and expression of human desmoglein: A cadherin-like component of the desmosome. *J. Cell Sci.* **99,** 809–821.

Novikoff, A. B., Novikoff, P. M., Rosen, O. M., and Rubin, C. S. (1980). Organelle relationships in cultured 3T3-L1 preadipocytes. *J. Cell Biol.* **87,** 180–196.

O'Keefe, E. J., Erickson, H. P., and Bennett, V. (1989). Desmoplakin I and desmoplakin II: Purification and characterization. *J. Biol. Chem.* **264,** 8310–8318.

Ouyang, P., and Sugrue, S. P. (1992). Identification of an epithelial protein related to the desmosome and intermediate filament network. *J. Cell Biol.* **118,** 1477–1488.

Papamarcaki, T., Kouklis, P., Kreis, T. E., and Georgatos, S. D. (1991). The "Lamin B-fold." *J. Biol. Chem.* **266,** 21247–21251.

Parry, D. A. D., Crewther, W. G., Fraser, R. D. B., and MacRae, T. P. (1977). Structure of α-keratin: Structural implication of the amino acid sequences of the type I and II chain segments. *J. Mol. Biol.* **113,** 449–454.

Peifer, M., McCrea, P., Green, K. J., Wieschaus, E., and Gumbiner, B. (1992). The vertebrate adhesive junction proteins b-catenin and plakoglobin and the Drosophila segment polarity gene armadillo form a multigene family with similar properties. *J. Cell Biol.* **118,** 681–691.

Pelliniemi, L. J., Dym, M., and Paranko, J. (1982). Cytoplasmic filaments in fetal Leydig cells. *Ann. N.Y. Acad. Sci.* **383,** 486–487.

Plancha, C. E., Carmo-Fonseca, M., and David-Ferreira, J. F. (1989). Cytokeratins filaments are present in golden hamster oocytes and early embryos. *Differentiation (Berlin)* **42,** 1–9.

Raats, J. M. H., Gerards, W. L. H., Schreuder, M. I., Grund, C., Henderick, J. B. J., Hendriks, I. L. A. M., Ramaekers, F. C. S., and Bloemendal, H. (1992). Biochemical and structural aspects of transiently and stably expressed mutant desmin in vimentin-free and vimentin-containing cells. *Eur. J. Cell Biol.* **58,** 108–127.

Rafferty, N. S. (1985). Lens morphology. *In* "The Ocular Lens, Structure, Function, and Pathology" (H. Maisel, ed.), pp. 1–160. Dekker, New York.

Ramaekers, F. C. S., and Bloemendal, H. (1981). Cytoskeletal and contractile structures in lens cell differentiation. *In* "Molecular and Cellular Biology of the Eye Lens" (H. Bloemendal, ed.), pp. 85–136. Wiley, New York.

Ramaekers, F. C. S., Dunia, I., Dodemont, H. J., Benedetti, E. L., and Bloemendal, H. (1982). Lenticular intermediate-sized filaments: Biosynthesis and interaction with plasma membrane. *Proc. Natl. Acad. Sci. U.S.A.* **79,** 32208–3212.

Rayns, D. G., Simpson, L. O., and Ledingham, J. M. (1969). Ultrastructure of desmosomes in mammalian intercalated discs: Appearances after lanthanum treatment. *J. Cell Biol.* **42,** 322–326.

Remington, S. G. (1993). Chicken filensin: A lens fiber cell protein that exhibits sequence similarity to intermediate filament proteins. *J. Cell Sci.* **105,** 1057–1068.

Richardson, F. L., Stromer, M. H., Huiatt, T. W., and Robson, R. M. (1981). Immunoelectron and immunofluorescence localization of desmin in mature avian muscles. *Eur. J. Cell Biol.* **26,** 91–101.

Riemer, D., Dodemont, H., and Weber, K. (1992). Analysis of the cDNA and gene encoding a cytoplasmic intermediate filament (IF) protein from the cephalochordate Branchiostoma lanceolatum; implications for the evolution of the IF protein family. *Eur. J. Cell Biol.* **58,** 128–135.

Ris, H. (1991). The three dimensional structure of the nuclear pore complex as seen by high voltage electron microscopy and high resolution low voltage scanning electron microscopy. *EMSA Bull.* **21,** 54–56.

Rungger-Brandle, E., Achtstatter, T., and Franke, W. W. (1989). An epithelium-type cytoskeleton in a glial cell: Astrocytes of amphibian optic nerves contain cytokeratin filaments and are connected by desmosomes. *J. Cell Biol.* **109,** 705–716.

Sarria, A. J., Nordeen, S. K., and Evans, R. M. (1990). Regulated expression of vimentin cDNA in cells in the presence and abscence of a preexisting vimentin filaments network. *J. Cell Biol.* **111,** 553–565.

Sarria, A. J., Panini, S. R., and Evans, R. M. (1992). A functional role for vimentin-intermediate filaments in the metabolism of lipoprotein derived cholesterol in human SW-13 cells. *J. Biol. Chem.* **267,** 19455–19463.

Sarria, A. J., Leiber, J. G., Nordeen, S. K., and Evans, R. M. (1994). The presence or absence of a vimentin-type intermediate filament network affects the shape of the nucleus in human SW13 cells. *J. Cell Sci.* **107,** 1593–1607.

Schafer, S., Koch, P. J., and Franke, W. W. (1994). Identification of the ubiquitous human desmoglein, Dsg2, and the expression catalogue of the desmoglein subfamily of desmosomal cadherins. *Exp. Cell Res.* **211,** 391–399.

Schmelz, M., and Franke, W. W. (1993). Complexus adhaerentes, a new group of desmoplakin-containing junctions in endothelial cells. The syndesmos connecting retothelial cells of lymph nodes. *Eur. J. Cell Biol.* **61,** 274–289.

Schmelz, M., Duden, R., Cowin, P., and Franke, W. W. (1986). A constitutive transmembrane glycoprotein of M_r 165 000 (desmoglein) in epidermal and nonepidermal desmosomes: I. Biochemical identification of the polypeptide. *Eur. J. Cell Biol.* **42,** 177–183.

Schmidt, A., Heid, H. W., Schafer, S., Nuber, U. A., Zimbelmann, R., and Franke, W. W. (1994). Desmosomes and cytoskeletal architecture in epithelial differentiation: Cell type-specific plaque components and intermediate filament anchorage. *Eur. J. Cell Biol.* **65,** 229–245.

Schultheiss, T., Lin, Z., Ishikawa, H., Zamir, I., Stoeckert, C. J., and Holtzer, H. (1991). Desmin/vimentin filaments are dispensable for many aspects of myogenesis. *J. Cell Biol.* **114,** 953–966.

Schwarz, M. A., Owaribe, K., Kartenbeck, J., and Franke, W. W. (1990). Desmosomes and hemidesmosomes: Constitutive molecular components. *Annu. Rev. Cell Biol.* **6,** 461–491.

Senior, A., and Gerace, L. (1988). Integral membrane proteins specific to the inner nuclear membrane and associated with the nuclear lamina. *J. Cell Biol.* **107,** 2029–2036.

Sharpe, A. H., Chen, L. B., Murphy, J. R., and Fields, B. N. (1980). Specific disruption of vimentin filament organization in monkey kidney CV-1 cells by diphteria toxin, endotoxin A and cycloheximide. *Proc. Natl. Acad. Sci. U.S.A.* **77,** 7267–7271.

Shoeman, R. L., and Traub, P. (1990). The in vitro DNA-properties of purified nuclear lamins proteins and vimentin. *J. Biol. Chem.* **265,** 9055–9061.

Shyy, T. T., Asch, B. A., and Asch, H. L. (1989). Concurrent collapse of keratin filaments, aggregation of organelles and inhibition of protein synthesis during the heat shock response in mammary epithelial cells. *J. Cell Biol.* **108,** 997–1008.

Simos, G., and Georgatos, S. D. (1992). The inner nuclear membrane protein p58 associates in vitro with a p58 kinase and the nuclear lamins. *EMBO J.* **11,** 4027–4036.

Simos, G., and Georgatos, S. D. (1994). The lamin B receptor-associated protein p34 shares sequence homology and antigenic determinants with the splicing factor 2-associated protein. *FEBS Lett.* **346,** 225–228.

Skalli, O., and Goldman, R. D. (1991). Recent insights into the assembly, dynamics, and function of intermediate filament networks. *Cell Motil. Cytoskel.* **19,** 67–79.

Skalli, O., Jones, J. C. R., Gagescu, R., and Goldman, R. D. (1994). IFAP 300 is common to desmosomes and hemidesmosomes and is a possible linker of intermediate filaments to these junctions. *J. Cell Biol.* **125,** 159–170.

Skerrow, C. J., and Matolsy, A. G. (1974). Isolation of epidermal desmosomes. *J. Cell Biol.* **63,** 515–523.

Small, J. V., and Sobieszek, A. (1977). Studies on the function and composition of the 10-nm (100-A) filaments of vertebrate smooth muscle. *J. Cell Sci.* **23,** 243–268.

Small, J. V., and Sobieszek, A. (1980). The contractile apparatus of smooth muscle. *Int. Rev. Cytol.* **64,** 241–306.

Smith, S., and Blobel, G. (1993). The first membrane spanning region of the lamin B receptor is sufficient for sorting to the inner nuclear membrane. *J. Cell Biol.* **120,** 631–637.

Somlyo, A. V., and Franzini-Armstrong, C. (1985). New views of smooth muscle structure using freezing, deep-etching and rotary shadowing. *Experientia* **41,** 841–856.

Sonnenberg, A., Calafat, J., Janssen, H., Daams, H., van der Raaji-Helmer, L. M. H., Baker, J., Loizidou, M., and Garrod, D. R. (1991). Integrin $\alpha_6 \beta_4$ complex is located in hemidesmosomes suggesting major role in epidermal cell-basement membrane adhesion. *J. Cell Biol.* **113,** 907–917.

Spinardi, L., Ren, Y. L., Sanders, R., and Giancotti, F. G. (1993). The beta 4 subunit cytoplasmic domain mediates the interaction of alpha 6 beta 4 integrin with the cytoskeleton of hemidesmosomes. *Mol. Cell. Biol.* **4,** 871–884.

Staehelin, L. A. (1974). Structure and function of intercellular junctions. *Int. Rev. Cytol.* **39,** 191–283.

Stappenbeck, T. S., and Green, K. J. (1992). The desmoplakin carboxyl terminus coaligns with and specifically disrupts intermediate filament networks when expressed in cultured cells. *J. Cell Biol.* **116,** 1197–1209.

Stappenbeck, T. S., Bornslaeger, E. A., Corcoran, C. M., Luu, H. H., Viarata, M. L. A., and Green, K. J. (1993). Functional analysis of desmoplakin domains: Specification of the interaction with keratin versus vimentin intermediate filament networks. *J. Cell Biol.* **123,** 691–705.

Steinert, P. M., and Roop, D. R. (1988). Molecular and cellular biology of intermediate filaments. *Annu. Rev. Biochem.* **57**, 593–625.

Stepp, M. A., Spurr-Michaud, S., Tisdale, A., Elwell, J., and Gipson, I. K. (1990). a_6b_4 integrin heterodimer is a component of hemidesmosomes. *Proc. Natl. Acad. Sci. U.S.A.* **87**, 8970–8974.

Stick, R., Angres, B., Lehner, C., and Nigg, E. A. (1988). The fates of chicken nuclear lamin proteins during mitosis: Evidence for a reversible redistribution of lamin B2 between inner nuclear membrane and elements of the endoplasmic reticulum. *J. Cell Biol.* **107**, 397–406.

Stromer, M. H., and Bendayan, M. (1988). Arrangement of desmin intermediate filaments in smooth muscle cells as shown by high-resolution immunocytochemistry. *Cell Motil. Cytoskel.* **11**, 117–125.

Stromer, M. H., and Bendayan, M. (1990). Immunocytochemical identification of cytoskeletal linkages to smooth muscle cell nuclei and mitochondria. *Cell Motil. Cytoskel.* **17**, 11–18.

Summerhayes, I. C., Wong, D., and Chen, L. B. (1983). Effect of microtubules and intermediate filaments on mitochondrial distribution. *J. Cell Sci.* **61**, 87–105.

Suzuki, S., and Naitoh, Y. (1990). Amino acid sequence of a novel integrin β_4 subunit and primary expression of the mRNA in epithelial cells. *EMBO J.* **9**, 757–763.

Takemura, R., Okabe, S., Kobayashi, N., and Hirokawa, N. (1993). Reorganization of brain spectrin (fodrin) during differentiation of PC12 cells. *Neuroscience* **52**, 381–391.

Tamura, R. N., Rozzo, C., Starr, L., Chambers, J., Reichardt, L. F., Cooper, M. H., and Quaranta, V. (1990). Epithelial integrin $\alpha_6\beta_4$: Complete primary structure of α_6 and variant forms of β_4. *J. Cell Biol.* **111**, 1593–1604.

Tanaka, T., Parry, D. A. D., Klaus-Kovtun, V., Steinert, P. M., and Stanley, J. R. (1991). Comparison of molecularly cloned bullous pemphigoid antigen to desmoplakin I confirms that they define a new family of cell adhesion junction plaque proteins. *J. Biol. Chem.* **266**, 12555–12559.

Toh, B. H., Lolait, S. J., Mathy, J. P., and Baum, R. (1980). Association of mitochondria with intermediate filaments and of polyribosomes with cytoplasmic actin. *Cell Tissue Res.* **211**, 163–169.

Tokuyasu, K. T. (1983). Visualization of longitudinally-oriented intermediate filaments in frozen sections of chicken cardiac muscle by a new staining method. *J. Cell Biol.* **97**, 562–565.

Tokuyasu, K. T., Dutton, A. H., and Singer, S. J. (1983). Immunoelectron microscopic studies of desmin (skeletin) localization and intermediate filament organization in chicken skeletal muscle. *J. Cell Biol.* **96**, 1727–1735.

Tokuyasu, K. T., Maher, P. A., and Singer, S. J. (1984). Distributions of vimentin and desmin in developing chick myotubes in vivo. I. Immunofluorescence study. *J. Cell Biol.* **98**, 1961–1972.

Tokuyasu, K. T., Maher, P. A., and Singer, S. J. (1985a). Distributions of vimentin and desmin in developing chick myotubes in vivo. II. Immunoelectron microscopic study. *J. Cell Biol.* **100**, 1157–1166.

Tokuyasu, K. T., Maher, P. A., Dutton, A. H., and Singer, S. J. (1985b). Intermediate filaments in skeletal and cardiac muscle tissue in embryonic and adult chicken. *Ann. N.Y. Acad. Sci.* **455**, 200–212.

Traub, P., and Nelson, W. J. (1983). The interaction in vitro of the intermediate filament protein vimentin with synthetic polyribo- and polydeoxyribonucleotides. *Hoppe-Seyler's Z. Physiol. Chem.* **364**, 575–592.

Traub, P., Nelson, W. J., Kühn, S., and Vorgias, C. E. (1983). The interaction in vitro of the intermediate filament protein vimentin with naturally occuring RNAs and DNAs. *J. Biol. Chem.* **258**, 1456–1466.

Traub, P., Vorgias, C. E., and Nelson, W. J. (1985a). Interaction in vitro of the neurofilament triplet proteins from porcine spinal cord with natural RNA and DNA. *Mol. Biol. Rep.* **10**, 129–136.

Traub, P., Perides, G., Scherbarth, A., and Traub, U. (1985b). Tenacious binding of lipids to vimentin during its isolation and purification from Ehrlich ascites tumor cells. *FEBS Lett.* **193**, 217–221.

Traub, P., Perides, G., Kühn, S., and Scherbarth, A. (1986a). Interaction in vitro of nonepithelial intermediate filament proteins with histones. *Z. Naturforsch. C: Biosci.* **42C**, 47–63.

Traub, P., Perides, G., Schimmel, H., and Scherbarth, A. (1986b). Interaction in vitro of nonepithelial intermediate filament proteins with total cellular lipids, individual phospholipids, and a phospholipid mixture. *J. Biol. Chem.* **261**, 10558–10568,

Traub, P., Perides, G., Kuhn, S., and Scherbarth, A. (1987). Efficient interaction of nonpolar lipids with intermediate filaments of the vimentin type. *Eur. J. Cell Biol.* **43**, 55–64.

Traub, P., Mothes, E., Shoeman, R. L., Schröder, R., and Scherbarth, A. (1992). Binding of nuclei acids to intermediate filaments of the vimentin type and their effects on filament formation and stability. *J. Biomol. Struct. Dyn.* **10**, 505–531.

Trevor, K. T., McGuire, J. G., and Leonova, E. V. (1995). Association of vimentin intermediate filaments with the centrosome. *J. Cell Sci.* **108**, 343–356.

Troyanovsky, S. M., Eshkind, L. G., Troyanovsky, R. B., Leube, R. E., and Franke, W. W. (1993). Contributions of cytoplasmic domains of desmosomal cadherins to desmosome assembly and intermediate filament anchorage. *Cell (Cambridge, Mass.)* **72**, 561–574.

Tsukita, S., and Tsukita, S. (1985). Desmocalmin: A calmodulin-binding high molecular weight protein isolated from desmosomes. *J. Cell Biol.* **101**, 2070–2080.

Uehara, Y., Campbell, G. R., and Burnstock, G. (1971). Cytoplasmic filaments in developing and adult vertebrate smooth muscle. *J. Cell Biol.* **50**, 484–497.

Unwin, P. N. T., and Milligan, R. A. (1982). A large particle associated with the perimeter of the nuclear pore complex. *J. Cell Biol.* **93**, 63–75.

Vorgias, C. E., and Traub, P. (1986). Nucleic acid-binding activities of the intermediate filament subunit proteins desmin and glial fibrillary acidic protein. *Z. Naturforsch, C: Biosci.* **41C**, 897–909.

Wang, K., and Ramirez-Mitchell, R. (1983). A network of transverse and longitudinal intermediate filaments is associated with sarcomeres of adult vertebrate skeletal muscle. *J. Cell Biol.* **96**, 562–570.

Wang, X., and Traub, P. (1991). Resinless section immunogold electron microscopy of karyo-cytoskeletal frameworks of eukaryotic cells cultured in vitro. Absence of a salt-stable nuclear matrix from mouse plasmacytoma MPC-11 cells. *J. Cell Sci.* **98**, 107–122.

Wang, X., Willingale-Theune, J., Shoeman, R. L., Giese, G., and Traub, P. (1989). Ultrastructural analysis of cytoplasmic intermediate filaments and the nuclear lamina in the mouse plasmacytoma cell line MPC-11 after the induction of vimentin synthesis. *Eur. J. Cell Biol.* **50**, 462–474.

Wedrychowski, A., Schmidt, W. N., and Hnilica, L. S. (1986a). The in vitro cross-linking of proteins and DNA by heavy metals. *J. Biol. Chem.* **261**, 3370–3376.

Wedrychowski, A., Schmidt, W. N., Ward, W. S., and Hnilica, L. S. (1986b). Cross-linking of cytokeratins to DNA in vivo by chromium salt and cis-diamminechloroplatinum (II). *Biochemistry* **25**, 1–9.

Weinstein, D. E., Shelanski, M. L., and Liem, R. M. (1991). Suppression by anti-sense mRNA demonstrates a requirement for the glial fibrillary acidic protein in the formation of stable astrocytic processes in response to neurons. *J. Cell Biol.* **112**, 1205–1213.

Wessels, G. M., and Chen, S. W. (1993). Transient localized accumulation of alpha spectrin during sea urchin morphogenesis. *Dev. Biol.* **155**, 161–171.

Wiche, G., Becker, B., Luber, K., Weitzer, G., Castanon, M. J., Hauptmann, R., Stratowa, C., and Stewart, M. (1991). Cloning and sequencing of rat plectin indicates a 466 kD polypeptide chain with a three-domain structure based on a central alpha-helical coiled coil. *J. Cell Biol.* **114**, 83–99.

Woodcock, C. L. F. (1980). Nucleus-associated intermediate filaments from chicken erythrocytes. *J. Cell Biol.* **85,** 881–889.

Worman, H. J., Yuan, J., Blobel, G., and Georgatos, S. D. (1988). A lamin B receptor in the nuclear envelope. *Proc. Natl. Acad. Sci. U.S.A.* **85,** 8531–8534.

Worman, H. J., Evans, C. D., and Blobel, G. (1990). The lamin B receptor of the nuclear envelope inner membrane: A polytopic protein with eight potential transmembrane domains. *J. Cell Biol.* **110,** 1535–1542.

Xu, Z., Cork, L. C., Griffin, J. W., and Cleveland, D. M. (1993). Increased expression of neurofilament subunit NF-L produces morphological alterations that resemble the pathology of human motor neuron disease. *Cell (Cambridge, Mass.)* **73,** 23–33.

Ye, Q., and Worman, H. J. (1994). Primary structure analysis and lamin B and Dna binding of human LBR, an integral protein on the nuclear envelope inner membrane. *J. Biol. Chem.* **269,** 11306–11311.

Yuan, J., Simons, G., Blobel, G., and Georgatos, S. D. (1991). Binding of lamin A to polynucleosomes. *J. Biol. Chem.* **266,** 9211–9215.

Zackroff, R. V., and Goldman, R. D. (1979). In vitro assembly of intermediate filaments from baby hamster kidney (BHK-21) cells. *Proc. Natl. Acad. Sci. U.S.A.* **76,** 6226–6230.

Zerban, H., and Franke, W. W. (1977). Structures indicative of keratinization in lactating cells of bovine mammary gland. *Differentiation (Berlin)* **7,** 127–131.

Developmental Failure in Preimplantation Human Conceptuses

Nicola J. Winston

Laboratoire de Physiologie du Developpement, Institut Jacques Monod,
CNRS-Université Paris VII, 75005 Paris, France

The majority of human conceptuses fertilized normally *in vitro* fail to establish a pregnancy following their replacement *in utero*. However, since conceptuses are usually transferred after only one or two cell divisions, their developmental outcome is not known. It has been found that a significant number of human oocytes which can be fertilized carry chromosomal abnormalities, even in the absence of ovarian stimulation. After fertilization, preimplantation-stage conceptuses developing *in vitro* display a high incidence of cellular abnormalities. Similar disruptions of cellular organization have also been noted in conceptuses fertilized *in vivo*. Thus, developmental abnormalities and the demise of the conceptus prior to the stage of implantation may stem from the poor quality of the oocyte. The conditions encountered *in vitro* have also been proposed to cause or contribute to the early demise of human conceptuses.

KEY WORDS: Human conceptus, Oocyte, Fertilization *in vitro*, Preimplantation development, Nuclear-cytoplasmic abnormalities.

I. Introduction

Infertility or subfertility has been estimated to influence the reproductive potential of as many as one in ten couples. One approach to overcoming human infertility is *in vitro* fertilization (IVF) therapy. However, while 25 years have elapsed since the first human oocytes were fertilized *in vitro* (Edwards *et al.*, 1969), although IVF in most cases produces conceptuses for transfer, pregnancy does not occur for the majority of couples (Interim Licensing Authority, 1991; American Fertility Society for Assisted Reproductive Technologies, 1994). The treatment of infertility by IVF and related techniques has also permitted the study of early events in human reproduc-

139

tion and the contribution of oogenesis, fertilization, and early embryogenesis to subsequent development. Examination of fresh and aged oocytes and conceptuses surplus to therapeutic needs has shown a high incidence of abnormality. Understanding the causes of conceptus wastage is complicated not only by the many variables associated with the treatment of individuals and our limited knowledge of gamete maturation and embryonic requirements *in vitro,* but also by the limited availability of material and by ethical constraints. Abnormalities at both the cellular and chromosomal level in oocytes and preimplantation conceptuses *in vivo* (Hertig *et al.,* 1956) and *in vitro* (Plachot *et al.,* 1986b; Tesarik *et al.,* 1987; Van Blerkom and Henry, 1992) suggest that the developmental process is frequently disrupted in human oocytes and conceptuses. To determine those factors most vital for normal conceptus development, it is necessary to establish whether the high level of human conceptus wastage results primarily from the production of poor quality gametes naturally, or whether inappropriate signals experienced during ovarian stimulation, culture *in vitro,* and conceptus transfer *in utero* are the principal cause or make an additional contribution to reduced conceptus viability. The effects of IVF therapy upon conceptus development are difficult to quantify, but the available data indicate that developmental failure is not exclusive to oocytes and conceptuses cultured *in vitro.* In this chapter, the nomenclature of early development by Johnson and Selwood (1996) will be used.

II. Gamete Competence

A. Oocyte Maturation

The failure of a conceptus to implant may stem from the quality of the oocyte. The cytoplasmic content of the fertilized zygote, which includes the organelles and protein synthetic apparatus required by the early conceptus, is almost exclusively of oocyte origin and, if incompetent, might reduce the potential for fertilization and early conceptus development.

The precise mechanisms governing oocyte maturation and the faithful production of a haploid cell containing a correct genetic complement remain unclear (Angell *et al.,* 1993). The situation is further complicated following ovarian stimulation, which may expose maturing follicular oocytes to abnormally high levels of gonadotrophic hormones. Many studies have been conducted to determine whether superovulation has detrimental consequences; however, to date there is no clear-cut answer. Furthermore, there are few oocytes available for study which are produced in the absence of exogenous hormone stimulation. However, those available illustrate that abnormal oocytes do arise in the absence of ovarian superovulation (Martin *et al.,* 1986; Wramsby and Fredga, 1987).

Follicular maturation is an intricate procedure which depends upon changes in the ovarian environment, appropriate to the developmental stages of the oocyte. The oocytes of most mammals, including the human, reach the diffuse diplotene or dictyate stage of the first meiotic division at the time of birth, or shortly thereafter, and remain arrested at this point until puberty. Prior to the discharge of a mature fertilizable oocyte in each ovulatory cycle, a carefully integrated developmental sequence must be completed before the oocyte is fully able to support fertilization (Edwards *et al.,* 1980a; Fritz and Speroff, 1982; Trounson *et al.,* 1982; Johnson and Everitt, 1995). The effects of gonadotrophic hormones, steroid hormones, and intrafollicular molecules on the oocyte are directed by the somatic follicular cells to which the oocyte is coupled metabolically. The exit from prophase and entry into metaphase of the first meiotic cycle is a critical phase of development which may be particularly susceptible to disruption during oocyte maturation. The developing oocyte must then halve its chromosome content just prior to ovulation to ensure euploidy following fertilization. Damage to the meiotic spindle may cause chromosomes to become lost from the oocyte, or left behind, following division of the cytoplasm.

During the menstrual cycle in the human, only one follicle usually acquires the competence to mature in response to preovulatory gonadotrophin stimulation (Hodgen, 1982). Pregnancies have been achieved after the transfer of a single *in vitro* fertilized oocyte recovered from an unstimulated follicular cycle (Steptoe and Edwards, 1978), although the recovery of several oocytes increases the chances of recovering at least one fertilizable oocyte in each IVF treatment cycle (Edwards *et al.,* 1980b). Since it became possible to retrieve multiple fertilizable oocytes which were able to establish pregnancies (Johnston *et al.,* 1981; Trounson *et al.,* 1981), several methods of artificially raising the levels of circulating gonadotrophic hormones have been developed to recruit and maintain the development of multiple follicles. However, the response of each follicle seems to be determined uniquely (Adashi, 1994, and references therein) and depends upon several factors. These include the quality of the ovarian stroma, the surrounding developing oocytes, the effect of other organ systems on the immediate environment of the follicle, and the heterogeneity among the follicles themselves, which influences their sensitivity to gonadotrophins. As a result, the sensitivity and response within and between individual women to gonadotrophins is highly variable (Scholtes *et al.,* 1988). Thus, the quality of the oocytes collected will vary both among patients and between oocytes from any one patient, and the potential for oocyte development may be highly unpredictable, even at this early stage. The poor success rate of IVF treatment might indicate that a significant proportion of meiotically mature oocytes recovered after ovarian stimulation are developmentally compromised, perhaps due to pathological changes in both cytoplasmic organization and function induced during preovulatory maturation (Van Blerkom, 1989; Van Blerkom and Henry, 1992). For example, elevated concentrations of estrogen (E_2) in stimulated cycles can induce premature luteinization in even the largest follicle by causing the normal follicular phase association between follicular diameter and the E_2 level to break down (Fleming and Coutts, 1986), thereby reducing oocyte potential (Suchanek *et al.,* 1994).

To overcome some of the problems of variable follicle response within a woman, gonadotrophin-releasing hormone (GnRH) agonists have been developed for therapeutic use. GnRH induces hypogonadotrophism by reducing the tonic level of luteinizing hormone (LH) and blocking the LH surge. Follicular ovulation can then be induced artificially to produce several synchronously developing and ovulatory follicles under the influence of exogenous gonadotrophins (Conaghan *et al.,* 1989; Dimitry *et al.,* 1991). The results show an improvement in the quality of oocytes and conceptuses as well as in the pregnancy rate (Rutherford *et al.,* 1988; Khan *et al.,* 1989). Thus, failure to manage ovarian stimulation carefully can increase the proportion of developmentally compromised or defective oocytes and abnormal conceptuses arising from them.

The exposure of follicular oocytes to high levels of gonadotrophins during ovarian stimulation has been associated with damage to oocytes in other mammals (mouse, Takagi and Sasaki, 1976; rabbit, Fujimoto *et al.,* 1974). It is difficult to determine whether, for example, the high degree of chromosomal abnormality in human oocytes is a direct consequence of ovarian hyperstimulation. Van Blerkom (1990a) observed that only a small proportion of oocytes matured *in vitro* displayed chromosomal abnormalities, suggesting that follicular stimulation regimes might be detrimental. However, other studies have proposed that superovulation procedures which precede IVF do not necessarily add to the naturally high aneuploidy rate in the human oocyte (Delhanty and Penketh, 1990). Whether high doses of gonadotrophic hormones have an effect on human oocyte quality which is not apparent at the structural level, how these influences are manifested, and what proportion of oocytes might be affected remain unclear.

B. Chromosomal Abnormalities

It is clear that chromosome anomalies make a considerable contribution to errors of reproduction in humans, with many cytogenetic syndromes resulting from even the slightest imbalance of chromosomes. Chromosomal aberrations occurring during gametogenesis and the early stages of development play a significant role in fetal loss (Chard, 1991). Boué *et al.* (1975b) reported that about 60% of spontaneously aborted fetuses developing from oocytes fertilized *in vivo* have an abnormal karyotype, most of which were trisomies or polyploidies. After *in vitro* fertilization, the rate of chromosomal anomalies in spontaneous abortions was assumed to be 62% (21/34, Plachot, 1989).

1. Oocyte

Chromosomal anomalies have been found to occur in oocytes and conceptuses fertilized *in vitro* and *in vivo*. The proportion of chromosomal abnormalities is greater in human oocytes than in spermatozoa (Pellestor, 1991b) and increases with advancing age (Pellestor and Sèle, 1988; Plachot *et al.*, 1988b; Delhanty and Penketh, 1990). However, the effect of losing or gaining chromosomes or single chromatids (Angell, 1991) might not be detectable morphologically in the oocyte, and can be unimportant during fertilization and early development (Jacobs *et al.*, 1978). Only a few studies have carried out cytogenetic analyses on freshly retrieved noninseminated oocytes and the incidence of reported chromosomal abnormalities differs greatly, ranging from 2/12 oocytes (16.7%; Edirisinghe *et al.*, 1992), 34% in a study of 50 oocytes (Martin *et al.*, 1986), to nearer 50% in a group of 23 oocytes (Wramsby and Fredga, 1987). Many more aged failed fertilized oocytes have been available for examination, although the rate of chromosomal abnormalities reported also varies considerably (from 7.4%, Michelmann and Hinney, 1988, to 65.4%, Wramsby *et al.*, 1987) These discrepancies may be attributed to several factors, including the method of chromosome prepararation preferred by individual authors, interclinic variations in the handling and culture of the oocytes during IVF treatment, the randomization of oocytes selected for analysis, the inclusion of oocytes that failed to fertilize due to sperm-related factors, the number of oocytes analyzed, and the difficulty in analyzing the chromosomes of mature oocytes in which the chromatids are very condensed. Three large studies found that a similar proportion of aged nonfertilized oocytes examined were chromosomally abnormal (26%, Plachot *et al.*, 1990; 24%, Pellestor, 1991a; 21.1%, Bongso *et al.*, 1988), which is similar to the proportion of cytogenetically abnormal fresh human oocytes, but significantly higher than in other mammalian oocytes following ovarian stimulation (1% to 10%, Asakawa and Dukelow, 1982; Asakawa *et al.*, 1988). Differences in the stimulation regimes used have also been proposed to contribute to variations in the reported frequency of chromosomal anomalies (Angell *et al.*, 1991), although other studies have contradicted this hypothesis (Plachot *et al.*, 1988b; Van Blerkom and Henry, 1988; Pellestor *et al.*, 1989; Edirisinghe *et al.*, 1992).

Whether aneuploidy is a primary cause of developmental failure, or a secondary consequence of other altered cellular or molecular processes of developmental significance is not clear. Oocyte aging to just a few hours after ovulation has been correlated with an increase in the proportion of abnormal oocytes (Plachot *et al.*, 1990) and associated with spindle instability (Szollosi, 1975), which may provide a cytological basis for chromosome nondisjunction and loss. However, some oocytes appear grossly normal and show no signs of spindle instability when examined up to 72 hr after collection (Eichenlaub-Ritter *et al.*, 1988). This observation might suggest that oocyte abnormality is restricted to a subpopulation of oocytes which contributes disproportionately to the relatively high frequency of aneuploidy observed after ovarian stimulation and ovulation induction (Van Blerkom and Henry, 1992).

Mouse oocytes in which the spindle was disrupted following freezing and thawing were also found to have chymotrypsin-resistant zonae and would thus be less likely to become fertilized (George *et al.*, 1994). Fertilization failure may, therefore, in some cases, be a direct consequence of oocyte abnormality and hence disproportionally increase the number of aneuploid oocytes in a group of aged oocytes that failed to be fertilized. However, in the same study George *et al.* (1994) also observed that although the implantation rate of frozen-thawed mouse oocytes that fertilized and were transferred on day two after insemination was equivalent to the control group, the incidence of spontaneous abortions was higher. Thus, in some cases, damage to components of the cytoskeletal system may be discrete from the overall function of the oocyte and only become apparent following fertilization, during cleavage or at later stages of development.

2. Spermatozoa

The fertilizing spermatozoon may also affect conceptus development (Lu *et al.*, 1994), although a certain amount of selection against abnormal spermatozoa may occur at the zona pellucida (Liu and Baker, 1992). However, alterations in spermatozoal DNA have been associated with fertilization failure (Peluso *et al.*, 1992) and other data also suggest associations with poor conceptus quality and delayed cleavage (Yovich and Stanger, 1984; Ron-El *et al.*, 1991; Parinaud *et al.*, 1993). In addition, Rosenbusch and Sterzik (1991) have reported that men from couples experiencing repeated spontaneous abortions have a higher incidence of sperm chromosome abnormalities then men from couples with no history of miscarriage. Furthermore, it has also been suggested recently that the human zygote inherits its mitotic potential from the male gamete (Palermo *et al.*, 1994; Schatten, 1994), whereby the sperm centriole controls the first mitotic division after fertilization. In some cases simply removing the zonal barrier (subzonal sperm injection; SUZI), or in addition the oolemma (intracytoplasmic sperm injection; ICSI), allows spermatozoa from patients with poor fertilization after standard IVF to fertilize oocytes at a rate equivalent to spermatozoa from patients with previously good or acceptable fertilization under standard IVF conditions (Nagy *et al.*, 1993). In other cases, though fertilization may occur, the conceptus is functionally abnormal (Cohen *et al.*, 1991; Van Steirteghem *et al.*, 1994). *In vitro* aging has been demonstrated to have a detrimental effect on the chromosomes of mouse spermatozoa (Munné and Estop, 1991) and may contribute to the extremely poor pregnancy outcome following reinsemination, or ICSI, of aged, failed fertilized human oocytes (Winston *et al.*, 1993; Tsirigotis *et al.*, 1994), which frequently uses spermatozoa collected on the day of oocyte retrieval.

Together, these data show that in the human there is a high incidence of errors at the chromosomal level that can originate from either gamete, although the effect of abnormal oocytes seems to be proportionately greater. These chromosomal abnormalities in the gametes can be compatible with fertilization, early cleavage, and in some cases implantation. This naturally occurring phenomenon will greatly influence the potential of any assisted reproductive techniques to alleviate problems of infertility.

C. Nongenetic Factors

Cytoplasmic immaturity has been observed by several authors in those oocytes which show no signs of fertilization as scored by the formation of pronuclei. The reported frequency of fertilized oocytes in which the sperm chromosomes remain condensed in the ooplasm (premature chromosome condensation, PCC) ranges from 3 to 28% (3–4%, Schmiady *et al.*, 1986; 8.35%, Michaeli *et al.*, 1990; 9%, Schmiady and Kentenich, 1989; 15.4%, Pieters *et al.*, 1989; 16.9%, Ma *et al.*, 1989; 28%, Janny *et al.*, 1990; 29.5%, Almeida and Bolton, 1993). In some instances, further culture allows the ooplasm to complete maturity and the oocytes can respond subsequently to the fertilizing spermatozoon. Recent studies on mouse oocytes show that the capacity of oocytes to become activated depends on the completion of cytoplasmic maturation and not nuclear meiotic maturation (Eppig *et al.*, 1994; McConnell *et al.*, 1995). Thus, although nuclear and cytoplasmic maturation usually occur more or less simultaneously, the data indicate that the competence to undergo maturation is acquired independently by each of the cellular components. Hence, while the development of oocytes with an immature cytoplasm is blocked, once the cytoplasm has completed maturation, development can commence in fertilized oocytes despite the condition of the nuclear component. Other nongenetic but as-yet undefined factors may act on subpopulations of oocytes produced by an individual patient. The incidence of fertilization of mature oocytes from couples with an unexplained cause of infertility was found to be significantly lower than that in an otherwise matched group of patients with damaged fallopian tubes (MacKenna et al., 1992). However, once fertilized, the cleavage and clinical pregnancy rates did not differ statistically between the two groups. Boatman et al. (1994) concluded that prefertilization exposure to oviductal fluid and not ovulation or the time post-HCG alters the morphology and fertilizability of oocytes. This would tally with the higher percentage of implantations and pregnancies reported following gamete intrafallopian transfer (G1FT) treatment, where mature human oocytes at the preovulatory follicular stage are retrieved and placed directly into the oviduct, than after standard IVF and conceptus transfer. Furthermore, the higher than natural concentrations of spermatozoa used for oocyte insemination in vitro may subject the oocytes and early zygotes to a potentially toxic environment high in hydrolases, peroxides, and free radicals. The extent to which individual oocytes are affected by conditions in vitro may differ according to their ability to resist damage (Johnson and Nasr-Esfahani, 1994).

III. Developmental Potential of the Human Conceptus

In general, the majority of human conceptuses fail (Munné et al., 1994). This is due primarily to numerical chromosome anomalies occurring either before or after fertilization. Overall, at least 30% of normally developing, and 75% of arrested conceptuses are abnormal at the chromosomal level (Plachot et al., 1988b). With the exception of polypronucleate zygotes, abnormal conceptuses, be they polyploid, aneuploid, or carrying gene defects, cannot be distinguished at the level of the light microscope. An abnormal genome may not affect conceptus morphology until gene activation is completed, but this usually occurs after the time of conceptus transfer (Braude et al., 1988). Furthermore, even grossly imbalanced conceptuses from tripronuclear zygotes are capable of forming blastocysts in vitro (Dokras et al., 1991; Balakier, 1993).

In order to minimize the duration of culture in vitro, human preimplantation conceptuses are usually transferred to the uterus at the 2- to 4-cell stage (day 2 postinsemination). As a result, the developmental potential of a large number of human conceptuses cannot be determined. The successful formation of blastocysts in vitro is generally taken as a measure of the adequacy of culture conditions and likely developmental competence for both mouse and human conceptuses (Hillier et al., 1985; George and Doe, 1989). The premise that those human conceptuses capable of development to the blastocyst stage are those which would have been most likely to develop *in utero* if transferred to recipients at the 2- to 4-cell stage has been tested (Dawson *et al.,* 1988; Bolton *et al.,* 1991; Bolton, 1992). Apparently normally fertilized conceptuses were kept in culture until day 5 postinsemination before being transferred to the patient. However, the clinical pregnancy rate following the transfer of blastocysts or morulae on day 5 after insemination did not better that achieved in the control group of matched patients in whom the transfer procedure was performed on day 2. Therefore, "blastocyst" formation alone seems to be an inadequate indicator of conceptus quality and developmental potential.

During IVF treatment, human conceptus development is assessed mostly by light microscopy. The viability of a conceptus is then assigned from its morphological appearance, combined with the developmental stage reached after a given period in culture. The development of conceptuses can also be assessed noninvasively upon the basis of their metabolic activity. Measurement of the nutrition and metabolism of individual human conceptuses has been made possible by the development of microchemical techniques (Gardner and Leese, 1993). Using this approach, Hardy *et al.* (1989b) demonstrated that the uptake of pyruvate by arrested conceptuses is lower than by those which develop normally. The measurement of pyruvate uptake by individual human conceptuses on successive days of culture prior to transfer to the patient on day 3 after insemination revealed that the metabolic activity of a conceptus on day 2 of development is correlated with its metabolic activity the following day (Gardner *et al.,* 1991). These data indicate that the metabolic activity of individual human conceptuses may provide an additional basis upon which more viable conceptuses may be selected for transfer (Conaghan *et al.,* 1991; Gardner and Leese, 1993; Turner *et al.,* 1994). Furthermore, the study of conceptus metabolism will increase our understanding of their physiological requirements and hence permit more suitable culture conditions to be established (Conaghan *et al.,* 1993; and see later discussion).

The cytogenetic status of a conceptus can be examined by biopsy and the removal of one or two blastomeres (Handyside *et al.,* 1989). Despite being invasive to the conceptus, this procedure is compatible with continued development, implantation, and live births (Handyside *et al.,* 1990; Hardy *et al.,* 1990). However, in general, conceptuses are selected for transfer on the basis of their gross morphology alone. A positive correlation between conceptus morphology and the implantation rate after IVF treatment has been reported (Shulman *et al.,* 1993). Chromosomal constitution has also been related to conceptus quality, whereby the incidence of chromosomal abnormalities in good-quality conceptuses was lowest (23.5%, Edirisinghe *et al.,* 1992), although still significant. However, Jamieson *et al.* (1994) found no difference between conceptus morphology and growth rate in apparently normal and aneuploid conceptuses. Furthermore, multinucleate blastomeres (MNBs) have been demonstrated in 50% of morphologically normal-looking conceptuses at stages from 2 to 16 cells (Lopata *et al.*, 1983) and in nearly 70% of all conceptuses developing *in vitro* (Tesarik, 1988). Winston *et al.* (1991a) also found that although the proportion of conceptuses showing abnormalities was higher in grades II to IV, a significant proportion of grade I conceptuses contained abnormally nucleated cells. Plachot *et al.* (1986a) reported NMBs in 69% of normally fertilized 4- to 8-cell conceptuses and in another study showed aneuploidy and mosaicism in 21.4% of healthy conceptuses 3 days after insemination, rising to 32.6% in fragmented and degenerate conceptuses (Plachot *et al.*, 1988a). The lack of a relationship between the parameters of conceptus morphology and chromosome constitution implies that conceptuses selected for transfer to the patient carry a genetic load (Jamieson *et al.*, 1994). From the data available, this is equivalent to about 30% of conceptuses with abnormalities. Thus, 9% of patients who have two conceptuses replaced will receive no normal conceptuses; 42% will receive only one normal conceptus; and only half (49%) will receive two conceptuses with no chromosomal abnormalities. Morphology is thus not a reliable criterion upon which to predict normal cellular development in human conceptuses.

A. Assessment of Cell Number and Cellular Integrity during Preimplantation Development

To establish the extent of abnormal development, the stage at which development becomes disrupted and the relationship between conceptus morphology and the proportion of multinucleate cells, the cell number and cellular integrity and nuclear number and integrity were recorded in normally fertilized (two pronucleate) conceptuses developing *in vitro*. The conceptuses were cultured to days 2, 3, or 5 after insemination. The number of oocytes inseminated initially in each group differed (325, 112, and 76 respectively); however, the proportion of normally fertilized oocytes was very similar (78%, 75%, and 77%, respectively).

Of 159 conceptuses cultured until day 5 (120–124 hr postinsemination, hpi), 62 (39%) formed blastocysts (defined as large, cavitated, fluid-filled structures). These blastocysts were fixed in 4% formaldehyde in PBS for 20 to 30 min and the cellular chromatin was stained with DAPI ([4′,6 diamidino-2-phenylindole, Sigma; 50 μg/ml diluted in H6 culture medium (Howlett *et al.*, 1987; Nasr-Esfahani *et al.*, 1990b)] supplemented with 4 mg/ml bovine serum albumin (lyophylized BSA; Sigma) for 30 min. In five blastocysts the chromatin staining was too poor to make an assessment of the number of nuclei. Of the remainder, the number of nuclei varied

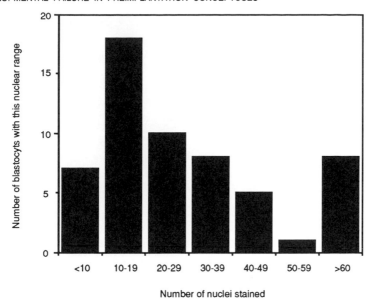

FIG. 1   Total number of nuclei in 57 human "blastocysts" cultured *in vitro* to day 5 (120–124 hr postinsemination).

widely from <10 to in excess of 60, with the majority (61%) having <30 nuclei (Fig. 1 and Fig. 2, a–f).

When the nuclear number in each blastocyst on day 5 was compared with the subjective assessment of its quality as a conceptus made on day 2, while conceptuses of all grades produced blastocysts, the majority had fewer than 50 cells, particularly in the less good groups, II to IV (Fig. 3). Apparently morphologically normal conceptuses produced 79% of the blastocysts with <50 nuclei. Thus overall, only 51% of those conceptuses graded I on day 2 cavitated and only 10% of grade I conceptuses cavitated with >50 nuclei.

To determine whether abnormal development was evident at earlier cleavage stages, a group of 59 conceptuses was cultured to day 3 (72 to 75 hpi), when an accurate cell count was made using phase contrast microscopy. Nuclear counts by DAPI staining were possible in 47 of these conceptuses (Table I). The median cell number was between 4 and 6 cells (range 2 to 10 cells) and the majority of conceptuses contained multinucleate and anucleate cells (Fig. 4, c and d). Only 14 (30%) were nucleated normally (Fig. 4, a and b), and seven of these had fewer than 5 cells. These data suggested that the anomalies of cell division occur as early as day 3 and possibly even earlier. Indeed, in a third group of 48 conceptuses examined

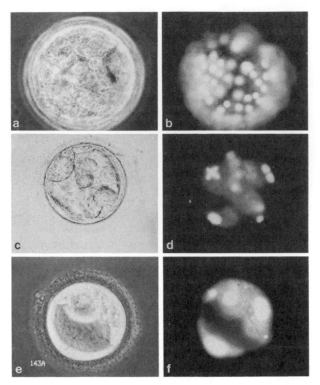

FIG. 2   Pairs of photomicrographs shown using phase contrast (a, c and e) and UV fluorescence (b, d and f), respectively, of DAPI-stained human conceptuses on day 5 of development *in vitro*. (a) and (b) "Normal" blastocyst stage conceptus showing >75 nuclei after DAPI staining. (c) and (d) "Blastocyst" at 120 hpi with only 22 nuclei. (e) and (f) Conceptus with a "blastocyst" appearance on day 5, showing only 4 nucleated cells after DAPI staining. ×292. Winston, N. J., Braude, P. R., Pickering, S. J., George, M. A., Cant, A., Currie, J., and Johnson, M. H. (1991a). The incidence of abnormal morphology and nucleocytoplasmic ratios in 2, 3 and 5 day human pre-embryos. *Hum. Reprod.* **6,** 17–24, by permission of Oxford University Press.

at 47 to 49 hpi, 23 conceptuses had between 2 and 4 cells and even at this early stage in development most conceptuses contained multinucleate and anucleate cells (Table II; Fig. 4, e and f). Only 17 conceptuses (35%) had a full complement of cells with a single nucleus.

The proportion of conceptuses which had formed "blastocyst" structures by day 5 in this study (39%) fell within the range of previous reports (45%, Fehilly *et al.,* 1985; 57%, Dawson *et al.,* 1988; 23%, Hardy *et al.,* 1989a; 25%, Bolton and Braude, 1987; 32%, Hardy *et al.,* 1990). The low incidence of blastocyst formation may be intrinsic to human conceptuses, a result of *in vitro* handling procedures (Pickering *et al.,* 1990), or a reflection of the use of "spare" conceptuses that are not selected as "best" for replacement to the patient (Braude *et al.,* 1988; Bolton *et al.,* 1989). Furthermore, since

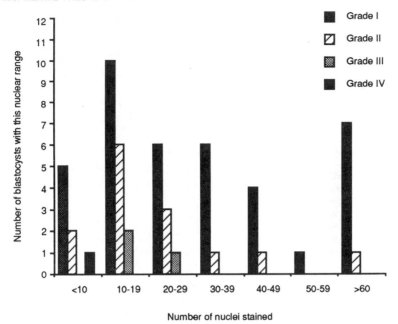

FIG. 3   Bar chart showing the range of cell numbers in conceptuses with a "blastocyst" appearance on day 5 after insemination and graded according to their morphological appearance on day 2.

TABLE I

Total Number of Cells (CN) and the Distribution of Nuclei in the Cells of 47 Human Conceptuses Cultured to Day 3 (72–75 hr Postinsemination)[a, b]

Cell number (CN)	Number of conceptuses with this cell number	Proportion of cells with a single nucleus (% of total)	Proportion of conceptuses in which all cells had a single nucleus (% of total)
2	2	4/4 (100)	2/2 (100)
3	6	5/18 (28)	1/6 (17)
4	9	30/36 (83)	4/9 (44)
5	10	30/50 (60)	2/10 (20)
6	11	40/66 (60)	1/11 (9)
7	5	31/35 (88)	3/5 (60)
8	3	16/24 (66)	1/3 (33)
10	1	9/10 (90)	0/1 (0)

[a] Mean number of cells per conceptus = 5.2.

[b] Mean number of nuclei per conceptus = 4.7.

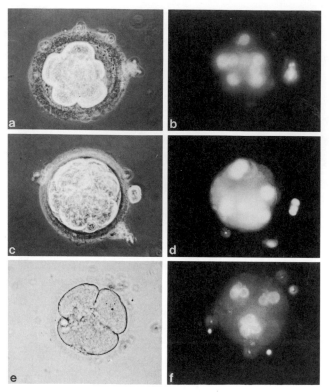

FIG. 4   Pairs of photomicrographs shown using phase contrast (a, c and e) and UV fluorescence (b, d and f), respectively, of DAPI-stained human conceptuses at various stages of development *in vitro*. (a) and (b) Eight-cell conceptus (day 3; 72 hpi) with regular blastomeres and a single nucleus in each cell. (c) and (d) Day 3 conceptus with asymmetrical fragments showing binucleate and anucleate cells after staining. (e) and (f) Four-cell conceptus (day 2; 48 hpi) with three multinucleate and one anucleate blastomere. ×250. Winston, N. J., Braude, P. R., Pickering, S. J., George, M. A., Cant, A., Currie, J., and Johnson, M. H. (1991a). The incidence of abnormal morphology and nucleocytoplasmic ratios in 2, 3 and 5 day human pre-embryos. *Hum. Reprod.* **6,** 17–24, by permission of Oxford University Press.

human conceptuses for research are generally only available in small numbers, part of the variation may be accounted for by the wide range in sample size among studies, interpatient variation, and the integrity of the starting material. The formation of a blastocoelic cavity in mouse conceptuses appears to be independent of the actual cell number present (Smith and McLaren, 1977; Chisholm *et al.,* 1985), or their ploidy (Snow, 1973). Blastocyst-like vesicles with only 4 cells can hatch and outgrow *in vitro* and implant *in vivo,* but they lack developmental competence because they have insufficient cells to form an appropriate inner cell mass (Pratt *et*

TABLE II

Total Number of Cells (CN) and the Distribution of Nuclei in the Cells of 48 Human Conceptuses Cultured to Day 2 (47–49 hr Postinsemination)[a, b]

Cell number (CN)	Number of conceptuses with this cell number	Proportion of cells with a single nucleus (% of total)	Proportion of conceptuses in which all cells had a single nucleus (% of total)
2	16	16/32 (50)	6/16 (38)
3	9	17/27 (63)	3/9 (33)
4	15	37/60 (62)	5/17 (29)
5	1	5/5 (100)	1/1 (100)
6	5	14/30 (47)	1/5 (20)
7	1	2/7 (28)	0/1 (0)
8	1	8/8 (100)	1/1 (100)

[a] Mean number of cells per conceptus = 3.5.
[b] Mean number of nuclei per conceptus = 4.1.

al., 1981). Thus, it can be inappropriate to assume the presence of full developmental competence from the gross morphological appearance of a "blastocyst." Other criteria, such as cell number and distribution are needed.

There are very few studies of cell number in human blastocysts which have developed *in vivo*. Hertig *et al.* (1954) recorded counts of 158 and 107 cells in two blastocysts estimated to be 5 days old when recovered at hysterectomy. Croxatto *et al.* (1972) obtained a blastocyst from uterine flushing which was estimated by serial sectioning to contain 186 cells, although the age of the blastocyst was not defined clearly. Steptoe *et al.* (1971) recorded nuclear counts of 112 and 110 in two blastocysts grown *in vitro* for 7 days following IVF. The most comprehensive study of blastocysts grown *in vitro* (Hardy *et al.*, 1989a) used nuclear fluorescent staining to report an average cell number of 58 ± 8.1 in nine day-5 blastocysts, although the number of cells varied widely (24 to 90). Thus, on the basis of these estimates *in vivo* and *in vitro*, a human blastocysts on day 5 would be expected to have completed six cleavage divisions and contain in excess of 60 cells. However, the relationship between the number of cell cycles elapsed and the onset of cavitation has not been established clearly in human conceptuses. Moreover, the assessment of conceptus development involves regular daily observation and is thus based on a chronological time course.

Studies in the mouse have revealed a considerable variation of about 15 hr in the absolute time at which conceptuses start cavitation (Chisholm *et al.*, 1985). Heterogeneity in the rate of development is also evident among human conceptuses (Fishel, 1985) and may account in part for the range in cell numbers observed following electron microscopic examination of serially sectioned cavitated conceptuses fixed at 120 to 124 hpi (Winston *et al.,* 1991a). The estimates made from fluorescently stained nuclei suggested that two thirds of blastocysts on day 5 have fewer than 30 cells, and that the cell number varied greatly, which is in agreement with Hardy *et al.* (1989a). By any criteria, these nuclear numbers seem low. Since only 39% of fertilized oocytes formed blastocysts, and it seems that only one third of these can be considered "normal," then only 13% of fertilized oocytes can be said to achieve an apparently normal blastocyst status.

There were, however, other morphological criteria observed which might question the "normality" of even the 13% of blastocysts with a seemingly appropriate cell number. The "cell number" count made following nuclear staining assumes that the number of nuclear structures reflects the number of cells present. However, some cells may be anucleate or multinucleate, or the nucleus might fail to stain with the dye. The results from serial sections of mammalian blastocysts (Long and Williams, 1982; Winston *et al.,* 1991a) show that the majority have a significant proportion of multinucleate cells. Hence, nuclear counts are likely to overestimate the cell number, although MNBs and mixiploidy seem to be a normal part of mammalian blastocyst development, including humans, both *in vivo* and *in vitro* (Iwasaki *et al.,* 1992; Benkhalifa *et al.,* 1993) and may arise as a result of cell fusion. These observations indicate that the majority of *in vitro* fertilized conceptuses fail to complete sufficient cell cycles to produce a "blastocyst" with adequate cell numbers to ensure normal differentiation of an inner cell mass. If implantation should occur, they are unlikely to be competent, although implanted blastocoelic vesicles could produce symptoms and signs of a pregnancy which fails very early (a biochemical pregnancy), or could even progress further to produce a missed abortion or an anembryonic pregnancy.

The examination of cell and nuclear number at precavitation stages showed a high incidence of MNBs, which is in agreement with previous reports (Lopata *et al.,* 1983; Plachot *et al.,* 1986b, 1988a; Tesarik *et al.,* 1987). The significance of multinucleated cells is unclear. Hardy *et al.* (1989a) reported large numbers of nuclei in fragmenting conceptuses. Cytoplasmic fragmentation is a significant feature of the early human conceptus, which is not evident in mouse conceptuses at the same stages. More heavily fragmented conceptuses are less likely to progress to the blastocyst stage (Fig. 5).

Lopata *et al.* (1982) noticed discarded cells in the blastoceole cavity and perivitelline space of human blastocysts, which might be the result of corrective measures by developing conceptuses. The surgical removal of cellular fragments from conceptuses is compatible with development and was observed to improve the hatching and pregnancy rates (Cohen *et al.,* 1992a; Alikani *et al.,* 1993). These data suggest that the conceptus may be sensitive to an imbalance in the nucleocytoplasmic ratio. However, the presence of multiple nuclear structures within individual blastomeres did

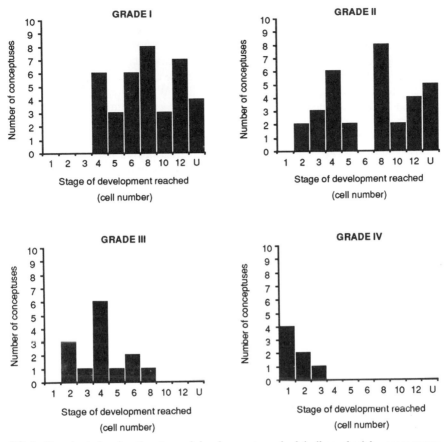

FIG. 5   Bar chart showing the stage of development reached (cell number) by conceptuses which failed to cavitate by day 5 after insemination and graded according to their morphological appearance on day 2. U = uncavitated conceptus.

not prevent some of these cells from dividing during overnight culture (Pickering *et al.,* 1995). Furthermore, as these authors point out, a negative correlation between the presence of MNBs and potential for development after transfer has yet to be confirmed.

B. Cellular DNA Content of Individual Blastomeres

The cellular DNA content of day 2 and day 3 conceptuses was assessed quantitatively in an attempt to determine the level at which normal development becomes disrupted. Twenty-eight cleaving human conceptuses were fixed in formaldehyde and stained with DAPI (3.12 μg/ml for 15 min precisely) on day 2 or day 3 after insemination (Table III). The conceptuses were then mounted on glass slides in citifluor mounting fluid to prevent bleaching (City University, London) and the discrete clusters of chromosomes and interphase nuclei in individual cells were assessed photocytometrically for their DNA content as described in Winston *et al.* (1991b). This method of quantification permits the DNA-specific fluorescent emission from the DAPI-stained chromosomes to be compared directly (Coleman and Goff, 1985) and is sufficiently sensitive to detect the order of the DNA content within a cell. Twenty-one of these conceptuses were examined on day 2 of development and the incidence of abnormal conceptuses (66%) was similar to that observed previously (see Table II). A single intact nucleus was observed in 39/60 (65%) embryonic blastomeres; however, the majority of conceptuses (14/21) had at least one multinucleate or anucleate cell. Only one of the 7 conceptuses examined on day 3 of development was fully normal, while 21/30 (70%) of the cells had one intact nucleus.

The DNA content could be measured in 82 of the 95 cells. No DAPI-stained chromatin was seen in 11 cells (anucleate), and in 2 cells the chromatin was dispersed throughout the cell and could not be measured (Table III). The readings showed that all the blastomeres, regardless of their nuclear phenotype or the period in culture, contained between 2C and 4C DNA (Fig. 6), which is appropriate for diploid cells traversing a mitotic cell cycle.

When the DNA readings were grouped according to the number of cells in the conceptus at the time of fixation (Fig. 7), most 1-cell and 2-cell

TABLE III

Distribution of Conceptus Development at the Time of Fixation and the Number of Cells in Which the DNA Content Could Be Measured Photocytometrically

	Number of cells at the time of fixation							
	1	2	3	4	5	6	7	Total
Day 2 conceptuses	3	7	3	7	—	1	—	21
Day 3 conceptuses	—	—	—	2	4	—	1	7
Total	3	7	3	9	4	1	1	28
Total number of cells	3	14	9	36	20	6	7	95
Number of anucleate cells	—	1	2	1	3	3	1	11
Number of cells with dispersing chromatin	—	—	—	1	—	—	1	2
Total number of cells analyzed	3	13	7	34	17	3	5	82

From Winston, N. (1993). *Zygote.* **1**, 1, 17–25, by permission of Cambridge University Press.

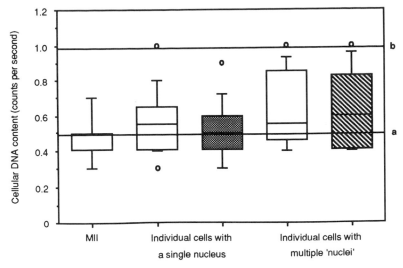

FIG. 6   Box plots* showing the range and distribution of values measured photocytometrically from (1) individual chromosome sets in unfertilized human oocytes arrested in metaphase II (MII, 2C DNA), (2) blastomeres with a single interphase nucleus or cluster of chromosomes (single nucleus) in conceptuses fixed on day 2 (empty box) or day 3 (filled box) of development, and (3) blastomeres with two or more interphase nuclei or clusters of chromosomes fixed on day 2 (empty box) or day 3 (filled box) of development. Line "a" represents the mean 2C DNA content and line "b" shows the calculated mean 4C DNA content. *Represents the 95% confidence band (filled areas) and individual points lying outside the 10th (lower) and 90th (upper) percentiles. From Winston, N. (1993). *Zygote*, **1**, 1, 17–25, by permission of Cambridge University Press.

conceptuses had between 2C and 4C DNA, as did the larger cell in 3-cell conceptuses. The smaller "one-quarter" size cells in 3-cell conceptuses had a 2C DNA content, suggesting that the parent cell had completed DNA replication prior to cell division. Four-cell conceptuses fixed on day 3 had, overall, a lower cellular DNA content than 4-cell conceptuses fixed on day 2. The reason for the reduced readings is not clear. The results from this study indicate, however, that DNA replication does not continue unchecked in arrested or slowly dividing conceptuses.

In a study of morphologically very poor (grade IV) conceptuses, 90% of those that were karyoptyped successfully were found to have abnormal or aberrant chromosome complements (Pellestor *et al.*, 1994). Twelve (10%) of these conceptuses contained cells with more than two sets of chromosomes. Multiple sets of chromosomes may result following the entry of multiple spermatozoa, the fertilization of diploid oocytes, or replication of the chromatin without cell division. Endomitosis is a mitotic modification which causes the duplication of chromosome number and is characterized by a defective mitotic spindle. Indeed, a further 71% of the conceptuses analyzed

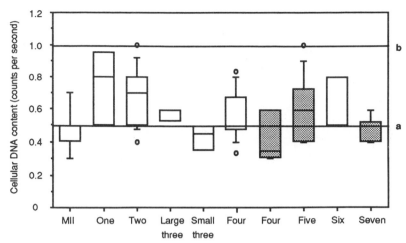

FIG. 7   Box plots* showing the range and distribution of values measured photocytometrically from individual chromosome sets in unfertilized human oocytes arrested in metaphase II (MII, 2C DNA) and individual blastomeres grouped according to the number of cells in the whole conceptus (range 1–7 cells). The distribution of readings from conceptuses fixed on day 2 are shown in empty boxes, and from conceptuses fixed on day 3 are shown in filled boxes. The blastomeres from "3-cell" conceptuses are grouped according to cell size, where "large" are "1/2 sized" cells and "small" are "1/4 sized" cells. *Represents the 95% confidence band (filled areas) and individual points lying outside the 10th (lower) and 90th (upper) percentiles.

were either noninterpretable or lacked chromosomes, which may indicate developmental failure due to factors other than chromosomal anomalies, such as cytoskeletal defects, mitochondrial DNA mutations, or metabolic abnormalities.

Varying amounts of cellular DNA, including polyploid cells apparently containing more than 4C DNA, have been found in conceptuses examined on day 2 or day 3 of development (Angell *et al.,* 1987). However, all except one of these polyploid cells arose in either multipronucleate conceptuses, or conceptuses which had been exposed to toxic culture conditions. Furthermore, these conceptuses were stained with ethidium to quantitate the cellular DNA, which may be subject to variations in the fluorescent emission among samples (Angell *et al.,* 1987). Pickering *et al.* (1995) found that 7 of the 131 (5.3%) individual blastomeres they examined had a DNA content > 4C. Six of these were multinucleate prior to overnight culture and failed to divide further. All normally fertilized conceptuses surplus to the requirements of the patients were included in this study, providing they contained at least one intact blastomere. Overall, the results from all these studies indicate that DNA replication rarely continues unchecked, the proportion of blastomeres scored as polyploid being very low (Tesarik *et al.,* 1987). Furthermore, the incidence of polyploidy does not appear to be correlated with the morphological quality of the conceptus, since a reading of $\leq 4C$ DNA was found even in MNBs and those blastomeres from very poor quality, degenerate conceptuses.

Cells containing two or more micronuclei could be distinguished morphologically into two types. Some multinucleate cells contained two equivalently sized nuclei positioned adjacent to one another (Fig. 8,a). Other cells contained three or more micronuclei which differed in size and shape and were frequently superimposed in the middle of the cell (Fig. 8,b). A few single nuclei were irregular in shape and appeared to have lobes bulging from the surface of the nuclear membrane (Fig. 8,c). Anucleate cells were not accompanied by multinucleate cells within the same conceptus in 3 out of 8 cases (Fig. 8,d), and only 4 of the 22 "companion cells" in these conceptuses contained more than 2C DNA at the time of analysis. In those zygotes which failed to divide at all, the nuclear material showed a variable phenotype: condensed and dispersed, decondensed and dispersed through-

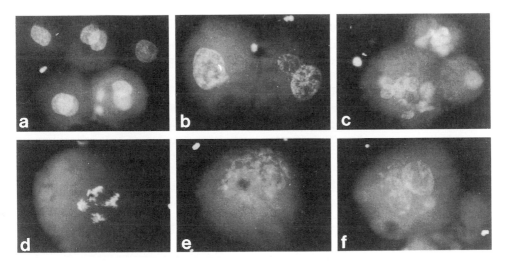

FIG. 8   Photomicrographs shown using UV fluorescence of DAPI-stained human conceptuses on day 2 (48 hpi; b–f) or day 3 (72 hpi; a) of development *in vitro* (magnification is ×312). (a) Five-cell conceptus with one or more binucleate cells. (b) Two-cell conceptus with irregularly shaped single nuclei, including one lobed nucleus. (c) Four-cell conceptus composed of a variety of abnormal cells (one anucleate, one multinucleate, and two binucleate). (d, e and f) Fertilized zygotes that failed to divide and in which the chromosomes became dispersed while remaining condensed (d), decondensed without a nuclear membrane (e), and (f) decondensed and packaged into several micronuclei. ×312.

out the cytoplasm or distributed in multiple micronuclear structures (Fig. 8,e, f, and g). Thus, from this and other studies (Tesarik *et al.*, 1987; Hardy *et al.*, 1993; Munné and Cohen, 1993), it seems likely that multinucleation of cells arises by several mechanisms of cellular failure, such as mitotic replication without cytokinesis, partial fragmentation of nuclei, or defective migration of the chromosomes at mitotic anaphase. However, in some cases the cell nucleation observed cannot be explained by these mechanisms either singularly or in combination and demonstrates the need for studies in which the division process in human conceptuses is recorded by serial observation.

C. Role of the Cytoskeleton in the Formation of Multinuclear Blastomeres

The observation of multinucleate cells in precavitation human conceptuses suggests that errors of karyo- and cytokinesis may be occurring *in vitro* at a high frequency, even as early as the 2- to 4-cell stage. In the mouse, the division of the cytoplasmic and nuclear material of a cell to produce two daughter cells, each containing a diploid set of chromosomes, requires the coordinated activity of both microfilaments and microtubules (Maro *et al.*, 1984; Schatten *et al.*, 1985; Maro *et al.*, 1986). These cytoskeletal elements are rearranged dramatically at mitosis and may be especially susceptible to disruption. Nuclear division is generally followed by cell division, which in the mouse requires intact spindle microtubules and functional microfilaments, respectively (Howlett *et al.*, 1985). However, it seems that once anaphase onset has been triggered, cells are committed to progress toward the interphase state, regardless of their ability to complete mitosis correctly (Mullins and Snyder, 1981; Snyder *et al.*, 1982; Vandre and Borisy, 1989).

The apparently different types of multinucleated cells that we observed (see Fig. 8) might reflect disruption of one or both of the cytoskeletal elements required for the completion of mitosis, without causing immediate arrest of the cell. The completion of nuclear division is dependent on proper assembly of the mitotic spindle, which itself is a complex structure requiring several components to reorganize the highly dynamic microtubules into a stable bipolar spindle. The molecules involved in spindle assembly and disassembly are not well defined; however, the activity of two motor molecules has been associated with cross-linkage of the spindle microtubules both at the polar region of the spindle (dynein-like, Pfarr *et al.*, 1990; Steuer *et al.*, 1990) and in the equatorial domain (kinesin-like, Vale and Goldstein, 1990). Chromosome arrangement and movement also rely upon the proper attachment of the spindle microtubules to the chromosomes. Thus, micronuclear structures may be formed if chromosomes become detached and stray from the spindle during anaphase. Any group of chromosomes which remains separate from the remainder will become enclosed in a nuclear membrane and form a micronucleus (Tesarik *et al.*, 1987). Meiotic spindle disruption and chromosomal dispersal from the metaphase plate are also associated frequently with oocyte aging (Pickering *et al.*, 1988). In addition, aged human oocytes show a diminished intensity of actin staining with time and an uncontrolled microtubule proliferation (M. H. Johnson, personal communication of data in preparation). Aged oocytes which are induced to mature by ovarian stimulation during therapeutic IVF may therefore contribute to abnormal conceptus development.

The cytoskeletal elements of the cell exist in a dynamic equilibrium between free tubulin and actin and polymerized microtubules and microfilaments, respectively. Even the microtubules within the metaphase spindle are turned over continually (de Pennart *et al.*, 1988; Gorbsky *et al.*, 1990), which may increase the chances of the spindle apparatus becoming disrupted. In addition, the extent of spatial organization and polymerization of the microtubules depends upon several parameters, including the ambient internal milieu of the cell and temperature. Adverse effects of transient cooling on the cytoskeleton of the human and mouse oocyte have been reported (Sathananthan *et al.*, 1988; Pickering *et al.*, 1990; Vincent *et al.*, 1990), which may not be reversed sufficiently quickly or completely prior to oocyte insemination during an IVF procedure (Pickering *et al.*, 1990). Conceptuses developed from frozen oocytes show clear evidence of micronuclear formation and incomplete incorporation of chromosomal material into the main nuclei (Sathananthan *et al.*, 1988). Thus, the highly complex functioning of the cellular cytoskeleton can be disrupted by external influences upon the cell. Furthermore, the temporal control of assembly and disassembly of cytoskeletal components during the cell cycle is itself regulated by enzymatic events which are centered around the activation of cyclin-dependent kinase complexes (Karsenti, 1991).

D. Regulation of the Cell Cycle and Cell Division

The progression of a cell through various phases of the cell cycle depends upon nuclear and cytoplasmic interaction. If cell division is to proceed efficiently, it should occur only after the DNA synthesis phase (S phase) of the cell cycle has been completed. The entry of a cell into mitosis involves coordinated and simultaneous events, including cytoskeletal rearrangements, disassembly of the nuclear envelope and nucleoli, and condensation of the chromatin into chromosomes. DNA synthesis is normally coupled tightly to mitosis to maintain the correct complement of chromosomes within cells. Regulators of downstream mitotic events have been identified

(Norbury and Nurse, 1992), and these "checkpoints" cannot be passed until previous cell cycle events have been completed (Hartwell and Weinert, 1989). However, DNA synthesis can become uncoupled from mitosis in cells exposed to mutational or chemical agents. For example, the *Drosophila* mutant *gnu* (Freeman *et al.*, 1986) can carry out periodic synthesis of DNA in the absence of mitosis. Conversely, entry into mitosis without the completion of DNA synthesis has been observed in vertebrate cells treated with caffeine (Schlegel and Pardee, 1986) and certain yeast mutants (Weinert and Hartwell, 1988).

1. Genetic Information

Much of our current knowledge of cell cycle control has been derived from complementation studies examining the effect of mutated genes upon progression through the cell cycle. The key events of the cell cycle seem to be governed by a growing cast of proteins encoded by cell-division cycle (*cdc*) genes. Although first isolated from yeast, these *cdc* genes appear to be highly conserved among species, including man (Dunphy and Newport, 1988). Several *cdc* genes and their products have been identified, one of which is the *cdc2* gene originally found in fission yeast (*Schizosaccharomyces pombe*) (Nurse and Bissett, 1981). The *cdc2* gene product is a 34-kDa protein-serine/threonine kinase (Simanis and Nurse, 1986) which can complex with a B-type cyclin protein to form a cell-cycle cytoplasmic control element, identified as maturation promoting factor (MPF) (Masui and Markert, 1971; Gerhart *et al.*, 1984; Dunphy *et al.*, 1988; Gautier *et al.*, 1988; Lohka *et al.*, 1988), that causes the entry of a cell into the mitotic M-phase. Cyclin proteins were first reported in marine organisms (Evans *et al.*, 1983) and show a gradual accumulation and rapid fall in their level during the cell cycle, being highest in the M-phase when MPF activity is also greatest. It seems that at different points in the cell cycle, different forms of cdc2-like molecules can associate with different classes of cyclins to form cyclin-dependent kinases, which are required for cell cycle transitions (summarized in Lock and Wickramasinghe, 1994).

The cdc2 protein kinase is in turn regulated at mitosis by the protein products of other genes which control cdc2 kinase activity by changing the level of its phosphorylation state. Phosphorylation of the cdc2 protein kinase by the gene products of *wee1* (Russel and Nurse, 1987) and *mik1* (Lundgren *et al.*, 1991) blocks its kinase activity and prevents the cell from entering mitosis. Conversely, cdc2 kinase activity is promoted following its dephosphorylation by cdc25 phosphatase (Gautier *et al.*, 1991; Millar *et al.*, 1991; Stausfield *et al.*, 1994), causing the cell to enter the M-phase. Mutation of the *cdc2* gene, or genes which act on *cdc2,* has been shown to cause nuclear abnormalities in human cells. Thus, disruption of the cdc2 cyclin-dependent kinase and/or its regulation can result in uncoupling of the cell cycle events and abnormal cellular development.

2. Biochemical Components

Genetic, biochemical, and morphological evidence implicates the cdc2 cyclin-dependent kinase as the master regulator controlling the induction of the events which take place as the cell cycle progresses into mitosis (Murray, 1989; Draetta, 1990; Nurse, 1990). As mitosis proceeds, cdc2 kinase appears to trigger a cascade of downstream mitotic events that complete the mitotic process, such as metaphase alignment of chromosomes, segregation of sister chromatids in anaphase, and cleavage furrow formation. Some of the proteins phosphorylated by cyclin-dependent kinases have been identified. These include those porteins involved in changes in nuclear structure during mitotic entry, such as the nuclear lamins responsible for assembly and disassembly of the nuclear envelope; histones, the hyperphosphorylation of which accompanies chromosome condensation (although chromosome condensation has been demonstrated in the absence of histone H1 (Ohsumi et al., 1993); intermediate filament proteins; and the centrosomes (Nigg, 1993). Centrosomal function must be regulated precisely for normal cell division to occur, otherwise mono- or multipolar spindles can result, with subsequent aneuploidy. The level of kinase activity has also been correlated directly with changes in the mean length of microtubules (Verde et al., 1990). Hence, the phosphorylation state of individual spindle components may be involved in the regulation of the spindle structure and function during mitosis. Andreassen and Margolis (1991) observed that the chromatin in cells exposed to the protein kinase inhibitor 2-aminopurine, which alters the phosphorylation of a very restricted subset of proteins (Farrel et al., 1977; Mahaderan et al., 1990), failed to enter mitosis and generate functioning chromosomes. Thus, post-translational modifications, particularly changes in the phosphorylation state of proteins during the cell cycle, are involved in regulating the function of nuclear and cytoskeletal structures during mitosis. The regulation of MPF activity also involves synthesis of the cyclin subunit and its periodic degradation at the metaphase–anaphase transition. Thus, if protein synthesis is impaired, progression of the cell cycle might be either inhibited or proceed abnormally. It is possible that the errors seen in cleavage of the early conceptus reflect an underlying problem with the quality or quantity of protein synthesis.

Analysis of the cellular DNA content in humans conceptuses showed that DNA replication can occur even when multiple nuclear structures are present within the cell (Fig. 6). DNA replication has been shown previously to occur in micronuclei (Tesarik et al., 1987), which also seem to resemble "normal" nuclei at the ultrastructural level (Sathananthan and Trounson, 1982; Trounson and Sathananthan, 1984). However, all inidividual embryonic blastomeres studied contained between 2C and 4C DNA (Fig. 6), suggesting that multiple nuclei were not arising as a result of successive rounds of DNA synthesis and karykinesis without cell division. The time of the last cell division was not recorded; thus cells with between 2C and 4C DNA might have been actively in the process of DNA replication, or may have failed to complete DNA replication and been arrested, as may be the case for the larger cells from 3-cell conceptuses (see Fig. 7).

In the human conceptuses studied, developmental arrest seems frequently to follow the formation of multiple micronuclei, rather than cytokinetic arrest preceding abnormal nuclear development. RNA synthesis has been shown to be low in multinucleate cells (Tesarik et al., 1987) and fits with the idea that multinucleation may often lead to irreparable arrest of blastomere development. Thus, it seems that abnormally nucleated human conceptus cells can enter but fail to complete S-phase and arrest. A human homolog of the cdc2-like gene (cdc2Hs) shows clear regulation of its activity at the boundary between G2 and entry into mitosis (Draetta and Beach, 1988; Draetta et al., 1988), and may be involved in the developmental arrest of some abnormal cell. Interestingly, anucleate cells were not always accompanied by multinucleate cells (Table IV), which might suggest that in some instances anucleation is the result of cell death rather than an error of cell division. Furthermore, the majority of companion cells in conceptuses with one or more anucleate cells contained only 2C DNA, regardless of their

TABLE IV

Details of the Companion Cells in Individual Conceptuses with One or More Anucleate Cell(s)[a]

Day of culture when fixed	Total number of cells	Number of anucleate cells	Number of cells with a single nucleus		Number of multinucleate cells	
			2C	4C	2C	4C
2	2	1	—	—	—	1
2	3	1	—	—	2	—
2	3	1	2	—	—	—
2	4	1	—	—	2	—[b]
2	6	3	1	—	1	1
3	5	1	3	1	—	—
3	5	2	—	—	2	1
3	7	1	5	—	—	—[b]

[a] Single and multinucleate companion cells are listed according to their DNA content, being either 2C or 4C.

[b] These conceptuses also contained one cell with dispersing chromatin.

nuclear phenotype. Whether the cytoplasmic component of a blastomere can continue through the cell cycle and divide in the absence of DNA replication is not known. Pickering *et al.* (1995) observed three incidences where an anucleate blastomere cleaved and formed anucleate daughter cells. However, from the data available, it is still not possible to identify conclusively those parameters, which will eventually lead to developmental arrest.

E. Effect of *in Vitro* Culture Conditions on the Zona Pellucida and Blastocyst Hatching

The function of the zona pellucida during conceptus cleavage seems to be mostly of a physical nature, preventing the dispersal of the blastomeres and direct contact between the conceptus and foreign cells (Moore *et al.*, 1968; Bronson and McLaren, 1970; Modlinski, 1970). The process by which conceptuses shed their zonae is probably complex, involving embryonically induced degradation as well as the effect of a uterine lysing agent (Parr, 1973; Pinkster *et al.*, 1974). There are a number of findings that support the hypothesis that some human IVF conceptuses lack the ability to produce zona lysins, or that their zonae pellucidae are too resilient to allow for normal dissolution. *In vitro* culture conditions may cause excessive hardening of the zona pellucida beyond that which would occur normally *in vivo*, as seen in mouse oocytes (Gianfortoni and Gulyas, 1985). Only one in four fully expanded blastocysts derived from *in vitro* fertilized ovarian oocytes will hatch (Fehilly *et al.*, 1985; Lindenberg *et al.*, 1989). This provides evidence that human blastocysts are capable of hatching in the absence of uterine lysins, but it also suggests that their zonae are often resilient to dissolution. The incidence of blastocyst formation and hatching was reported to increase when human conceptuses were cultured on monolayers of reproductive tract cells (Lindensberg *et al.*, 1989; Wiemer *et al.*, 1989). This finding supports the hypothesis that many *in vitro* fertilized conceptuses are deficient in the amount of zona lysins required to initiate and complete hatching.

Cohen *et al.* (1989) and Wright *et al.* (1990) observed that cleaved conceptuses with a good prognosis for implantation have a reduced, or nonuniform zona thickness. It was found subsequently that the proportion of blastocysts capable of hatching and the pregnancy rate were improved when a small hole was made in the zona pellucida prior to conceptus transfer (Cohen *et al.*, 1990a). These data indicate that *in vitro* culture conditions cause hardening of the zona pellucida and impairment of the hatching process. Some clinics routinely practice assisted hatching manipulations on all transferred conceptuses; however, patients whose conceptuses already have thin zonae may be jeopardized by the procedure (Cohen *et al.*, 1992b). In addition, Depypere and Leybaert (1994) showed that zona drilling with acid tyrodes causes a transient reduction in the intracellular pH of aged human oocytes and may be related to the high incidence of cytoplasmic degeneration after zona drilling of human oocytes. More recently, the zona has been slit mechanically or by using a laser. These techniques appear compatible with conceptus development, blastocyst formation, and improved pregnancy rates (Dokras *et al.*, 1994; Obruca *et al.*, 1994), although the size of the gap may be important in preventing twinning at the hatching stage. The higher implantation and pregnancy rates from zona-drilled conceptuses suggest that the conceptus does not suffer from any immunological influences *in utero*. However, the implantation rate was increased in patients receiving low-dose immunosuppression during the 4 days following oocyte retrieval compared with those patients without immunosuppression to whom drilled conceptuses were transferred (Cohen *et al.*, 1990b). Finally, the hatching (Malter and Cohen, 1989) and implantation (Liu *et al.*, 1993) of human conceptuses has been observed to occur significantly earlier following assisted hatching manipulation. Thus, assisted hatching may enhance conceptus implantation not only by mechanically facilitating the hatching process but also by allowing earlier conceptus–endometrium contact. This may optimize the synchronization between the conceptus and the endometrium, resulting in improved pregnancy rates.

F. Endometrial Synchronization

Studies of conceptus development assume frequently that the uterine environment is supportive of all conceptuses transferred *in utero* and is receptive to implanting conceptuses. The use of assays for either human chorionic gonadotropin (hCG) or other trophoblastic markers indicates that many "pregnancies" are short lived and fail around the time of the next menstruation, even with conceptuses fertilized *in vivo* (Leridon, 1977; Edmonds *et al.*, 1982). It is not clear whether this reflects a uterine environment frequently hostile to the implanting conceptus, or that many conceptuses are abnormal and unable to establish a pregnancy. However, the studies of Hertig *et al.* (1959) imply that implantation presents a major hurdle for abnormal conceptuses. The development of endometrial receptiveness is a major factor determining the outcome of assisted reproductive technology treatments. The primary function of the endometrium is to provide a substrate for the implantation and maintenance of blastocyst development. During the follicular phase, the glands and stroma grow rapidly due to estrogen-stimulated proliferation. After ovulation, progesterone secreted by the corpus luteum promotes secretory transformation and decidualization of the endometrium, which in turn provides essential support for implantation and early pregnancy. Abnormalities in endometrial development during nonstimulated cycles are associated with infertility and early and repetitive miscarriage.

In IVF treatment cycles, the administration of pituitary antagonists and/or high levels of exogenously administered gonadotrophins might affect the functioning or development of the uterine tissues, changing the period of their receptiveness to implantation (see Schoolcraft *et al.*, 1994). High follicular-phase LH concentrations have been associated with a reduced probability of pregnancy (Regan *et al.*, 1990). Benadiva and Metzger (1994) found that in biopsies of late luteal phase endometrium, the glandular epithelium and the stroma did not develop synchronously following hyperstimulation. These changes did not correlate with differences in preovulatory or midluteal serum estrogen or progesterone levels, or with the administration of exogenous progesterone, but did occur significantly more often in stimulated cycles than in nonstimulated cycles in the same patient. However, there may also be an individual variation in endometrial sensitivity to circulating steroid hormone concentrations. The authors suggest that if gland development is retarded sufficiently, the implantation window (Navot *et al.*, 1991) may be shifted late enough in the luteal phase to make rescue of the corpus luteum by embryonic hCG difficult, resulting in nonconception cycles or early pregnancy losses.

Culture *in vitro* has been correlated frequently with retarded development in human and mouse conceptuses compared with conceptuses of a similar age developing *in vivo* (Bowman and McLaren, 1970; Harlow and Quinn, 1982; Dokras *et al.*, 1991). A "developmental lag" might reduce the chances of implantation due to asynchrony between the maturation status of the uterine endometrium and the stage of conceptus development. The viable pregnancy rates reported following GIFT and tubal conceptus transfer (TET) procedures (20% to 35%) are higher than those reported for IVF (13% to 20%) per treament cycle (Asch *et al.*, 1985; Wong *et al.*, 1988; Yovich and Matson, 1988). Thus, 1 or 2 days of culture *in vitro* might be sufficient to reduce conceptus viability. Furthermore, the replacement of cryopreserved conceptuses following standard IVF caused a more dramatic increase in the implantation and pregnancy rates per oocyte collection cycle than following the GIFT procedure (Wang *et al.*, 1994).

These results suggest that conceptuses replaced immediately after fertilization *in vitro* may be further compromised as a result of development desynchronization. In rodents, sychronization between embryonic development and endometrial development is known to be critical to successful implantation (Fisher and Smithberg, 1973). However, Huisman *et al.* (1994) found equivalent pregnancy rates when human conceptuses were transferred after 2, 3, or 4 days in culture, suggesting that either the rate of conceptus development *in vitro* is not retarded relative to that of the uterus, or that strict synchronization between embryonic and endometrial development may be unneccesary in the human.

It is difficult to define to what extent conceptus viability and the capacity of the uterine endometrium to support a pregnancy are at fault in the failure of both natural and assisted reproduction. Several studies in which conceptuses from donated oocytes were transferred have indicated that endometrial function is independent of the age of the female (Navot *et al.*, 1991; Cohen *et al.*, 1992a; Check *et al.*, 1994), which would point to failure at the level of conceptus. There are, however, other reports which conclude conversely that the endometrium is subject to an age-related decline in receptivity (Yaron *et al.*, 1993). It is likely that hyperstimulation regimes, culture *in vitro,* and the high level of chromosomal abnormalities which occur in human oocytes and conceptuses play some role in determining the success or failure of pregnancy and this is compounded further by the varying the responsiveness of individual patients.

IV. Extrinsic Determinants of Embryonic Development

Successful culture of *in vitro* fertilized human oocytes to the blastocyst stage was reported by Steptoe in 1971 (Steptoe *et al.*, 1971). Initially, human oocytes and conceptuses were cultured in simple defined media which supported the development of mouse conceptuses. However, it became clear that human conceptus development was improved by culture in a more complex medium supplemented with serum, as well as a slight increase in the osmotic pressure (Edwards *et al.*, 1970). Human preimplantation conceptus normally move into the uterine cavity at the morula or blastocyst stage and were hence retained in culture for several days before being transferred *in utero.* However, Marston *et al.* (1977) observed that recently fertilized rhesus monkey oocytes, unlike those of mice, survived to initiate implantation when transferred directly back to the uterus. Shortly after, a pregnancy was achieved from a human conceptus transferred *in utero* at the 8-cell stage (Steptoe and Edwards, 1978). It is now clear that culture conditions *in vitro* are unable to support the optimal development of all human conceptuses (Bolton and Braude, 1987) and are presumably responsible for some of the anomalies observed in oocytes and conceptuses cultured *in vitro.*

A. Influence of Culture Media upon Conceptus Development

The composition of most culture media used for human IVF may not be sufficiently comprehensive to support optimum development of human conceptuses. Some of the ultrastructural changes observed, including swelling of the nuclear envelope (Trounson and Sathananthan, 1984), may reflect the inadequacy of culture conditions. Irregular cleavage was observed when human conceptuses were cultured in medium with a low osmotic pressure (270 mosm/kg), inducing an increase in apparent cell number without nuclear division (Edwards *et al.*, 1970). Pickering *et al.* (1995) examined the effect of three different media on the incidence of multinucleation in human conceptuses. The proportion of mononucleated blastomeres did not differ among the three groups, but differences did occur in the proportion of blastomeres that were multinucleated and anucleate. However, overall, the incidence of MNBs correlated better with the period in culture, being generally higher in conceptuses examined on day 3 after insemination compared with conceptuses on day 2, suggesting that MNB formation was progressive. Indeed, 24% of mononucleated blastocyts that cleaved in culture overnight formed abnormally nucleated daughter cells. Can the progressive disorganization of cellular nucleation and arrest in development of normally fertilized human conceptuses be attributed directly to culture conditions *in vitro;* is it simply an integral problem of conceptus development; or do both these factors contribute to the early demise of many human conceptuses?

Many studies have been conducted to examine the effects of *in vitro* culture conditions upon the developmental potential of human conceptuses. The levels of glucose and pyruvate available in traditional culture media have been shown to meet the requirements of developing human conceptuses (Leese *et al.*, 1986; Hardy *et al.*, 1989b). Even at the highest rate of turnover, less than 15% of the glucose available was utilized (Wales *et al.*, 1987). The role of vitamins and amino acids is less clear. Media supplemented with a full complement of amino acids and vitamins supported better development of surplus conceptuses to the blastocyst stage, compared with unsupplemented media (Lopata and Hay, 1989; Taylor *et al.*, 1991). However, short-term exposure of conceptuses for 24 to 30 hr to amino acids and vitamins showed no improvement in morphological quality or a higher pregnancy rate after transfer (Staessen *et al.*, 1991).

More recent attempts to improve culture conditions have introduced epithelial cells into the culture environment (coculture). Improvements in conceptus morphology, development, and pregnancy rates have been reported, including those patients with previous implantation failures (Bongso *et al.*, 1989; Wiemer *et al.*, 1993a; Schillaci *et al.*, 1994). The beneficial effect of such feeder cells may be through the release of embryotrophic factors, such as growth factor ligands (Gardner and Leese, 1990). These ligands have been isolated from bovine oviductal epithelial cells (Watson *et al.*, 1992) and can reduce the imbalance of certain constituents and detoxify the culture medium. The presence of the cumulus cells, which in natural cycles are normally taken up by the fallopian tube, either during and/ or following fertilization, has also been correlated with enhanced conceptus development (Magier et al., 1990; Mansour et al., 1994; Saito et al., 1994).

However, the incidence of abnormal nuclear development was not reduced (Sathananthan et al., 1990) and the reported rates of blastocyst expansion, hatching, and pregnancy are also not improved consistently (Sathananthan et al., 1990; Plachot et al., 1993; Sakkas et al., 1994). In a detailed prospective study, Van Blerkom (1993) found no significant improvement in the development of early human conceptuses cultured from day 2 after insemination to the blastocyst stage in the presence of Vero cells compared with the development of conceptuses cultured in medium alone. The actual advantages of coculture compared with conventional culture methods for increasing the frequency of developmentally viable conceptuses remain unclear (Bavister, 1987).

B. Other Factors Affecting Conceptuses Cultured in Vitro

Embryonic stress induced by exposure to chemical components and conditions not normally encountered in vivo may also reduce conceptus viability or cause developmental arrest in vitro. The culture of animal cells in the presence of various chemicals can induce the formation of a series of small nuclei (karyomeres). There is now considerable evidence that oxygen radical-induced cellular dysfunction is reponsible for suboptimal conceptus development in vitro in a variety of mammalian species (Nasr-Esfahani et al., 1990a). Oxygen toxicity is known to be elicited by reactive oxygen substances that are generated within cells (Jones, 1985). Key factors contributing to oxygen toxicity are high oxygen tension, transition metals, and illumination. Atmospheric oxygen tension is several times higher than in the uterus or the oviduct and may lead to nonphysiologically high concentrations of oxygen in conceptuses. Several reports have illustrated clearly that the development of conceptuses protected from oxidative stress was enhanced (Nasr-Esfahani et al., 1990b, 1992; Noda et al., 1994). The protection of oocytes and conceptuses from excessive illumination may also assist in reducing the action of free radical species. Deleterious factors might be introduced by the addition of serum to the culture media (Hemmings et al., 1994). Protein supplementation of the medium can be achieved effectively using follicular fluid or a plasma protein fraction containing albumin (Pool and Martin, 1984). In addition, the support of oocyte and conceptus development in this way will mimic more closely physiochemical properties in vivo.

However, recent studies propose that the best pregnancy results are obtained by using a combination of culture medium akin to tubal fluid, coculture with oviductal or other epithelial cells, and micromanipulation of the zona pellucida (Wiemer *et al.*, 1993a,b; Tucker *et al.*, 1994). It seems likely that different culture conditions are required to maintain optimal conceptus viability on successive days of development to mimic the changing environment normally encountered by the conceptus as it passes along the fallopian tube and into the uterine cavity (Conaghan *et al.*, 1993; Gardner and Sakkas, 1993).

The similar incidence of multinucleate blastomeres reported by different authors (Lopata *et al.*, 1983; Plachot *et al.*, 1986b; Tesarik, 1988) *in vitro* and observations of fragmented and multinucleate cells in conceptuses developed *in vivo* (Hertig *et al.*, 1954; Alvarez *et al.*, 1988; Pereda *et al.*, 1989) suggest that this characteristic of human conceptuses may not depend primarily on the culture environment (Pickering *et al.*, 1995). Furthermore, the nuclei and cytoplasm in multinucleate cells do not appear to be altered noticeably (Dvorak *et al.*, 1982). Monks *et al.* (1993) observed in a group of natural cycle patients that conceptus quality and pregnancy rate were not correlated with the type of culture medium used, but rather the type of female infertility. The cause of developmental abnormalities appears to lie in the inherent quality of the conceptus, although abnormal cells may be more susceptible to environmental damage.

V. Concluding Remarks

The development of *in vitro* fertilization has made possible the study of mechanisms of fertilization and early conceptus development. To date, many studies of mammalian oocytes and early conceptuses, particularly in the mouse, have described architectural rearrangements of the cytoskeleton and the nuclear transformations that accompany them. It is clear from the data that the processes of fertilization and subsequent development are very complex and in the human frequently fail to produce conceptuses with the potential to reach the stage of implantation and to establish a viable pregnancy (Hertig *et al.*, 1959; Boué *et al.*, 1975b; Jacobs *et al.*, 1978).

From our results and the data avilable, it is not possible to determine conclusively the functional level at which the developmental process fails in the human conceptus. There are probably other factors in addition to those discussed in this review which remain undefined and which are most likely frequently classified under the category "unexplained." There are also still many gaps in our knowledge of the precise functioning of the human reproductive system; for example, the period during which the oocyte can be fertilized and how long it can remain developmentally viable are poorly defined. Oocytes inseminated 48 hr after retrieval with donor spermatozoa were still fertilizable but gave rise to lower quality conceptuses with a high percentage of chromosomal abnormalities (Bongso *et al.*, 1992). Likewise, 24-hr aged oocytes inseminated intracytoplasmically can become fertilized and procedure cleaving conceptuses; however, their potential for implantation appears limited (Tsirigotis *et al.*, 1994). The number of pregnancies following the replacement of oocytes fertilizing after reinsemination was also extremely small (Winston, *et al.*, 1993). Various naturally occurring features have also been proposed as possible explanations for the low success rate of assisted conception therapies, for examples, seasonality in fecundity (Stolwijk *et al.*, 1994).

The relative importance of each gamete for development and the nature of the interactions between the gametes remain basic questions in developmental biology. The data indicate strongly that the status of the oocytes is likely to affect subsequent development, although abnormal oocytes might not be precluded from fertilization (Almeida and Bolton, 1994) and development to postimplantation stages (Boué *et al.*, 1975a). Further, it seems that in the mammalian oocytes, nuclear and cytoplasmic maturation can occur as two separate processes causing abnormalities following fertilization (Schmiady *et al.*, 1986; Kubiak, 1989; Schmiady and Kentenich, 1989; Eppig *et al.*, 1994; McConnell *et al.*, 1995).

A. Influence of the Nucleocytoplasmic Ratio on Development

In the mouse, it is clear that an appropriate ratio of cytoplasmic and nuclear material is essential for development to occur (McGrath and Solter, 1986; Surani *et al.*, 1986; Howlett *et al.*, 1987). However, some tripronucleate human oocytes develop to the blastocyst stage and this raises an intriguing question concerning the regulation of development in genetically abnormal human conceptuses. How some polyspermic oocytes can continue to develop while normal fertilization is associated with such a high rate of developmental failure is far from clear (Van Blerkom and Henry, 1991; Balakier, 1993). It has been suggested that multispermic human conceptuses have some ability to restore a diploid state, and hence an appropriate nucleocytoplasmic ratio in their cells (Van Blerkom *et al.*, 1984; Tesarik, 1988). This may be achieved by extrusion of the extra complement of chromosomes into one nondeveloping cell, or perhaps these oocytes divide directly to the 3- or 4-cell stage, explaining the mosaic nature of some multispermic conceptuses (Kola and Trounson, 1989; Van Blerkom and Henry, 1991).

B. Activation of the Embryonic Genome

Many human conceptuses arrest early in development, particularly between the 4 and 8-cell stages, at which time the embryonic genome becomes activated (Braude *et al.*, 1988). Multinucleated blastomeres have also been observed in human parthenotes (Winston, 1992), the majority of which arrest prior to the 8-cell stage (Winston, 1992; Taylor and Braude, 1994) and might lend support to the idea that developmental failure results from the inability to undergo genomic activation. However, developmental failure might not be caused simply by the failure of embryonic gene expression. Fertilized conceptuses have been shown to transcribe RNAs which are sensitive to α-amanitin inhibition, without necessarily undergoing cleavage. Furthermore, fertilized conceptuses which have multiple micronuclei and/ or have arrested show the presence of transcription products associated with the embryonic genome (Artley *et al.*, 1992). In contrast, cells which have not undergone the transition from a maternally encoded developmental program to an embryonic encoded one have been found in conceptuses which have developed beyond the 8-cell stage (Tesarik *et al.*, 1986). It seems, however, that for conceptuses to cavitate and continue development to the blastocyst stage, a certain proportion of cells must have undergone the transition to a "progressed-cleavage transcription pattern." Therefore, embryonic development to later stages may be associated partly with the number of competent blastomeres.

C. Future Developments

The finding of abnormal cellular and nuclear development at very early cleavage stages has important clinical consequences. First, many conceptuses on day 2 of development that are graded as suitable for transfer under a dissecting microscope clearly have gross structural abnormalities, which may not be compatible with further development. This finding alone would be sufficient to account for the poor success of IVF. Second, the successful development of fertilized oocytes in culture to the blastocyst stage cannot be taken as indicative of normal development and adequacy of culture conditions. Third, single cells removed as a biopsy from early cleavage stages for early diagnosis of genetic disease (Handyside *et al.*, 1989, 1990) cannot be relied on to be representative of the whole conceptus, and methods need to be developed for the assessment of the "normality" of the conceptus and its biopsies. The ability to conduct accurately the kind of observations required to recognize the developmental potential of an oocyte or conceptus is extremely limited in the clinical context.

A variety of techniques for assessing conceptus quality have now been developed (Handyside *et al.*, 1989, 1990; Penketh *et al.*, 1989; Pickering *et al.*, 1992). These data provide invaluable information both for the current treatment cycle and toward our understanding of human conceptus development. However, the use of specific embryonic products as markers of development potential may be unreliable, since these conceptuses might still suffer developmental arrest (Artley *et al.*, 1992). Intrafollicular factors might be more suitable markers of developmental potential (Nayudu *et al.*, 1989b; Harika *et al.*, 1991). Recently, specific stages of human oocyte cytoplasmic structure and organization have been detected at the light microscope level and were shown to be predictive of fertilization failure and early developmental arrest (Nayudu *et al.*, 1989a; Van Blerkom, 1990a,b). This approach to assessing the potential for development of human oocytes and conceptuses during IVF therapy may still be the most accessible currently.

Major differences in the distribution and orientation of cytoskeletal elements have been reported between the oocytes and zygotes of humans and rodents (Pickering *et al.*, 1988; Wiker *et al.*, 1990; Santella *et al.*, 1992). Variations in the responses of mouse and human oocytes when exposed to compromised culture conditions (Pickering *et al.*, 1990; Winston and Johnson, 1992) and from studies of their parthenogenetic activability (Kaufman, 1983; Abramczuk and Lopata, 1990; Winston *et al.*, 1991b) also illustrate the danger of relying on results extrapolated from another species (Bolton *et al.*, 1986). These data underline the extreme importance of carrying out carefully planned and interpreted studies on the limited amount of human oocyte and embryonic material available and within the restrictions currently imposed in many countries upon the use of gametes and conceptuses beyond the direct treatment of the patient.

Finally, it has been calculated that only 30% of human oocytes fertilized *in vivo* go to term (Biggers, 1981). This would suggest that the human reproductive system is naturally highly fallible and/or highly selective and that the selection of developmental viability can occur at stages beyond fertilization and early cleavage. The possibility of increasing the success of assisted conception therapies might therefore be limited by the human reproductive system itself. However, the reported average birth rate when a single conceptus is replaced after fertilization *in vitro* is even lower (10%) (Steptoe *et al.*, 1986). Thus, fertilization and culture *in vitro* do appear to accentuate developmental failure. Additional factors such as maternal age and male infertility alone or in combination further increase the likelihood of failure of individual treatment cycles (American Fertility Society for Assisted Reproductive Technologies, 1994). Many effects of assisted conception treaments remain undefined and comparisons drawn from natural cycle treatment regimes are limited by the problems of optimally timing oocyte collection in unstimulated cycles and relying on the status of a single oocyte (Ritchie, 1985; MacDougal *et al.*, 1994). Furthermore, the possibility of defining fixed parameters suitable for different individuals or even one person in different treatment cycles seems unlikely. The success of IVF will greatly rely upon a better understanding of the developmental requirements of early conceptuses, and the resilience of those normal oocytes and conceptuses to support conditions *in vitro* and produce viable pregnancies.

Acknowledgments

I would like to thank Martin Johnson, Bernard Maro, and Jeremy Garwood for their helpful suggestions and critical reading of the manuscript. The work was supported by a program grant from the Medical Research Council to Professor M. H. Johnson, Department of Anatomy, University of Cambridge, UK and Professor P. R. Braude, Assisted Conception Unit, St. Thomas' Hospital, London, UK.

References

Abramczuk, J. W., and Lopata, A. (1990). Resistance of the human follicular oocytes to parthenogenetic activation: DNA distrubution and content in oocytes maintained *in vitro. Hum. Reprod.* **5,** 578–581.

Adashi, E. Y. (1994). Endocrinology of the ovary. *Hum. Reprod.* **9,** 815–827.

Alikani, M., Olivennes, F., and Cohen, J. (1993). Microsurgical correction of partially degenerate mouse embryos promotes hatching and restores their viability. *Hum. Reprod.* **8,** 1723–1728.

Almeida, P. A., and Bolton, V. N. (1993). Immaturity and chromosomal abnormalities in oocytes that fail to develop pronuclei following insemination *in vitro. Hum. Reprod.* **8,** 229–232.

Almeida, P. A., and Bolton, V. N. (1994). The relationship between chromosomal abnormalities in the human oocyte and fertilization *in vitro. Hum. Reprod.* **9,** 343–346.

Alvarez, F., Barache, V., Fernandez, F., Guerreri, B., Grulott, E., Hess, R., Salvatierra, A. M., and Zacharias, S. (1988). New insight on the mode of action of intrauterine contraceptive devices in women. *Fertil. Steril.* **49,** 768–773.

American Fertility Society for Assisted Reproductive Technologies (1994). Assisted reproductive technology in the United States and Canada: 1992 results generated from The American Fertility Society/Society for Assisted Reproductive Technology Registry. *Fertil. Steril.* **62,** 1121–1128.

Andreassen, P. R., and Margolis, R. L. (1991). Induction of partial mitosis in BHK cells by 2-aminopurine. *J. Cell Sci.* **100,** 299–310.

Angell, R. R. (1991). Predivision in human oocytes at meiosis I: A mechanism for trisomy formation in man. *Hum. Genet.* **86,** 383–387.

Angell, R. R., Summer, A. T., West, J. D., Thatcher, S. S., Glasier, A. F., and Baird, D. T. (1987). Post-fertilization polyploidy in human preimplantation embryos fertilized *in vitro. Hum. Reprod.* **2,** 721–727.

Angell, R. R., Ledger, W., Yong, E. L., Harkness, L., and Baird, D. T. (1991). Cytogenetic analysis of unfertilized human oocytes. *Hum. Reprod.* **6,** 568–573.

Angell, R. R., Xian, J., and Keith, J. (1993). Chromosome anomalies in human oocytes in relation to age. *Hum. Reprod.* **8,** 1047–1054.

Artley, J. K., Braude, P. R., and Johnson, M. H. (1992). Gene activity and cleavage arrest in human pre-embryos. *Hum. Reprod.* **7,** 1014–1021.

Asakawa, T., and Dukelow, W. R. (1982). Chromosomal analysis after in vitro fertilisation of squirrel monkey (*Saimiri sciureus*) oocytes. *Biol. Reprod.* **26,** 579–583.

Asakawa, T., Ishikawa, M., Shimizu, M., and Dukelow, W. R. (1988). The chromosomal normality of in vitro-fertilised rabbit oocytes. *Biol. Reprod.* **38,** 292–295.

Asch, R. H., Ellsworth, R. L., Balmeceda, J. P., and Wong, P. C. (1985). Birth following intrafallopian tube transfer. *Lancet* **2,** 163.

Balakier, H. (1993). Tripronuclear human zygotes: The first cell cycle and subsequent development. *Hum. Reprod.* **8,** 1892–1897.

Bavister, B. (1987). Co-culture for embryo development: Is it really necessary? *Hum. Reprod.* **7,** 1339–1341.

Benadiva, C. A., and Metzger, D. A. (1994). Superovulation with human menopausal gonadotrophins is associated with endometrial gland-stroma dyssynchrony. *Fertil. Steril.* **61,** 700–704.

Benkhalifa, D., Janny, L., Vye, P., Malet, P., Boucher, D., and Menezo, Y. (1993). Assessment of polyploidy in human morulae and blastocysts using co-culture and fluorescent in-situ hybridization. *Hum. Reprod.* **8,** 895–902.

Biggers, J. D. (1981). In vitro fertilization and embryo transfer in human beings. *N. Engl. J. Med.* **304,** 336.

Boatman, D. E., Felson, S. E., and Kimura, J. (1994). Changes in the morphology, sperm penetration and fertilization of ovulated hamster eggs induced by oviductal exposure. *Hum. Reprod.* **9,** 519–526.

Bolton, V. N. (1992). Controversies and opinions in embryonic culture: Two- to four-cell transfer vs blastocyt. *J. Assist. Reprod. Genet.* **9,** 506–508.

Bolton, V. N., and Braude, P. R. (1987). Development of the human preimplantation embryo *in vitro. In* "Recent Advances in Mammalian Development" (A. McLaren and G. Siracusa, eds.), pp. 93–114. Academic Press, London.

Bolton, V. N., Johnson, M. H., Maro, B., Howlett, S. K., Pickering, S., and Webb, M. (1986). The mouse embryo as a model for human IVF failure. *In* "Proceedings of the Fourth World Conference on In Vitro Fertilization." Plenum, New York.

Bolton, V. N., Hawes, S. M., Taylor, C. T., and Parsons, J. H. (1989). Development of spare human preimplantation embryos in vitro: An analysis of the correlations among gross morphology, cleavage rates and development to the blastocyst. *J. In Vitro Fertil. Embryol. Transfus.* **6,** 30–35.

Bolton, V. N., Wren, M. E., and Parsons, J. H. (1991). Pregnancies after in vitro fertilization and transfer of human blastocysts. *Fertil. Steril.* **55,** 830–832.

Bongso, A., Soon-Chye, N., Ratnam, S., Sathananthan, H., and Wong, P. C. (1988). Chromosome anomalies in human oocytes failing to fertilize after insemination *in vitro. Hum. Reprod.* **3,** 645–649.

Bongso, A., Soon-Chye, N., Sathananthan, H., Lian, N. P., Rauff, M., and Ratnam, S. (1989). Improved quality of human embryos when co-cultured with human ampullary cells. *Hum. Reprod.* **4,** 706–713.

Bongso, A., Fong, C. Y., Ng, S., and Ratnam, S. (1992). Fertilization, cleavage and cytogenetics of 48-hour zona-intact and zona-free human unfertilized oocytes reinseminated with donor sperm. *Fertil. Steril.* **57,** 129–133.

Boué, J., Boué, A., and Lazar, P. (1975a). The epidemilogy of human spontaneous abortions with chromosomal anaomalies. *In* "Aging Gametes" (C. J. Bandau, ed.), p. 330. Karger, Basel.

Boué, J., Boué, A., and Lazar, P. (1975b). Retrospective and prospective epidemiological studies of 1500 karyotyped spontaneous human abortions. *Teratoloy* **12,** 11–26.

Bowman, P., and McLaren, A. (1970). Cleavage rate of mouse embryos in vivo and *in vitro. J. Embryol. Exp. Morphol.* **24,** 203–207.

Braude, P. R., Bolton, V. N., and Moore, S. (1988). Human gene expression first occurs between the four- and eight-cell stages of preimplantation development. *Nature (London)* **332,** 459–461.

Bronson, R. A., and McLaren, A. (1970). Transfer to the mouse oviduct of eggs with and without zona pellucida. *J. Reprod. Fertil.* **22,** 129–136.

Chard, T. (1991). Frequency of implantation and pregnancy loss in natural cycles. *In* "Factors of Importance for Implantation" (M. Seppälä, ed.), p. 179. Baillière Tindall, London.

Check, J. H., Askari, H. A., Fisher, C., and Vanaman, L. (1994). The use of a shared donor oocyte program to evaluate the effect of uterine senescence. *Fertil. Steril.* **61**, 252–256.

Chisholm, J. C., Johnson, M. H., Warren, P. D., Fleming, T. P., and Pickering, S. J. (1985). Developmental variability within and between mouse expanding blastocysts and their ICMs. *J. Embryol. Exp. Morphol.* **86**, 311–336.

Cohen, J. Inge, K. L., Suzman, M., Wiker, S., and Wright, G. (1989). Videocinematography of fresh and cryopreserved embryos: A retrospective analysis of embryonic morphology and implantation. *Fertil. Steril.* **51**, 820–827.

Cohen, J., Elsner, C., Kort, H., Malter, H., Massey, J., Mayer, M. P., and Wiemer, K. (1990a). Impairment of the hatching process following IVF in the human and improvement of implantation by assisted hatching using micromanipulation. *Hum. Reprod.* **5**, 7–13.

Cohen, J., Malter, H., Elsner, C., Kort, H., Massey, J., and Mayer, M. P. (1990b). Immunosuppression supports implantation of zona pellucida dissected human embryos. *Fertil. Steril.* **53**, 662–665.

Cohen, J., Talansky, B. E., Malter, H., Alikani, M., Alder, A., Reing, A., Berkeley, A., Graf, M., Davis, O., Liu, H., Bedford, M., and Rosenwaks, Z. (1991). Microsurgical fertilization and teratozoospermia. *Hum. Reprod.* **6**, 118–123.

Cohen, J., Alikani, M., Reing, A. M., Ferrara, T. A., Trowbridge, J., and Tucker, M. (1992a). Selective assisted hatching of human embryos. *Ann. Acad. Med. Singapore* **21**, 565–570.

Cohen, J., Alikani, M., Trowbridge, J., and Rosenwaks, Z. (1992b). Implantation enhancement by selective assisted hatching using zona drilling of human embryos with poor prognosis. *Hum. Reprod.* **7**, 685–691.

Coleman, A. W., and Goff, L. J. (1985). Applications of fluorochromes to pollen biology. I. Mithramycin and 4′,6-diamidino-2-phenylindole (DAPI) as vital stains for quantitation of nuclear DNA. *Stain Technol.* **60**, 145–154.

Conaghan, J., Dimitry, E. S., Mills, M., Margara, R. A., and Winston, R. M. L. (1989). Delayed human chorionic gonadotrophin administration for in vitro fertilisation. *Lancet* **1**, 1323–1324.

Conaghan, J., Martin, K., Hardy, K., Winston, N. J., Dawson, K. J., Winston, R. M. L., and Leese, H. (1991). Prospects for pre-transfer assessment of human embryo viability following in vitro fertilization. *Hum. Reprod.* **6**, 232.

Conaghan, J., Handyside, A. H., Winston, R. M. L., and Leese, H. J. (1993). Effects of pyruvate and glucose on the development of human preimplantation embryos *in vitro. J. Reprod. Fertil.* **99**, 87–95.

Croxatto, H. B., Diaz, S., Fuentealba, B., Croxatto, H., Carrillo, D., and Farbes, C. (1972). Studies on the duration of egg transport in the human oviduct. 1. The time interval between ovulation and egg recovery from the uterus in normal women. *Fertil. Steril.* **23**, 447–458.

Dawson, K. J., Rutherford, A. J., Winston, N. J., Subak-Sharpe, R., and Winston, R. M. L. (1988). Human blastocyst transfer, is it a feasible proposition. *Hum. Reprod. Suppl.* p. 44.

Delhanty, J. D., and Penketh, R. J. A. (1990). Cytogenetic analysis of unfertilized oocytes retrieved after treatment with the LHRH analogue, buserelin. *Hum. Reprod.* **5**, 699–702.

de Pennart, H., Houliston, E., and Maro, B. (1988). Post translational modification of tubulin and the dynamics of microtubules in mouse oocytes and zygotes. *Biol. Cell.* **64**, 375.

Depypere, H. T., and Leybaert, L. (1994). Intracellular pH changes during zona drilling. *Fertil. Steril.* **61**, 319–323.

Dimitry, E. S., Oskarsson, T., Conaghan, J., Margara, R., and Winston, R. M. L. (1991). Beneficial effects of a 24h delay in human chorionic gonadotrophin administration during in-vitro fertilization treatment cycles. *Hum. Reprod.* **6**, 944–946.

Dokras, A., Sargent, I. L., Ross, C., Gardner, R. L., and Barlow, D. H. (1991). The human blastocyst: Morphology and human chorionoc gonadotrophin secretion *in vitro. Hum. Reprod.* **6**, 1143–1151.

Dokras, A., Ross, C., Gosden, B., Sargent, I. L., and Barlow, D. H. (1994). Micromanipulation of human embryos to assist hatching. *Fertil. Steril.* **61**, 514–520.

Draetta, G. (1990). Cell cycle contol in eukaryotes, molecular mechanisms of cdc 2 activation. *Trends Biochem. Sci.* **15,** 378–383.

Draetta, G., and Beach, D. (1988). Activation of cdc2 protein during mitosis in human cells: Cell cycle-dependent phosphorylation and subunit rearrangement. *Cell (Cambridge, Mass.)* **54,** 17–26.

Draetta, G., Piwnica-Worms, H., Morrison, D., Druker, B., Roberts, T., and Beach, D. (1988). cdc2 is a major cell-cycle regulated tyrosine kinase substrate. *Nature (London)* **336,** 738–743.

Dunphy, W. G., and Newport, J. W. (1988). Unravelling of mitotic control mechanisms. *Cell (Cambridge, Mass.)* **55,** 925–928.

Dunphy, W. G., Brizuela, L., Beach, D., and Newport, J. (1988). The *Xenopus* cdc2 protein is a component of MPF, a cytoplasmic regulator of mitosis. *Cell (Cambridge, Mass.)* **54,** 423–431.

Dvorak, M., Tesarik, J., Pilka, L., and Travnik, P. (1982). Fine structure of human two-cell ova fertilized and cleaved *in vitro. Fertil. Steril.* **37,** 661–667.

Edirisinghe, W. R., Murch, A. R., and Yovich, J. L. (1992). Cytogenetic analysis of human oocytes and embryos in an in vitro fertilization programme. *Hum. Reprod.* **7,** 230–236.

Edmonds, P. K., Lindsay, K. S., Miller, J. F., Williamson, E., and Wood, P. J. (1982). Early embryonic mortality in women. *Fertil. Steril.* **38,** 447–453.

Edwards, R. G., Bavister, B. D., and Steptoe, P. C. (1969). Early stages of fertilization *in vitro* of human oocytes matured in vitro. *Nature (London)* **211,** 632–635.

Edwards, R. G., Steptoe, P. C., and Purdy, J. M. (1970). Fertilization and cleavage in vitro of preovulatory human oocytes. *Nature (London)* **227,** 1307–1309.

Edwards, R. G., Steptoe, P. C., Fowler, R. E., and Baillie, J. (1980a). Observation on preovulatory human ovarian follicles and their aspirates. *Br. J. Obstet. Gynaecol.* **87,** 769–779.

Edwards, R. G., Steptoe, P. C., and Purdy, J. M. (1980b). Establishing full-term pregnancies using cleaving embryos grown *in vitro. Br. J. Obstet. Gynaecol.* **87,** 737–756.

Eichenlaub-Ritter, U., Stahl, A., and Luciani, J. M. (1988). The microtubular cytoskeleton and chromosomes of unfertilised human oocytes aged *in vitro. Hum. Genet.* **80,** 259–264.

Eppig, J. J., Schultz, R. M., O'Brien, M., and Chesnel, F. (1994). Relationships between the developmental programs controlling nuclear and cytoplasmic maturation of mouse oocytes. *Dev. Biol.* **164,** 1–9.

Evans, T., Rosenthal, E. T., Youngblom, J., Distel, D., and Hunt, T. (1983). Cyclin: A protein specified by maternal mRNA in sea urchin eggs that is destroyed at each cleavage division. *Cell (Cambridge, Mass.)* **33,** 389–396.

Farrel, P. J., Balkow, K., Hunt, T., and Jackson, R. J. (1977). Phosphorylation of initiation factor eIF-2 and the control of reticulocyte protein synthesis. *Cell (Cambridge, Mass.)* **11,** 187–200.

Fehilly, C. B., Cohen, J., Simons, R. F., Fishel, S. B., and Edwards, R. G. (1985). Cryopreservation of cleaving and expanded blastocysts in the human: A comparative study. *Fertil. Steril.* **44,** 638–644.

Fishel, S. (1985). Time-dependent motility changes of human spermatozoa after preparation for in vitro fertilization. *J. In Vitro Fertil. Embryo. Transf.* **2,** 233–235.

Fisher, D. L., and Smithberg, M. (1973). Host-transplant relationship of cultured mouse embryos. *J. Exp. Zool.* **183,** 263–266.

Fleming, R., and Coutts, J. R. T. (1986). Induction of multiple follicular growth in normally menstruating women with endogenous gonadotrophon suppression. *Fertil. Steril.* **45,** 226–230.

Freeman, M., Nusslein-Volhard, C., and Glover, D. M. (1986). The dissociation of nuclear and centrosomal division in gnu, a mutant causing giant cell formation in Drosophila. *Cell (Cambridge, Mass.)* **46,** 457–468.

Fritz, M. A., and Speroff, L. (1982). The endocrinology of the menstrual cycle: The interaction of folliculogenesis and neuroendocrine mechanisms. *Fertil. Steril.* **38,** 509.

Fujimoto, S., Pahlavan, N., and Dukelow, W. R. (1974). Chromosome abnormalities in rabbit preimplantation blastocysts induced by superovulation. *J. Reprod. Fertil.* **40,** 177–181.

Gardner, D. K., and Leese, H. J. (1990). Concentrations of nutrients in mouse oviduct fluid and their effects on embryo development and metabolism *in vitro. J. Reprod. Fertil.* **88,** 361–368.

Gardner, D. K., and Leese, H. J. (1993). Assessment of embryo metabolism and viability. *In* "Handbook of In Vitro Fertilization" (A. Trounson and D. K. Gardner, eds.), pp. 195–211. CRC Press, London.

Gardner, D. K., and Sakkas, D. (1993). Mouse embryo cleavage, metabolism and viability: Role of medium composition. *Hum. Reprod.* **8,** 288–295.

Gardner, D. K., Spitzer, A., and Osborn, J. C. (1991). Development of a complex serum-free medium and its effects on the development and metabolism of the preimplantation human embryo. *Proc. Am. Fertil. Soc.* **164,** 164.

Gautier, J., Norbury, C., Lohka, M., Nurse, P., and Maller, J. (1988). Purified maturation-promoting factor contains the product of a *Xenopus* homolog of the fission yeast cell cycle control gene cdc2+. *Cell (Cambridge, Mass.)* **54,** 433–439.

Gautier, J., Solomon, M. J., Booher, R. N., Bazan, J. F., and Kirschner, M. W. (1991). cdc25 is a specific tyrosine phosphatase that directly activates p34cdc2. *Cell (Cambridge, Mass.)* **67,** 197–211.

George, M. A., and Doe, B. G. (1989). The influence of handling procedures during mouse oocyte and embryo recovery on viability and subsequent development in vitro. *J. In Vitro Fertil. Embryol. Transfus.* **6,** 69–72.

George, M. A., Johnson, M. H., and Howlett, S. K. (1994). Assessment of the developmental potential of frozen-thawed mouse oocytes. *Hum. Reprod.* **9,** 130–136.

Gerhart, J., Wu, M., and Kirschner, M. (1984). Cell cycle dynamics of an M-phase specific cytoplasmic factor in *Xenopus* laevis oocytes and eggs. *J. Cell Biol.* **98,** 1247–1255.

Gianfortoni, J. G., and Gulyas, B. J. (1985). The effect of short-term incubation (aging) of mouse oocytes on in vitro fertilization, zona solubility and embryonic development. *Gamete Res.* **11,** 59–68.

Gorbsky, G. J., Simerly, C., Schatten, G., and Borisy, G. G. (1990). Microtubules in the metaphase-arrested mouse oocyte turn over rapidly. *Proc. Natl. Acad. Sci. U.S.A.* **87,** 6049–6053.

Handyside, A. H., Pattinson, J. K., Penketh, R. J. A., Delhanty, J. D. A., Winston, R. M. L., and Tuddenham, E. G. D. (1989). Biopsy of human pre-embryos and sexing by DNA amplification. *Lancet* **1,** 347–349.

Handyside, A. H., Kontogianni, E. H., Hardy, K., and Winston, R. M. L. (1990). Pregnancies from human preimplantation embryos sexed by Y-specific DNA amplification. *Nature (London)* **344,** 768–770.

Hardy, K., Handyside, A. H., and Winston, R. M. L. (1989a). The human blastocyst: cell number, death and allocation during late preimplantation development *in vitro. Development (Cambridge, UK)* **107,** 597–604.

Hardy, K., Hooper, M. A., K., Handyside, A. H., Rutherford, A. J., Winston, R. M. L., and Leese, H. J. (1989b). Non-invasive measurement of glucose and pyruvate uptake by individual human oocytes and preimplantation embryos. *Hum. Reprod.* **4,** 188–191.

Hardy, K., Martin, K. L., Leese, H. J., Winston, R. M. L., and Handyside, A. H. (1990). Human preimplantation development in vitro is not adversely affected by biopsy at the 8-cell stage. *Hum. Reprod.* **5,** 708–714.

Hardy, K., Winston, R. M. L., and Handyside, A. H. (1993). Binucleate blastomeres in preimplantation human embryos in vitro: Failure of cytokinesis during early cleavage. *J. Reprod. Fertil.* **98,** 96–103.

Harika, G., Zeinoun, R., Pigeon, F., Palot, M., Quereux, C., and Whal, P. (1991). Failure of cleavage: The guilty party. *Hum. Reprod.* **6,** 235.

Harlow, G. M., and Quinn, P. (1982). Development of mouse preimplantation embryos *in vitro* and *in vivo*. *Aust. J. Biol. Sci.* **35,** 187–193.

Hartwell, L. H., and Weinert, T. A. (1989). Checkpoints: Controls that ensure the order of cell cycle events. *Science* **246,** 629–634.

Hemmings, R., Lachapelle, M.-H., Falcone, T., Miron, P., Ward, L., and Guyda, H. (1994). Effect of follicular fluid supplementation on the in vitro development of human pre-embryos. *Fertil. Steril.* **62,** 1018–1021.

Hertig, A. T., Rock, J., Adams, E. C., and Mulligan, W. J. (1954). On the preimplantation stages of the human ovum: A description of four normal and four abnormal specimens ranging from the second to fifth day of development. *Contrib. Embryol.* **35,** 199–220.

Hertig, A. T., Rock, J., and Adams, C. E. (1956). A description of 34 human ova within the first 17 days of development. *Am. J. Anat.* **98,** 435–459.

Hertig, A. T., Rock, J., and Adams, E. C. (1959). Thirty-four fertilized human ova, good, bad and indifferent, recovered from 210 women of known fertility. *Pediatrics* **23,** 202–211.

Hillier, S. G., Dawson, K. J., Afnan, M., Margara, R. A., Ryder, T. A., Wickings, E. J., and Winston, R. M. L. (1985). Embryonic culture: Quality control in clinical in vitro fertilization. *In* "Proceedings of the Twelfth Study Group of the Royal College of Obstetritians and Gynaecologists" (E. Thompson, D. N. Joyce, and J. R. Newton, eds.), p. 125. Royal College of Obstetritians and Gynaecologists, London.

Hodgen, G. D. (1982). The dominant follicle. *Fertil. Steril.* **38,** 281.

Howlett, S. K., Webb, M., Maro, B., and Johnson, M. H. (1985). Meiosis II, mitosis I and the linking interphase: A study of the cytoskeleton in the fertilized mouse egg. *Cytobios* **43,** 295–305.

Howlett, S. K., Barton, S. C., and Surani, M. A. (1987). Nuclear cytoplasmic interactions following nuclear transplantation in mouse embryos. *Development (Cambridge, UK)* **101,** 915–923.

Huisman, G. J., Alberda, A. T., Leerentveld, R. A., Verhoeff, A., and Zeilmaker, G. H. (1994). A comparison of in vitro fertilization results after embryo transfer after 2, 3, and 4 days of embryo culture. *Fertil. Steril.* **61,** 970–971.

Interim Licensing Authority (1991). "The Sixth Report of the Interim Licensing Authority for Human In Vitro Fertilisation and Embryology." Interim Licensing Authority, London.

Iwasaki, S., Hamano, S., Kuwayama, M., Yamashita, M., Ushijima, H., Nagaoka, S., and Nakahara, T. (1992). Developmental changes in the incidence of chromosomal anomalies of bovine embryos fertilized *in vitro*. *J. Exp. Zool.* **261,** 79–85.

Jacobs, P. A., Angell, R. R., Buchanan, J. M., Hassold, T. J., Matsuyama, A. M., and Manuel, B. (1978). The origin of human triploids. *Hum. Genet.* **42,** 49–57.

Jamieson, M. E., Coutts, J. R. T., and Connor, J. M. (1994). The chromosome constitution of human preimplantation embryos fertilized in vitro. *Hum. Reprod.* **9,** 709–715.

Janny, L., Benkhalifa, M., Pouly, J. L., Vye, P., Canis, M., Geneix, A., Malet, P., and Boucher, D. (1990). La condensation prématurée des chromosomes: Facteur fréquent d'échec en fécondation in vitro? *Contracep. Fertil. Sexual.* **18,** 536–537.

Johnson, M. H., and Everitt, B. J. (1995). "Essential Reproduction." Blackwell, Oxford.

Johnson, M. H., and Nasr-Esfahani, M. H. (1994). Radical solutions and cultural problems: Could free oxygen radicals be responsible for the impaired development of preimplantion mammalian embryos *in vitro*. *BioEssays* **16,** 31–38.

Johnson, M. H., and Selwood, L. (1996). The nomenclature of early development in mammals. *Reprod. Fertil. Dev.* (in press).

Johnston, I., Lopata, A., Spiers, A., Hoult, I., Kellow, G., and DuPlessis, Y. (1981). In vitro fertilization: The challenge of the eighties. *Fertil. Steril.* **36,** 699.

Jones, D. P. (1985). The role of oxygen concentration in oxidative stress: Hypoxic and hyperoxic models. *In* "Oxidative Stress" (H. Sies, ed.), pp. 151–195. Academic Press, London.

Karsenti, E. (1991). Mitotic spindle morphogenesis in animal cells. *Semin. Cell Biol.* **2,** 251–260.

Kaufman, M. H. (1983). "Early Mammalian Development: Parthenogeneti Studies." Cambridge Univ. Press, Cambridge, UK.

Khan, I., Staessen, C., Van den Abbeel, E., Camus, M., Wisanto, A., Smitz, J., Derroey, P., and Van Stertegham, A. C. (1989). Time of insemination and its effects on in vitro fertilization, cleavage and pregnancy rates in GnRH agonist/HMG-stimulated cycles. *Hum. Reprod.* **4,** 921–926.

Kola, I., and Trounson, A., eds. (1989). Dispermic fertilization: Violation of expected cell behaviour. *In* "The Cell Biology of Fertilization" (G. Schatten and H. Schatten, eds.), p. 277. Academic Press, New York.

Kubiak, J. Z. (1989). Mouse oocytes gradually develop the capacity for activation during the Metaphase II arrest. *Dev. Biol.* **136,** 537–545.

Leese, H. J., Hooper, M. A. K., Edwards, R. G., and Ashwood-Smith, M. J. (1986). Uptake of pyruvate by early human embryos determined by non-invasive technique. *Hum. Reprod.* **1,** 181–182.

Leridon, H. (1977). "Human Fertility: The Basic Components." Univ. of Chicago Press, Chicago.

Lindenberg, C., Hyttel, P., Sjögren, A., and Greve, T. (1989). A comparative study of attachment of human, bovine and mouse blastocysts to uterine epithelial monolayer. *Hum. Reprod.* **4,** 446–456.

Liu, D. Y., and Baker, H. W. G. (1992). Morphology of spermatozoa bound to the zona pellucida of human oocytes that failed to fertilize in vitro. *J. Reprod. Fertil.* **94,** 71–84.

Liu, H.-C., Cohen, J., Alikani, M., Noyes, N., and Rosenwaks, Z. (1993). Assisted hatching facilitates earlier implantation. *Fertil. Steril.* **60,** 871–875.

Lock, L. F., and Wickramasinghe, D. (1994). Cycling with CDKs. *Trends Cell Biol.* **4,** 404–405.

Lohka, M. J., Hayes, M., and Maller, J. (1988). Purification of maturation promoting factor, an intracellular regulator of early mitotic events. *Proc. Natl. Acad. Sci. U.S.A.* **85,** 3009–3013.

Long, S. E., and Williams, C. V. (1982). A comparison of the chromosome complement of inner cell mass and trophoblast cells in day 10 pig embryos. *J. Reprod. Fertil.* **66,** 645–648.

Lopata, A., and Hay, D. L. (1989). The surplus embryo: Its potential for growth, blastulation, hatching and human chorionic gonadotrophin production in culture. *Fertil. Steril.* **51,** 984–991.

Lopata, A., Kohlman, D. J., and Kellow, G. N. (1982). The fine structure of human blastocysts developed in culture. *In* "Embryonic Development, Part B: Cellular Aspects" (M. M. Burger, ed.), pp. 69–85. Alan R. Liss, New York.

Lopata, A., Kohlman, D., and Johnston, I. (1983). The fine structure of normal and abnormal embryos developed in culture. *In* "Fertilization of the Human Egg In Vitro" (H. M. Beier and H. R. Linder, eds.), pp. 189–209. Springer-Verlag, Berlin.

Lu, Y., Hammitt, D. G., Zinmeister, A. R., and Dewald, G. W. (1994). Dual color fluorescence in situ hybridization to investigate aneuploidy in sperm from 33 normal males and a man with a t(2;4;8)(q23;q27;p21). *Fertil. Steril.* **62,** 394–399.

Lundgren, K., Walworth, N., Booher, R., Debski, M., Kirschner, M., and Beach, D. (1991). mik1 and wee1 cooperate in the inhibitory tyrosine phosphorylation of cdc2. *Cell (Cambridge, Mass.)* **64,** 1111–1122.

Ma, S., Kalousek, D. K., Zouves, C., Ho Yeun, B., Gomel, V., and Moon, Y. S. (1989). Chromosome analysis of human oocytes failing to fertilize in vitro. *Fertil. Steril.* **51,** 992–997.

MacDougal, M. J., Tan, S.-L., Hall, V., Balen, A., Mason, B., and Jacobs, H. S. (1994). Comparison of natural with clomiphene citrate-stimulated cycles in in vitro fertilization: A prospective, randomized trial. *Fertil. Steril.* **61,** 1052–1057.

MacKenna, A. I., Zegers-Hochschild, F., Fernandez, E. O., Fabres, C. V., Huidobro, C. A., Prado, J. A., Roblero, L. S., Alteri, E. L., Guadarrama, A. R., and Lopez, T. H. (1992). Fertilization rate in couples with unexplained infertility. *Hum. Reprod.* **7,** 223–226.

Magier, S., Van Der Van, H., Diedrich, K., and Krebs, D. (1990). Significance of cumulus oophorus in in-vitro fertilization and oocyte viability and fertility. *Hum. Reprod.* **5,** 847–852.

Mahaderan, L. C., Wills, A. J., Hirst, E. A., Rathjen, P. D., and Heath, J. K. (1990). 2-aminopurine abolishes EGF- and TPA-stimulated pp33 phosphorylation and c-fos induction without affecting the activation of protein kinase C. *Oncogene* **5,** 327–335.

Malter, H. E., and Cohen, J. (1989). Partial zona dissection of the human oocyte: A nontraumatic method using micromanipulation to assist zona pellucida penetration. *Fertil. Steril.* **51,** 139–148.

Mansour, R. T., Aboulghar, M. A., Serour, G. I., and Abbass, A. M. (1994). Co-culture of pronucleate oocytes with their cumulus cells. *Hum. Reprod.* **9,** 1727–1729.

Maro, B., Johnson, M. H., Pickering, S. J., and Flach, G. (1984). Changes in actin distribution during fertilization of the mouse egg. *J. Embryol. Exp. Morphol.* **81,** 211–237.

Maro, B., Johnson, M. H., Webb, M., and Flach, G. (1986). Mechanisms of polar body formation in the mouse oocyte: An interaction between chromosomes, the cytoskeleton and the plasma membrane. *J. Embryol. Exp. Morphol.* **92,** 11–32.

Marston, J. H., Penn, R., and Sivelle, P. C. (1977). Successful autotransfer of tubal eggs in the rhesus monkey (*Macaca mulatta*). *J. Reprod. Fertil.* **49,** 175–176.

Martin, R. H., Mahadevan, M. M., and Taylor, P. J. (1986). Chromosomal analysis of unfertilized human oocytes. *J. Reprod. Fertil.* **78,** 673–678.

Masui, Y., and Markert, C. L. (1971). Cytoplasmic control of nuclear behaviour during meiotic maturation of frog oocytes. *J. Exp. Zool.* **177,** 129–146.

McConnell, J. M. L., Campbell, L., and Vincent, C. (1995). Capacity of mouse oocytes to become activated depends on completion of cytoplasmic but not nuclear meiotic maturation. *Zygote* **3,** 45–55.

McGrath, J., and Solter, D. (1986). Nucleocytoplasmic interactions in the mouse. *J. Embryol. Exp. Morphol.* **97,** 277–289.

Michaeli, G., Fejgin, M., Ghetler, Y., Ben Nun, I., Beyth, Y., and Amiel, A. (1990). Chromosomal analysis of unfertilized oocytes and morphologically abnormal preimplantation embryos from an in vitro fertilization program. *J. In Vitro Fertil. Embryol. Transfus.* **7,** 341–346.

Michelmann, H. W., and Hinney, B. (1988). The influence of chromosomal factors on the fertilization rate during IVF. *Hum. Reprod.* **3,** 46.

Millar, J. B. A., McGowan, C. H., Lenaers, G., Jones, R., and Russell, P. (1991). p80[cdc25] mitotic inducer in the tyrosine phosphatase that activates p34[cdc2] kinase in fission yeast. *EMBO J.* **10,** 4301–4309.

Modlinski, J. A. (1970). The role of the zona pellucida in the development of mouse eggs *in vivo. J. Embryol. Exp. Morphol.* **23,** 539–551.

Monks, N. J., Turner, K., Hooper, M. A. K., Kumar, A., Verma, S., and Lenton, E. A. (1993). Development of embryos from natural cycle in-vitro fertilization: Impact of medium type and female infertility factors. *Hum. Reprod.* **8,** 266–271.

Moore, N. W., Adams, C. E., and Rowson, L. E. A. (1968). Development potential of single blastomeres of the rabbit egg. *J. Reprod. Fertil.* **17,** 527–533.

Mullins, J. M., and Snyder, J. A. (1981). Anaphase progression and furrow establishment in nocodazole-arrested PtK1 cells. *Chromosoma* **83,** 493–505.

Munné, S., and Cohen, J. (1993). Unsuitability of multinucleated human blastomeres for preimplantation genetic diagnosis. *Hum. Reprod.* **8,** 1120–1125.

Munné, S., and Estop, A. (1991). The effect of in vitro ageing on mouse sperm chromosomes. *Hum. Reprod.* **6,** 703–708.

Munné, S., Alikani, M., and Cohen, J. (1994). Monospermic polyploidy and atypical morphology. *Hum. Reprod.* **9,** 506–510.

Murray, A. W. (1989). The cell cycle as a cdc2 cycle. *Nature (London)* **342,** 14–15.

Nagy, Z. P., Joris, H., Liu, J., Staessen, C., Devroey, P., and Van Steirteghem, A. C. (1993). Intracytoplasmic single sperm injection of 1-day old unfertilized human oocytes. *Hum. Reprod.* **8,** 2180–2184.

Nasr-Esfahani, M. H., Aitken, J. R., and Johnson, M. H. (1990a). Hydrogen peroxide levels in oocytes and early cleavage stage embryos from blocking and non-blocking strains of mice. *Development (Cambridge, UK)* **109**, 501–507.

Nasr-Esfahani, M. H., Johnson, M. H., and Aitken, R. J. (1990b). The effect of iron and iron chelators on the in vitro block to development of the mouse preimplantation embryo: BAT6 a new medium for improved culture of mouse embryos *in vitro*. *Hum. Reprod.* **5**, 997–1003.

Nasr-Esfahani, M. H., Winston, N. J., and Johnson, M. H. (1992). The effect of glucose, glutamine, EDTA and oxygen on the level of reactive oxygen species and on the development of the mouse preimplantation embryo in vitro. *J. Reprod. Fertil.* **96**, 219–231.

Navot, D., Scott, R. T., Droesch, K., Veeck, L. L., Liu, H.-C., and Rosenwaks, Z. (1991). The window of embryo transfer and the efficiency of human conception *in vitro*. *Fertil. Steril.* **55**, 114–118.

Nayudu, P. L., Lopata, A., Jones, G. M., Gook, D. A., Bourne, H. M., Sheather, S. J., Brown, T. C. and Johnston, W. I. H. (1989a). An analysis of human oocytes and follicles from stimulated cycles: Oocyte morphology and associated follicular fluids. *Hum. Reprod.* **4**, 558–567.

Nayudu, P. L., Gook, D. A., Hepworth, G., Lopata, A., and Johnston, W. I. H. (1989b). Prediction of outcome in human in vitro fertilization based on follicular and stimulation response variables. *Fertil. Steril.* **51**, 117–125.

Nigg, E. A. (1993). Cellular substrates of p34^{cdc2} and its companion cyclin-dependent kinases. *Trends Cell Biol.* **3**, 296–301.

Noda, Y., Goto, Y., Umaoka, Y., Shiotani, M., Nakayama, T., and Mori, T. (1994). Culture of human embryos in alpha modification of Eagle's medium under low oxygen tension and low illumination. *Fertil. Steril.* **62**, 1022–1027.

Norbury, C., and Nurse, P. (1992). Animal cell cycles and their control. *Annu. Rev. Biochem.* **61**, 441–470.

Nurse, P. (1990). Universal control mechanism regulating onset of M-phase. *Nature (London)* **344**, 503–508.

Nurse, P., and Bissett, Y. (1981). Gene required in G_1 for commitment to cell cycle and in G_2 for control of mitosis in fission yeast. *Nature (London)* **292**, 558–560.

Obruca, A., Strohmer, H., Sakkas, D., Menezo, Y., Kogosowski, A., Barak, Y., and Feichtinger, W. (1994). Use of lasers in assisted fertilization and hatching. *Hum. Reprod.* **9**, 1723–1726.

Ohsumi, K., Katagiri, C., and Kishimoto, T. (1993). Chromosome condensation in Xenopus mitotic extracts without Histone H1. *Science* **262**, 2033–2035.

Palermo, G., Munné, S., and Cohen, J. (1994). The human zygote inherits its mitotic potential from the male gamete. *Hum. Reprod.* **9**, 1220–1225.

Parinaud, J., Mieusset, R., Vieitez, G., Labal, B., and Richoilley, G. (1993). Influence of sperm parameters on embryo quality. *Fertil. Steril.* **60**, 888–892.

Parr, E. L. (1973). Shedding of the zona pellucida by guinea pig blastocysts: An ultrastructural study. *Biol. Reprod.* **8**, 531–544.

Pellestor, F. (1991a). Frequency and distribution of aneuploidy in human female gametes. *Hum. Genet.* **86**, 283–288.

Pellestor, F. (1991b). Differential distribution of aneuploidy in human gametes according to their sex. *Hum. Reprod.* **6**, 1252–1258.

Pellestor, F., and Sèle, B. (1988). Assessment of aneuploidy in the human female by using cytogenetics of in vitro fertilization failures. *Am. J. Hum. Genet.* **42**, 274–283.

Pellestor, F., Sèle, B., Jalbert, H., and Raymond, L. (1989). Human oocytes chromosome analysis. *Am. J. Hum. Genet.* **44**, A104.

Pellestor, F., Dufour, M.-C., Arnal, F., and Humeau, C. (1994). Direct assessment of the rate of chromosomal abnormalities in grade IV human embryos produced by in vitro fertilization procedure. *Hum. Reprod.* **9**, 293–302.

Peluso, J. J., Luciano, A. A., and Nulsen, J. C. (1992). The relationship between alterations in spermatozoal deoxyribonucleic acid, heparin binding sites and semen quality. *Fertil. Steril.* **57,** 665–670.

Penketh, R. J. A., Delhanty, J. D. A., Van den Berghe, J. A., Finklestone, E. M., Handyside, A. H., Malcolm, S., and Winston, R. M. L. (1989). Rapid sexing of human embryos by non-radioactive in situ hybridization: Potential for preimplantation diagnosis of X-linked disorders. *Prenatal Diagn.* **9,** 489–500.

Pereda, J., Cheviakoff, S., and Croxatto, H. B. (1989). Ultrastructure of a 4 cell human embryo developed *in vivo. Hum. Reprod.* **4,** 680–688.

Pfarr, C. M., Coue, M., Grissom, P. M., Hays, T. S., Porter, M. E., and McIntosh, J. R. (1990). Cytoplasmic dynein is localized to kinetochores during miosis. *Nature (London)* **345,** 263–265.

Pickering, S. J., Johnson, M. H., Braude, P. R., and Houliston, E. (1988). Cytoskeletal organisation in fresh, aged and spontaneously activated human oocytes. *Hum. Reprod.* **3,** 978–989.

Pickering, S. J., Braude, P. R., Johnson, M. H., Cant, A., and Currie, J. (1990). Transient cooling to room temperature can cause irreversible disruption of the meiotic spindle in the human oocyte. *Fertil. Steril.* **54,** 102–108.

Pickering, S. J., McConnell, J. M., Johnson, M. H., and Braude, P. R. (1992). Reliability of detection by polymerase chain reaction of the sickle cell-containing region of the β-globin gene in single human blastomeres. *Hum. Reprod.* **7,** 630–636.

Pickering, S. J., Taylor, A., Johnson, M. H., and Braude, P. R. (1995). An analysis of multinucleated blastomere formation in human embryos. *Hum. Reprod.* **10,** In press.

Pieters, M. H. E. C., Geraedts, J. P. M., Dumoulin, J. C. M., Evers, J. L. H., Bras, M., Kornips, F. H. A. C., and Menheere, P. P. C. A. (1989). Cytogenetic analysis of in vitro fertilization (IVF) failures. *Hum. Genet.* **81,** 367–370.

Pinkster, M. C., Sacco, A. G., and Mintz, B. (1974). Implantation-associated proteinase in mouse uterine fluid. *Dev. Biol.* **38,** 285–290.

Plachot, M. (1989). Chromosome analysis of spontaneous abortions after in vitro fertilization. A European study. *Hum. Reprod.* **4,** 425–429.

Plachot, M., Junca, A.-M., Mandelbaum, J., Cohen, J., Salat-Baroux, J., and DaLage, C. (1986a). Timing of in vitro fertilization of cumulus-free and cumulus-enclosed human oocytes. *Hum. Reprod.* **1,** 237–242.

Plachot, M., Junca, A. M., Mandelbaum, J., De Grouchy, J., Salat-Baroux, J., and Cohen, J. (1986b). Cytogenetical analysis of human oocytes and embryos in an IVF program. *Hum. Reprod.* **1,** Abstr. 21.

Plachot, M., De Grouchy, J., Junca, A.-M., Mandelbaum, J., Salat-Baroux, J., and Cohen, J. (1988a). Chromosome analysis of human oocytes and embryos: Does delayed fertilization increase chromosome imbalance? *Hum. Reprod.* **3,** 125–127.

Plachot, M., Veiga, A., Montagut, J., de Grouchy, J., Calderon, G., Leprete, S., Junca, A.-M., Santalo, J., Carles, E., Mandelbaum, J., Barri, P., Degoy, I., Cohen, J., Egozcue, J., Sabatier, J. C., and Salat-Baroux, J. (1988b). Are clinical and biological IVF parameters correlated with chromosomal disorders in early life: A multicentric study. *Hum. Reprod.* **3,** 627–635.

Plachot, M., de Grouchy, J., Junca, A.-M., Mandelbaum, J., Cohen, J. and Salat-Baroux, J. (1990). Genetics of human oocytes. *In* Gamete Physiology" (R. H. Asch, J. P. Balmeceda, and I. Johnston, eds.), pp. 145–152. Serono Symposia, Norwell, MA.

Plachot, M., Antoine, J. M., Alvarez, S., Firmin, C., Pfister, A., Mandelbaum, J., Junca, A.-M., and Salat-Baroux, J. (1993). Granulosa cells improve embryo development *in vitro. Hum. Reprod.* **8,** 2133–2140.

Pool, T. B., and Martin, J. E. (1994). High continuing pregnancy rates after in vitro fertilization-embryo transfer using medium supplemented with a plasma protein fraction containing α- and β-globulins. *Fertil. Steril.* **61,** 714–719.

Pratt, H., Chakraborty, J., and Surani, M. (1981). Molecular and morphological differentiation of the mouse blastocyst after manipulations of compaction with Cytochalasin D. *Cell (Cambridge, Mass.)* **26**, 279–292.

Regan, L., Owen, E. J., and Jacobs, H. S. (1990). Hypersecretion of luteinising hormone, infertility and miscarriage. *Lancet* **336**, 1141–1144.

Ritchie, W. G. M. (1985). Ultrasound in the evaluation of normal and induced ovulation. *Fertil. Steril.* **43**, 167–180.

Ron-El, R., Nachum, H., Herman, A., Golan, A., Caspi, E., and Soffer, E. (1991). Delayed fertilization and poor embryonic development associated with impaired semen quality. *Fertil. Steril.* **55**, 338–344.

Rosenbusch, B., and Sterzik, K. (1991). Sperm chromosomes and habitual abortion. *Fertil. Steril.* **56**, 370–372.

Russel, P., and Nurse, P. (1987). Negative regulation of mitosis by wee1, a gene encoding a protein kinase homolog. *Cell (Cambridge, Mass.)* **49**, 559–567.

Rutherford, A. J., Subak-Sharpe, R. J., Dawson, K. J., Magara, R. A., Franks, S., and Winston, R. M. L. (1988). Improvement of in vitro fertilisation after treatment with buserilin, an agonist of luteinising hormone releasing hormone. *Br. Med. J.* **296**, 1765–1768.

Saito, H., Hirayama, T., Koike, K., Saito, T., Nohara, M., and Hiroi, M. (1994). Cumulus mass maintains embryo quality. *Fertil. Steril.* **62**, 555–558.

Sakkas, D., Jaquenoud, N., Leppens, G., and Campana, A. (1994). Comparison of results after in vitro fertilized human embryos are cultured in routine medium and in co-culture on vero cells: A randomized study. *Fertil. Steril.* **61**, 521–525.

Santella, L., Alikani, M., Talansky, B. E., Cohen, J., and Dale, B. (1992). Is the human oocyte plasma membrane polarized? *Hum. Reprod.* **7**, 999–1003.

Sathananthan, A. H., and Trounson, A. O. (1982). Ultrastructure of cortical granule release and zona interaction in monospermic and polyspemic human ova fertilised *in vitro. Gamete Res.* **6**, 225.

Sathananthan, A. H., Trounson, A., Freeman, L., and Brady, T. (1988). The effect of cooling human oocytes on the meiotic spindles and oocyte structure. *Hum. Reprod.* **3**, 968–977.

Sathananthan, H., Bongso, A., Ng, S.-C., Ho, J., Mok, H., and Ratnam, S. (1990). Ultrastructure of preimplantation human embryos co-cultured with human ampullary cells. *Hum. Reprod.* **5**, 309–318.

Schatten, G. (1994). The centrosome and its mode of inheritance: The reduction of the centrosome during gametogenesis and its restoration during fertilization. *Dev. Biol.* **165**, 299–335.

Schatten, G., Simerly, C., and Schatten, H. (1985). Microtubule configuration during fertilization, mitosis and early development in the mouse. *Proc. Natl. Acad. Sci. U.S.A.* **82**, 4152–4156.

Schillaci, R., Ciriminna, R., Cefalù, E. (1994). Vero cell effect on *in vitro* human blastocyst development: preliminary results. *Hum. Reprod.* **9**, 1131–1135.

Schlegel, R., and Pardee, A. B. (1986). Caffeine-induced uncoupling of mitosis from the completion of DNA replication in mammalian cells. *Science* **232**, 1264–1266.

Schmiady, H. and Kentenich, H. (1989). Premature chromosome condensation after in vitro fertilization. *Hum. Reprod.* **4**, 689–695.

Schmiady, H., Sperling, K., Kentich, H., and Stauber, M. (1986). Prematurely condensed human sperm chromosomes after in vitro fertilization. *Hum. Genet.* **74**, 441–443.

Scholtes, M. C. W., Hop, W. C. J., Alberda, A. T., Janssen-Caspers, H. A. B., Leerenveld, R. A., van Os, H. C., and Zeilmaker, G. H. (1988). Factors influencing the outcome of successive IVF treatment cycles in attaining a follicular puncture. *Hum. Reprod.* **3**, 755–759.

Schoolcraft, W. B., Schlenker, T., Gee, M., Jones, G. S., and Jones, H. W. J. (1994). Assisted hatching in the treatment of poor prognosis in vitro fertilization candidates. *Fertil. Steril.* **62**, 551–554.

Shulman, A., Ben-Nun, I., Ghetler, Y., Kaneti, H., Shilon, M., and Beyth, Y. (1993). Relationship between embryo morphology and implantation rate after in vitro fertilization treatment in conception cycles. *Fertil. Steril.* **60,** 123–126.

Simanis, V., and Nurse, P. (1986). The cell cycle control gene cdc2+ of fission yeast encodes a protein kinase potentially regulated by phosphorylation. *Cell (Cambridge, Mass.)* **45,** 261–268.

Smith, R., and McLaren, A. (1977). Factors affecting the time of formation of the mouse blastocoele. *J. Embryol. Exp. Morphol.* **41,** 79–92.

Snow, M. H. L. (1973). Tetraploid mouse embryos produced by cytochalasin B during cleavage. *Nature (London)* **244,** 513–514.

Snyder, J. A., Hamilton, R. T., and Mullins, J. M. (1982). Loss of mitotic centrosomal microtubule initiation capacity at the metaphase-anaphase transition. *Eur. J. Cell Biol.* **27,** 191–199.

Staessen, C., Van der Abbeel, E., Janssenswillen, C., Devroey, P., and Van Steirteghem, A. C. (1991). Is the addition of vitamins and amino acids to Earle's culture medium beneficial for early human embryo development *in vitro? Hum. Reprod.* **6,** 99.

Stausfield, U., Fernandez, A., Capony, J.-P., Girard, F., Lautredou, N., Derancourt, J., Labbe, J.-C., and Lamb, N. J. (1994). Activation of p34cdc2 protein kinase by microinjection of human cdc25C into mammalian cells. Requirement for prior phosphorylation of cdc25C by p34cdc2 on sites phosphorylated at mitosis. *J. Biol. Chem.* **269,** 5989–6000.

Steptoe, P. C., and Edwards, R. G. (1978). Birth after reimplantation of a human embryo. *Lancet* **2,** 366.

Steptoe, P. C., Edwards, R. G., and Purdy, J. M. (1971). Human blastocysts grown in culture. *Nature (London)* **229,** 132–133.

Steptoe, P. C., Edwards, R. G., and Walters, D. E. (1986). Observations on 767 clinical pregnancies and 500 births after human in vitro fertilization. *Hum. Reprod.* **1,** 89–94.

Steuer, E. R., Wordeman, L., Schroer, T. A., and Sheetz, M. P. (1990). Localization of cytoplasmic dynein to mitotic spindles and kinetochores. *Nature (London)* **345,** 266–268.

Stolwijk, A. M., Ruivers (?), J. C. M., Hamilton, C. J. C. M., Jongbloet, P. H., Hollanders, J. M. G., and Zielhuis, G. A. (1994). Seasonality in the results of IVF. *Hum. Reprod.* **9,** 2300–2305.

Suchanek, E., Simunic, V., Juretic, D., and Grizelj, V. (1994). Follicular fluid contents of hyaluronic acid, follicle-stimulating hormone and steroids relative to the success of in vitro fertilization of human oocytes. *Fertil. Steril.* **62,** 347–352.

Surani, M. A. H., Barton, S. C., and Norris, M. L. (1986). Nuclear transplantation in the mouse: Heritable differences between parental genomes after activation of the embryonic genome. *Cell (Cambridge, Mass.)* **45,** 127–136.

Szollosi, D. (1975). Mammalian eggs ageing in the fallopian tubes. *In* "Ageing Gametes" (R. J. Blandeau, ed.), pp. 98–121. Karger, Basel.

Takagi, N., and Sasaki, M. (1976). Digynic triploidy after superovulation in mice. *Nature (London)* **264,** 278–281.

Taylor, A. S., and Braude, P. R. (1994). The early development and DNA content of activated human oocytes and parthenogenetic embryos. *Hum. Reprod.* **9,** 2389–2397.

Taylor, C. T., Smullen, L. A., Ankers, L., and Kingsland, C. R. (1991). The effect of water-soluble vitamins on the development of human preimplantation embryos *in vitro. Hum. Reprod.* **6,** 53.

Tesarik, J. (1988). Developmental control of human preimplantation embryos: A comparative approach. *J. In Vitro Fertil. Embryol. Transfus.* **5,** 347–362.

Tesarik, J., Kopecny, V., Plachot, M., and Mandelbaum, J. (1986). Activation of nucleolar and extranucleolar RNA synthesis and changes in the ribosomal content of human embryos developing in vitro. *J. Reprod. Fertil.* **78,** 463–470.

Tesarik, J., Kopecny, V., Plachot, M., and Mandelbaum, J. (1987). Ultrastructural and autoradiographic observations on multinucleated blastomeres of human cleaving embryos obtained by I.V.F. *Hum. Reprod.* **2,** 127–136.

Trounson, A. O., and Sathananthan, A. H. (1984). The application of electron microscopy in the evaluation of 2-4 cell human embryos cultured in vitro for embryo transfer. *J. In Vitro Fertil. Embryol. Transfus.* **3,** 153–165.

Trounson, A. O., Leeton, J. F., Wood, C., Webb, J., and Wood, J. (1981). Pregnancies in humans by fertilization in vitro and embryo transfer in the controlled ovulatory cycle. *Science* **212,** 681.

Trounson, A. O., Mohr, L. R., Wood, C., and Leeton, J. F. (1982). Effect of delayed insemination on in-vitro fertilization, culture and transfer of human embryos. *J. Reprod. Fertil.* **64,** 285–294.

Tsirigotis, M., Redgment, C., and Craft, I. (1994). Late intracytoplasmic sperm injection (ICSI) in in vitro fertilization (IVF) cycles. *Hum. Reprod.* **9,** 1359.

Tucker, M. J., Ingargiola, P. E., Massey, J. B., Morton, P. C., Wiemer, K. E., Wiker, S. R., and Wright, G. (1994). Assisted hatching with or without bovine oviductal epithelial cell co-culture for poor prognosis in-vitro fertilization patients. *Hum. Reprod.* **9,** 1528–1531.

Turner, K., Martin, K. L., Woodward, B. J., Lenton, E. A., and Leese, H. J. (1994). Comparison of pyruvate uptake by embryos derived from conception and nonconception matural cycles. *Hum. Reprod.* **9,** 2362–2366.

Vale, R. D., and Goldstein, L. S. B. (1990). One motor, many tails: An expanding repertoire of force-generating enzymes. *Cell (Cambridge, Mass.)* **60,** 883–885.

Van Blerkom, J. (1989). Developmental failure in human reproduction associated with preovulatory oogenesis and preimplantation embryogenesis. *In* "Ultrastructure of Human Gametogenesis and Early Embryogenesis" (J. Van Blerkom and P. Motta, eds.), pp. 125–180. Kluwer Academic Publishers, Boston.

Van Blerkom, J. (1990a). Extrinsic and intrinsic influences on human oocyte and early embryo developmental potential. *In* "The Biology and Chemistry of Mammalian Fertilization" (P. M. Wassarman, cd.), pp. 81–109. CRC Press, London.

Van Blerkom, J. (1990b). Occurence and developmental consequences of abberant cellular organization in meiotically mature human oocytes after exogenous ovarian stimulation. *J. Electron Microsc. Tech.* **16,** 324–346.

Van Blerkom, J. (1993). Development of human embryos to the hatched blastocyst stage in the presence or absence of a monolayer of Vero cells. *Hum. Reprod.* **8,** 1525–1539.

Van Blerkom, J., and Henry, G. (1988). Cytogenetic analysis of living human oocytes: Cellular basis and developmental consequences of pertubations in chromosomal organization and complement. *Hum. Reprod.* **3,** 777–790.

Van Blerkom, J., and Henry, G. (1991). Dispermic fertilization of human oocytes. *J. Electron Microsc. Tech.* **17,** 437–449.

Van Blerkom, J. and Henry, G. (1992). Oocyte dysmorphism and aneuploidy in meioticaly mature human oocytes after stimulation. *Hum. Reprod.* **7,** 379–390.

Van Blerkom, J., Henry, G., and Porrecco, R. (1984). Preimplantation human embryonic development from polypronuclear eggs after in vitro fertilization. *Fertil. Steril.* **41,** 686–696.

Vandre, D. D., and Borisy, G. G. (1989). Anaphase onset and dephosphorylation of mitotic phosphoproteins occur concommitantly. *J. Cell Sci.* **94,** 245–258.

Van Steirteghem, A. C., Van der Elst, J., Van der Abbeel, E., Joris, H., Camus, M., and Devroey, P. (1994). Cryopreservation of supernumary multicellular human embryos obtained after intracytoplasmic sperm injection. *Fertil. Steril.* **62,** 775–780.

Verde, F., Labbé, J.-C., Doree, M., and Karsenti, E. (1990). Regulation of microtubule dynamics by cdc2 protein kinase in cell-free extracts of *Xenopus* eggs. *Nature (London)* **343,** 233–238.

Vincent, C., Pickering, S. J., Johnson, M. H., and Quick, S. J. (1990). Dimethylsulfoxide affects the organisation of microfilaments in the mouse oocyte. *Mol. Reprod. Dev.* **26,** 227–235.

Wales, R. G., Whittingham, D. G., Hardy, K., and Craft, I. L. (1987). Metabolism of glucose by human embryos. *J. Reprod. Fertil.* **79,** 289–297.

Wang, X. J., Ledger, W., Payne, D., Jeffrey, R., and Matthews, C. D. (1994). The contribution of embryo cryopreservation to in-vitro fertilization/gamete intra-fallopian transfer: 8 years experience. *Hum. Reprod.* **9,** 103–109.

Watson, A. J., Hogan, A., Hahnel, A., Wiemer, K. E., and Schultz, G. A. (1992). Expression of growth factor ligand and receptor genes in the preimplantation bovine embryo. *Mol. Reprod. Dev.* **31,** 87–95.

Weinert, T. A., and Hartwell, L. H. (1988). The RAD9 gene controls the cell cycle response to DNA damage in *Saccharomyces cerevisiae. Science* **241,** 317–322.

Wiemer, K. E., Cohen, J., Amborski, G. F., Munuyakani, L., Wiker, S., and Godke, R. A. (1989). In vitro development and implantation of human embryos following culture on fetal uterine fibroblast cells. *Hum. Reprod.* **4,** 595–600.

Wiemer, K. E., Hoffman, D. I., Maxson, W. S., Eager, S., Muhlberger, B., Fiore, I., and Cuervo, M. (1993a). Embryonic morphology and rate of implantation of human embryos following co-culture on bovine oviductal epithelial cells. *Hum. Reprod.* **8,** 97–101.

Wiemer, K. E., Hu, Y., Cuervo, M., Genetis, P., and Leibowitz, D. (1993b). The combination of co-culture and selective assisted hatching: results from their clinical application. *Fertil. Steril.* **61,** 105–110.

Wiker, S., Malter, H., Wright, G., and Cohen, J. (1990). Recognition of paternal pronuclei in human zygotes. *J. In Vitro Fertil. Embryol. Transfus.* **7,** 33–37.

Winston, N. J. (1992). Approaches to the study of early human embryo development. Ph.D. Thesis, Cambridge University, Cambridge, UK.

Winston, N. J., and Johnson, M. H. (1992). Can the mouse embryo provide a good model for the study of abnormal cellular development seen in human embryos? *Hum. Reprod.* **7,** 1291–1296.

Winston, N. J., Braude, P. R., Pickering, S. J., George, M. A., Cant, A., Currie, J., and Johnson, M. H. (1991a). The incidence of abnormal morphology and nucleocytoplasmic ratios in 2, 3 and 5 day human pre-embryos. *Hum. Reprod.* **6,** 17–24.

Winston, N. J., Johnson, M. H., Pickering, S. J., and Braude, P. R. (1991b). Parthenogenetic activation and development of fresh and aged human oocytes. *Fertil. Steril.* **56,** 904–912.

Winston, N. J., Braude, P. R. and Johnson, M. H. (1993). Are non-fertilized human oocytes useful? *Hum. Reprod.* **8,** 503–507.

Wong, P. C., Bongso, A., Ng, S. C., Chan, C. L. K., Hagglund, L., Anandakumar, C., Wong, Y. C., and Ratnam, S. S. (1988). Pregnancies after human tubal embryo transfer (TET): A new method of infertility treatment. *Singapore J. Obstet. Gynaecol.* **19,** 41–43.

Wramsby, H., and Fredga, K. (1987). Chromosome analysis of human oocytes failing to cleave after insemination in vitro. *Hum. Reprod.* **2,** 137–142.

Wramsby, H., Fredga, K., and Liedholm, P. (1987). Chromosome analysis of human oocytes recovered from preovulatory follicles in stimulated cycles. *N. Engl. J. Med.* **316,** 121–124.

Wright, G., Wiker, S., Elsner, C., Kort, H., Massey, J., Mitchell, D., Toledo, A., and Cohen, J. (1990). Observations on the porphology of human zygotes, pronuclei and nucleoli and implications for cryopreservation. *Hum. Reprod.* **5,** 109–115.

Yaron, Y., Botchen, A., Amit, A., Kogosowski, A., Yovel, I., and Lessing, J. B. (1993). Endometrial receptivity: The age-related decline in pregnancy rates and the effect of ovarian function. *Fertil. Steril.* **60,** 314–318.

Yovich, J. L., and Matson, P. L. (1988). Early pregnancy wastage after gamete manipulation. *Br. J. Obstet. Gynaecol.* **95,** 1120–1127.

Yovich, J. L., and Stanger, J. D. (1984). The limitations of in vitro fertilization from males with severe oligospermis and abnormal sperm morphology. *J. In Vitro Fertil. Embryol. Transfus.* **1,** 172–179.

G-Protein-Coupled Receptors in Insect Cells

Jozef J. M. Vanden Broeck
Laboratory for Developmental Physiology and Molecular Biology, Department of Biology, Zoological Institute K. U. Leuven, B-3000 Leuven, Belgium

The main classes of transmembrane signaling receptor proteins are well conserved during evolution and are encountered in vertebrates as well as in invertebrates. All members of the G-protein-coupled receptor superfamily share a number of basic structural and functional characteristics. In both insects and mammals, this receptor class is involved in the perception and transduction of many important extracellular signals, including a great deal of paracrine, endocrine, and neuronal messengers and visual, olfactory and gustatory stimuli. Therefore, most of the receptor subclasses appear to have originated several hundred million years ago, before the divergence of the major animal Phyla took place. Nevertheless, many insect-specific molecular interactions are encountered and these could become interesting tools for future applications, e.g., in insect pest control. Insect cell lines are well suited for large-scale expression and characterization of cloned receptor genes. Furthermore, novel methods for the production of stably transformed insect cells may form a major breakthrough for insect signal transduction research.

KEY WORDS: G-protein-coupled receptors, Insects, Amines, Peptides, Signal transduction, Gene expression, Molecular cloning.

I. Introduction

All organisms "communicate" with their environment. In metazoans, the activities of the multitude of cells which form together the individual animal are well coordinated by contact, paracrine, endocrine, and neuronal communication mechanisms. The physiological processes that occur in living cells can be adapted to meet external circumstances. In this way environmental and intercellular messages are translated into intracellular responses via signal-transduction pathways and signaling networks. Usually these responses are initiated by the interaction of an extracellular signaling molecule with a cellular receptor protein. This receptor is not a simple binding protein, but it has the properties of a signal transducer, which can elicit further physiological effects. It is the aim of this chapter to take a look at the signal-transducing receptor types which are encountered in insect cells. Molecular cloning, biochemical characterizations, physiological measurements, and genetic analyses indicate that the basic intracellular signaling events and the major components involved in this signaling are well conserved during evolution. Therefore, the main receptor types found to be important in mammals also occur in insect cells. In this chapter, the general features of different receptor families are discussed. Insect members of almost all these receptor families have already been discovered: ligand-gated ion channels, receptor protein kinases, receptor protein phosphatases, guanylate cyclases, cytokine receptors, cell contact receptors, nuclear receptors, and G-protein-coupled receptors. A number of intracellular regulators of signaling processes such as small (e.g., p21-Ras) and large (heterotrimeric) guanine nucleotide binding (G) proteins are known to play a "key" role in both mammalian and insect signal transduction.

Insect endocrine and nervous systems use a large number of distinct messenger molecules which act as hormones, transmitters, or modulators: a multitude of amino acid, amine, and peptide messengers are present in insect tissues. Novel methodologies have led to the purification and sequencing of insect neuropeptides. A single messenger may elicit pleiotropic responses, whereas a single effect can be controlled by a large number of distinct messenger molecules. It is therefore a major challenge for insect physiologists to explain this complex situation by characterizing the specific receptor proteins. Pharmacological structure and activity studies, radioligand binding assays, and molecular cloning of receptor genes have already contributed to our insight in the complexity of neuroendocrine signaling. Insect-specific features of molecular interactions can become important for pest control or for other applications.

In laboratories all over the world, insect baculoviral expression systems are extensively used for large-scale production of recombinant proteins. Novel methods for the creation of stably transformed insect cell lines have now been developed. These cell lines are well suited for the expression of correctly modified membrane proteins. In addition, several aspects of insect signal-transduction processes can be studied in these cell lines.

II. Features of Receptor Families and Examples

A. Ligand-Gated Ion Channels

1. Nicotinic Acetylcholine Receptors

In a wide variety of metazoans, acetylcholine (ACh) is an abundant neuronal signaling molecule. In general, it is formed from choline and acetyl-CoA by the activity of a choline-acetyltransferase enzyme (ChAT). ACh acts as a transmitter in the central nervous system (CNS) of both vertebrates and invertebrates. It is the most important transmitter at the vertebrate neuromuscular junction, but in most invertebrates this is not the case. Extracellular ACh is degraded by acetylcholinesterase (AChE), and choline, the product of this degradation, is taken up in the nerve endings by a choline carrier. In insects, AChE is the target protein for insecticides such as organophosphates and carbamates.

The effects of ACh can be attributed to the presence of two major classes of receptors. They are named after the agonists that allow us to distinguish these receptors: (1) the nicotinic ACh receptor (nAChR), a ligand-gated cation channel, is formed by the interaction of multiple integral membrane subunits and (2) the muscarinic ACh receptor (mAChR) exerts its effects through the activation of G-proteins and thus belongs to another superfamily of receptors (G-protein-coupled receptors). Both types are found in vertebrates as well as invertebrates. In the vertebrate nervous system, the mAChR is more abundant than the nAChR. In insects, the opposite is found: nAChR is highly abundant in the CNS, reaching almost the same concentrations as in electric eel tissue. Activation of insect nAChR by binding of ACh results in an increase of membrane permeability for cations. A high-affinity antagonist is the snake toxin α-bungarotoxin. The insect nAChRs have been investigated by electrophysiological methods (e.g., patch clamp), by radio-ligand binding studies, and by further biochemical analysis (purification and molecular cloning studies). Expression of nACh-RNAs in *Xenopus* oocytes allows the electrophysiological characterization of the receptor and its subunits. Insect nAChR is related to vertebrate nAChR, as shown by pharmacological and binding profiles (Breer and Sattelle, 1987). Molecular cloning of nicotinic receptor subunits revealed structural conservation through the presence of four putative hydrophobic, membrane-spanning regions, and sequence similarity with mammalian subunits was observed (Hermans-Borgmeyer *et al.*, 1986; Wadsworth *et al.*, 1988). In *Drosophila* as well as in *Schistocerca*, α-like subunits have been cloned (Bossy *et al.*, 1988; Sawruk *et al.*, 1990; Marshall *et al.*, 1990). Nicotinic cationic channels produced in *Xenopus* oocytes (e.g., SAD or AL1) can be blocked by α-bungarotoxin. Although the electrophysiological and pharmacological data obtained in *Xenopus* oocytes are consistent with analyses of insect nAChR receptors, it is not unlikely that *in vivo* additional subunit types are involved in nAChR subunit oligomerization. Non-α subunits have indeed been identified and are probably present *in vivo* in hetero-oligomeric nAChR complexes (e.g., ARD, SBD, Schloss *et al.,* 1988). In different *in vivo* studies on insect tissues, more than one type of nAChR is found. In preparations of cultured insect neurons, at least two major classes of nAChR were observed by single-channel recordings (Albert and Lingle, 1993). Such differences in conductance properties might be correlated with the existence of synaptic and extrajunctional receptor sites. In the CNS of adult locusts, nAChR localization was analyzed by immunocytochemistry. Immunopositive labeling was present in synapses and at extrajunctional sites; some antibody binding was also observed in the cytoplasm of neuronal cell bodies at structures associated with Golgi complexes and at the membranes of particular types of glial cells (Leitch *et al.,* 1993).

2. Amino Acid Receptors

a. L-Glutamate   L-Glutamate is a transmitter which is used for many excitatory responses in the vertebrate nervous system. Distinct receptor subclasses have been identified in the vertebrate brain and these are characterized by their sensitivity to a variety of ligands: kainate (KA), α-amino-3-hydroxy-5-methyl-isooxazol-4-propionate (AMPA), N-methyl-D-aspartate (NMDA), L-2-amino-4-phosphono-butyrate (APBu) and 1-amino-cyclopentyl-1,3-dicarboxylate (ACPD). Four of these subclasses are type I receptors (ligand-gated ion channels), one is type II (ACPD) and is linked to the inositol triphosphate pathway. In insects, L-Glu is an important transmitter of neuromuscular junctions. It can also induce inhibitory responses via an L-Glu-gated Cl^- channel which has no vertebrate counterpart. Electrophysiological and binding data indicate that there are indeed at least two major classes of glutamate receptor channels in insects: the cation-selective D-type channel (*d*epolarization) and the anion-selective H-type (*h*yperpolarization). L-Glu and ibotenate are the most effective agonists. There are probably several subtypes in the CNS, at neuromuscular junctions and on extrajunctional locations. Molecular cloning studies have indicated that at least some insect GluR subunits show clear sequence similarity to their vertebrate counterparts. In *Drosophila melanogaster,* certain GluR homologs were cloned and sequenced: DGluR-I, a kainate receptor type from the CNS; DGluR-II, a novel ionotropic subtype derived from developing muscle, and DNMDAR-I, a putative NMDA receptor expressed in the head of adult flies. Attempts to get an entirely functional NMDA receptor by expression of this DNMDAR-I subunit in *Xenopus* oocytes failed, indicating that other subunits may be needed to assemble a functional *Drosophila* NMDA receptor protein (Ultsch *et al.*, 1993). NMDA receptors are probably involved in long-term potentiation phenomena, learning, and memory consolidation as shown by Jaffe and Blanco (1994) in the cricket.

Xenopus oocytes have been used to monitor the expression of nAChR and amino acid receptor-channel mRNAs. The expression system is well suited for electrophysiological characterization and expression of channel proteins. Desert locust (*Schistocerca gregaria*) leg muscle mRNA expressed in *Xenopus* oocytes allowed the characterization of several distinct types of amino acid (especially L-Glu and GABA) receptors (Fraser *et al.*, 1990): receptor channels gated by the agonists L-Glu, L-quisqualate, DL-ibotenate, and gamma-aminobutyric acid (GABA) were identified. The pharmacology of the expressed receptors is similar to the *in vivo* locust muscle receptors.

The hindleg skeletal muscles of the locust receive excitatory, inhibitory, and modulatory inputs. At insect excitatory neuromuscular synapses L-Glu is probably the most commonly used transmitter substance. The postjunctional excitatory L-Glu receptor is a cation-selective channel of the D-type (depolarization) and in most cases it is quisqualate sensitive (although other subpopulations of ibotenate and Asp-sensitive receptors have been found). Extrajunctional receptors with distinct sensitivities have been found: D-type (quisqualate-sensitive), H-type (ibotenate-sensitive hyperpolarizing Cl⁻ channels) and probably also metabotropic L-Glu receptors (affecting second messenger levels). NMDA-sensitive cationic channels are present on presynaptic locations. Metabotropic receptors might be involved in some actions of L-Glu. L-Glu induces adenylate and guanylate cyclase activation in dipteran body-wall muscles and some L-Glu effects can be mimicked by cAMP (Usherwood, 1994).

b. GABA In both vertebrates and invertebrates, GABA is the most common inhibitory amino acid transmitter of the nervous system. Vertebrate GABA receptors are restricted to the CNS and belong to two major classes: GABA-A-Rs are GABA-gated ion channels composed of several subunits, and GABA-B-Rs are G-protein-coupled receptors. The multisubunit hetero-oligomeric proteins (GABA-A-Rs) are integral Cl⁻ channels which bind a variety of agonists [GABA, muscimol, piperidine-4-sulfonic acid (P4S), tetrahydroisoxazolopyridinol (THIP), 3-aminopropane sulfonic acid (3APS), isoguvacine] and antagonists (bicuculline, and noncompetitor picrotoxin). Modulatory effects are observed for barbiturates, benzodiazepines, and steroids. Multiple isoforms of the subunits exist (e.g., generated by alternative splicing of mRNAs) and this creates a large potential for receptor diversity. GABA-B-Rs act via G-protein activation and are coupled to cationic channels to change the conductance for Ca^{2+} and K^+. They are sensitive to baclofen.

Inhibitory neuromuscular synapses exist in insects and they generally contain GABA, which is the most common inhibitory signaling molecule. These GABA-A-like receptors are picrotoxin-sensitive (block) and in most cases they are insensitive to bicuculline and baclofen. This was also shown in the *Xenopus* oocyte expression system when locust hindleg mRNAs were introduced (Fraser *et al.,* 1990). Neuronal GABA responses have been widely studied in insect preparations or on cultured neurons. These have an agonist profile similar to that of vertebrate GABA-A-R-mediated effects and can be modulated by barbiturates and benzodiazepines. However, some differences with GABA-A-Rs have also been clearly demonstrated: larger anions can be gated; there is usually no block with bicuculline; and the pharmacological profile differs for 3APS, P4S, and THIP, which are less active or inactive (Lummis, 1990). Insect binding sites for tritiated GABA or muscimol have been shown to be similar in most preparations. Nevertheless, there may be more than one type of GABA receptor in the insect CNS, but it is not entirely clear whether GABA-B-like receptors occur in insects.

Benzodiazepine sites are in three-to-fourfold excess compared with GABA sites. Unlike vertebrate sites, tritiated flunitrazepam binding is displaced most potently by the agent Ro5-4864. Tritiated Ro5-4864 binding in vertebrates is not linked to a GABA-associated benzodiazepine site but to another type of peripheral benzodiazepine binding site located on mitochondrial membranes. In contrast, all the insect [^{3}H]benzodiazepine binding sites investigated so far can be modulated by GABA (binding is enhanced by GABA). Convulsant binding sites have also been studied by using radiolabeled DHP α-dihydropicrotoxinin (DHP), tetrabicyclophosphorothionate (TBPS), propylbicyclophosphate (PrBP), and butyl(cyanophenyl)-trioxabicyclo-octane (BuOB). In general, there are fewer convulsant sites than GABA and benzodiazepine sites. The sites are modulated by GABA in at least some preparations, but the effects seem to differ (from enhancement to inhibition of binding by GABA). It is not entirely clear whether these binding sites are present on the GABA receptor complex. Steroids and ecdysteroids are almost without effect on GABA-induced responses in insects, at least in the cases where this was tested.

GABA receptors are a target for some classes of insecticides: cyclodiene, avermectin, and others (Casida, 1993). Cyclodiene (CD) and lindane-resistant cockroaches are resistant to picrotoxin as well. CDs inhibit Cl^- uptake by cockroach muscle. DHP binding sites are fewer in resistant cockroaches. Chlorinated CDs and cyclohexanes inhibit tritiated DHP binding in cockroaches and TBPS binding in houseflies. Cyclohexane and the CD, endrin, can block GABA responses in motoneuron D1 of the cockroach. Selective screening of dieldrin-resistant, mutant strains of *Lucilia cuprina* (Diptera, Calliphoridae) after ethyl methanesulfonate (EMS) mutagenesis of a susceptible strain resulted in four strains that were phenotypically similar. The genetic basis was monogenic and results pointed toward a single locus, Rdl, determining the resistant status of each strain. This indicates that the number of monogenic options for resistance to a particular chemical can be very limited (Smyth *et al.*, 1992). The effects of avermectin appear to be more complex. In locust muscle, dihydroavermectin (DHAVM) acts on at least two sites, one of which is a GABA receptor Cl^- channel. AVM can act as a GABA mimetic on cockroach nerve cord, but the exact modes of action are not entirely known yet. Some effects of pyrethroid- and DDT-like compounds on insect GABA responses might be secondary effects, since voltage-gated Na^+ channels are the primary target (Lummis, 1990; Sattelle, 1990, 1992).

Molecular cloning of GABA receptor channel subunits has shown conservation of membrane-spanning regions in both vertebrates and insects. Henderson *et al.* (1994) identified three *Drosophila* members of the ligand-gated chloride channel family via molecular cloning. Cyclodiene resistance genes have been characterized as GABA receptor subunits (e.g., Rdl gene of *Drosophila melanogaster,* Lee *et al.*, 1993; Bloomquist *et al.*, 1992; ffrench-Constant, 1994). Expression of the Rdl subunit cRNA leads to the formation of highly functional GABA-gated Cl^- channels. The pharmacology of this receptor subunit (with its possible splice variants) expressed in a baculoviral system is the same as in cultured insect neurons. An Ala^{-320} to Ser replacement in the M2 transmembrane domain is responsible for the insensitivity to picrotoxin and cyclodiene insecticides. In cyclodiene-resistant yellow fever mosquitoes (*Aedes aegypti*), the identical Ala-Ser substitution was found in the M2 domain of the GABA receptor (Thompson *et al.*, 1993). This resistance mechanism, based on a single point mutation, was confirmed in a number of other insect species (ffrench-Constant, 1994).

B. Receptor Tyrosine Kinases (RTK)

1. General Properties

In mammals an important number of growth factor receptors combine the properties of transmembrane signaling receptor proteins with those of protein Tyr kinases which are able to phosphorylate "themselves" (autophosphorylation) as well as a number of crucial intracellular signaling proteins at certain tyrosine residues. These receptors are generally called receptor tyrosine kinases (RTKs) and are usually involved in certain developmentally important processes by controlling cell growth and mitogenesis, transformation or differentiation, although other functions may be encountered in more differentiated cell populations. The predominant role of most growth factor receptors is to stimulate cell growth and proliferation. Depending on the situation, RTKs can also be involved in the establishment of the opposite, i.e., growth arrest and differentiation. The RTKs have a similar molecular topology, consisting of three major domains. They possess a large glycosylated extracellular part which is responsible for ligand binding, a single membrane-spanning region, and a cytoplasmic domain that contains a tyrosine kinase catalytic domain. The cytoplasmic part also has distinct peptide sequence stretches with Tyr, Ser and/or Thr phorphorylation sites. RTKs can be classified into several subclasses as reviewed by Ullrich and Schlessinger (1990). The cytokine, growth hormone, prolactin, erythropoietin etc. receptor family members have a relatively short cytoplasmic domain and appear to mediate their functions via the activation of associated cytoplasmic PTKs (Kelly *et al.,* 1991).

Ligand-induced activation of kinase activity is mediated by receptor dimerization (oligomerization). Ligand binding stabilizes the interaction between receptor "monomers" and leads to the activation of the adjacent kinase domains. Some receptors, such as the insulin receptor, have a heterotetrameric structure stabilized by disulfide bonds. In other cases, the ligand itself is bivalent (e.g., growth hormone) or is a homo- or heterodimer (e.g., PDGF), thus facilitating the homo- or heterodimerization of its receptors. Receptor dimerization is most probably a general property of all growth factor receptors (Schlessinger and Ullrich, 1992). RTK dimerization appears to be essential for kinase activation. Autophosphorylation of ligand-activated receptors and transphosphorylation (of a related but associated receptor in heterodimers or -oligomers) is the result of an intermolecular reaction mechanism. It establishes a conformation that is competent to interact with and phosphorylate other cellular substrates: the phosphotyrosine sites in the phosphorylated receptors represent specific binding sites for other target proteins (enzymes, regulatory components of larger protein complexes, and "adaptor" proteins) which are involved in the initiation of several signal-transduction cascades. The regulation of the activity of small GTPases, such as p21-Ras, is an extremely important element in RTK signal transduction (Boguski and McCormick, 1993). Moreover, many protein Ser/Thr kinases have been shown to be activated via growth factor signaling: e.g., MAP/ERK kinases, c-Raf, PKC, ribosomal-S6-protein kinase. Ser/Thr phosphorylations/dephosphorylations of transcription factors have been correlated with eukaryotic gene regulation, induction, or repression. This regulation may mainly occur at three levels of interaction: (1) interaction of transcription factors with cytoskeletal anchor proteins or other conditions influencing cytoplasmic versus nuclear localization; (2) modulation of DNA binding by phosphorylation/dephosphorylation of transcription factors;

(3) modulation of the interaction with the transcriptional machinery (transcriptional activation) (Hunter and Karin, 1992). It is indeed very intriguing to observe that almost every eukaryotic transcription factor that has been analyzed in detail has been proved to contain phosphorylation sites.

2. Data on Insect RTKs

Members of this receptor family were encountered in insects, especially in *Drosophila,* which is a most interesting model species for developmental biologists and geneticists, and were found to be involved in the same kind of "key processes": the RTK family has an ancient origin and its functional importance has been conserved during animal evolution.

Thompson *et al.* (1985) described the possible presence of an insulin–EGF receptor type by performing immunoblots and affinity labeling studies. Petruzzelli *et al.* (1986) isolated a genomic sequence that showed homology to the kinase domain of the human insulin receptor. The phosphorylation of the protein at tyrosine residues, as detected with an antipeptide antibody, was dependent on insulin administration. Insulin homologs have been identified in insects (small PTTH of *Bombyx; Locusta* insulin-like factor), but the *in vivo* physiological effects of these endogenous peptide hormones are not entirely elucidated and cannot yet be linked to the activation of a certain receptor type. On the other hand, there is no evidence available for the presence of an EGF-like ligand, although a *Drosophila* EGF receptor homolog was cloned and identified (DER; Livneh *et al.,* 1985; Schejter *et al.,* 1986; Wadsworth *et al.,* 1986). EGF-like repeats are present in several larger proteins (such as Notch, Delta, Serrate, Slit) and are thought to act as adhesion molecules involved in developmental processes regulated by cell–cell contact (Wharton *et al.,* 1985; Knust *et al.,* 1987; Vässin *et al.,* 1987; Johansen *et al.,* 1989; Rothberg *et al.,* 1990). The DER receptor protein is preferentially expressed in certain mitotic cell populations during embryonic and larval development, and oogenesis (Kammermeyer and Wadsworth, 1987). This DER function in mitotic cell signaling has probably been well conserved during evolution. Indeed, DER mutations (torpedo gene) disrupt a variety of developmental processes in *Drosophila,* indicating that the receptor may have different activities at different developmental stages. A number of mutants and the resulting molecular lesions, tissue-specific effects, oligomerization of DER receptors, and the phosphorylation of multiple substrates are discussed by Clifford and Schüpbach (1994).

Another interesting example of an insect RTK member is the product of the *Drosophila* sevenless gene (Hafen *et al.,* 1987). This gene is involved in the signaling that induces the establishment of cell fate in the eye: R7 photoreceptor cells are specified by *sev* gene action. Cell fate in the developing eye is determined by a cascade of inductive interactions. The correct assembly of a visual unit, an ommatidium, requires a number of genes. Bride of sevenless (*boss*) and sevenless (*sev*) are involved as mutations in which R7 fails to develop correctly. Instead it becomes a non-neuronal cone cell. Boss, the membrane-bound ligand of Sev, is needed in R8 for proper R7 development and acts as an inductive signal (Basler *et al.*, 1991). It consists of seven putative transmembrane segments and has a large extracellular portion (Hart *et al.*, 1990). A number of other proteins are involved in producing a correct Sevenless response: E(sev) gene products are most probably involved in the intracellular signaling cascade down-stream of activated Sevenless receptor tyrosine kinase. Some of the 7 E(sev) genes are now identified and encode "key" proteins such as Ras1 GTPase, Sos ("Son of sevenless") protein, Drk SH3-SH2-SH3 domain protein, Dm Hsp83 stress protein, and Dm Cdc37 (Cutforth and Rubin, 1994). Thus, these elegant genetic studies have revealed that several signal-transduction molecules found in vertebrates to act downstream of RTKs are also involved in Sev signaling. Ligand binding at RTKs generally leads to autophosphory-lation and to binding and phosphorylation of a number of cellular signaling factors that have SH2 (Src-Homology region 2) domains. These domains are found in a number of important proteins, including phospholipases, phosphatases, GAP, cytoskeletal binding proteins, and cytoplasmic kinases. Sos functions downstream of Sev and also of DER. It is homologous to other exchange proteins that activate Ras by inducing loading of GTP. Ras proteins are small GTPases which play a most prominent role in RTK signaling: C-ras affects mitogenesis, neurite outgrowth, c-fos expression, and gene expression. Raf1-protein kinase probably acts as a protein kinase at steps subsequent to Ras activation (Pazin and Williams, 1992). The ETS domain protein Pointed-P2 is a nuclear target in the sevenless signaling cascade downstream of Rolled/mitogen-activated protein (MAP) kinase (Brunner *et al.*, 1994).

Other receptor tyrosine kinases have been described in *Drosophila*. These include a homolog of the mammalian FGF receptor (DFGFr) and Torso, a RTK that is needed for the correct induction of other genes that specify anterior or posterior terminals of the developing embryo (Casanova and Struhl, 1993). This receptor tyrosine kinase is activated only in the vicinity of the poles, although it is uniformly distributed. The product of the torso-like gene (tsl) is most probably the ligand for Torso. The Tsl protein is secreted and spatially restricted to form a concentration gradient (Martin *et al.*, 1994).

In fruit fly embryos, a homolog of the mammalian Ror1 and Ror2 RTKs, which are members of the Trk neurotrophin receptor family, is specifically expressed in the nervous system. This homolog may represent an insect neurotrophin receptor that functions in the early stages of neural develop-ment. Neurotrophins are a group of small secreted proteins that promote neuronal cell survival, proliferation, and differentiation. Examples of vertebrate members of this subfamily of growth factors are nerve growth factor (NGF), brain-derived neurotrophic factor (BDNF), and additional factors identified by molecular cloning such as the neurotrophins NT-3, NT-4, and NT-5. All these neurotrophins are capable of promoting survival and differentiation of sensory neurons, but their activities may affect markedly distinct, or overlapping, populations of neuronal cells. The neurotrophin receptors belong to the Trk receptor subfamily: e.g., NGF-Trk, TrkB, TrkC (Chao, 1992). Insect members of this receptor subfamily of protein tyrosine kinases have been identified: e.g., DTrk and DRor (Wilson *et al.*, 1993).

3. RTK Signaling Pathways

During the past few years, much information on specific and selective protein–protein interactions induced by receptor activation has been reported. The intracellular signal-transduction pathways and networks are still intensively studied through physiological, biochemical, molecular biological, and genetic analyses. Many growth and differentiation factors, hormonal polypeptides, cytokines, antigens, and extracellular matrix compounds exert their functions by binding to transmembrane receptors that signal through associated protein kinase domains that are present at the cytoplasmic side. Although the targets or substrates of these kinases (protein tyrosine kinases, PTKs) may mediate quite different cellular functions, they usually contain short regions of homology. These regions, e.g., SH2 (Src-homology 2), SH3 (Src-homology 3), or PH (pleckstrin homology) domains, are generally involved in the recognition of certain short (peptide) motifs and thus allow these proteins to interact specifically with (a number of) other signaling proteins or even to form larger protein complexes. The SH2 domains recognize certain peptide sequence motifs that contain phosphotyrosine residues. Therefore these domains are especially important in tyrosine kinase signaling pathways. SH3 domains interact with Pro-rich motifs, while PH domains are thought to be crucial for recruitment of the protein to a membrane-bound state. There are good indications that PH domains might interact with $\beta\gamma$ subunits of heterotrimeric G-proteins. Most of the signaling proteins contain several of these domains or their binding motifs, indicating that RTK activation initiates complicated protein–protein interactions. The establishment of such signaling networks leads to the regulation of a number of cellular processes which are linked and interdependent. An interesting achievement in this respect is the identification of adaptor proteins such as Sem 5 in *Caenorhabditis elegans,* Drk in *Drosophila melanogaster,* or Grb2 and Crk in mammals. These adaptors are formed by an SH2 domain flanked by two SH3 domains and their role is to link proteins together. Drk is thought to be involved in *Drosophila*

DER, Torso, and Sevenless signaling. Another important factor is Sos (son of sevenless), which is a guanine nucleotide release protein (GNRP) containing proline-rich motifs that bind to the Drk SH3 domains. Functioning as an exchange factor, Sos may then activate Ras GTP-ase. GTP-bound Ras triggers a protein Ser/Thr kinase cascade/network that eventually leads to the phosphorylation of transcription factors by MAP kinase. Not only Ras activation, but also its inactivation is regulated: RasGAPs (GTPase-activating proteins) enhance the rate of hydrolysis of GTP to GDP. P120-GAP contains SH2, SH3, and PH domains and binds to certain phospho-Tyr proteins, including growth factor receptors. The basic steps arising from RTK stimulation and the routes that lead to the (in)activation of the "small" G-protein Ras appear to be basically conserved throughout all metazoans. Moreover, the basic protein modules and motifs have been conserved during evolution (Pawson, 1995). Protein–protein interactions appear to play a major role in many signal-transduction events. Growth and differentiation factors can induce pleiotropic effects that lead to major cell physiological changes. These seem to be regulated via versatile networks of interactions and biochemical reactions, including the phosphorylation and dephosphorylation of proteins (Freeman and Donoghue, 1991). Oncoproteins span almost the entire network of intracellular signaling factors, ranging from growth factors over RTKs, cytoplasmic protein kinases and GTP-binding proteins (G-proteins) to nuclear transcription factors (Pouyssegur and Seuwen, 1992). "Key," "switch," or "integrator" proteins, which sense intra- and/or extracellular signaling events, connect different pathways and coordinate the establishment of cellular responses including mitogenesis, transcriptional control at the level of the genes, mRNA translation, protein trafficking and subcellular localization, control of cell architecture, cell–cell interactions, ionic and/or electrical changes, and release of extracellular signaling factors. Different responses can be obtained within distinct tissues.

C. Ser/Thr Kinase Receptors

The TGFβ superfamily of cytokines includes a number of differentiation and growth factors that are extremely important in developmental, especially inductive, processes. In vertebrates, this family includes TGFβ-like factors, activins, inhibins, Müllerian duct-inhibiting substance, several bone morphogenetic proteins (BMPs) and the products of the Vg-related genes. The corresponding receptors, which contain an extracellular part for ligand binding, a single membrane-spanning region, and a Ser/Thr kinase domain at their intracellular region, are divided into two families: type I and type II. Type I receptors require association with type II receptors for binding, while type II ones can independently bind the TGFβ or activin-like ligand

(Mathews and Vale, 1991; Donaldson *et al.*, 1992). The mechanism of TGFβ receptor activation has been proposed by Wrana *et al.* (1994). TGFβ binds directly to rec.II, then rec.I is recruited into the complex and becomes phosphorylated by rec.II. Rec.I propagates the signal to downstream signaling molecules. Therefore both receptors are required for proper signaling in order to produce any response. Rec.II is a constitutively (autophosphorylation) active kinase which uses the ligand to recruit, phosphorylate, and signal through rec.I.

In insects, certain differentiation and growth factors also belong to the TGFβ superfamily: Dpp (decapentaplegic), Scw (screw), and 60A factors in *Drosophila* are related to mammalian bone morphogenetic proteins (Padgett *et al.*, 1987). In metazoans, development seems to be regulated by a fundamental strategy and the many factors involved are functionally and structurally well conserved. Dpp is a secreted signaling protein that organizes cell fate in the dorsoventral axis of *Drosophila* ectoderm. It has also been shown to induce mesodermal gene expression in a later stage of development (Staehling-Hampton *et al.*, 1994).

Receptors that belong to the TGFβ receptor type I and II families have been identified in *Drosophila*. AtrII has been characterized as a fruitfly activin receptor II-like protein containing the typical protein Ser/Thr kinase domain. Maternal AtrII transcript and gene product are abundant in oocytes, and during development they reach their highest levels in the mesoderm and gut (Childs *et al.*, 1993). The AtrII receptor even binds vertebrate activin (type II receptor) with high specificity and affinity. The products of the genes saxophone (*sax*) and thickveins (*tkv*) have been identified as type I receptors (Nellen *et al.*, 1994; Penton *et al.*, 1994; Brummel *et al.*, 1994) and correlated with the Dpp ligand. Padgett *et al.* (1993) have shown that the vertebrate bone morphogenetic protein BMP4 can rescue *Drosophila* dpp mutants, indicating that a noninsect member of the BMP subfamily can exert the same functions as Dpp and can completely replace it *in vivo*.

D. Other Receptors Involved in Inductive Processes

Other interesting examples of evolutionarily conserved signaling have been discovered. One of the most striking is formed by the Toll receptor pathway. Toll protein is an integral membrane receptor that shows some similarity to interleukin-1 receptors. It is involved in the signal-transduction pathway that establishes dorsoventral polarity in the *Drosophila* embryo. The "ventralizing" signal for this receptor is found in the perivitelline space and is probably generated via a Ser-protease cascade (*snake* and *easter* genes; Smith and DeLotto, 1994) that leads to the spatially restricted production of ligand in the perivitelline space at the ventral side of the embryo (Stein *et al.,* 1991). This ligand is probably a derivative of the product of the *spätzle* gene (Schneider *et al.,* 1994). Through a series of intracellular events, Toll activation leads to the dissociation of Dorsal protein from the cytoskeletal Cactus–Dorsal complex. Dorsal is relocalized and forms a dorsoventral gradient in the nuclei. This protein is a Rel/NFκB-like transcription factor and is important for the establishment of dorsoventral polarity in the embryo (St. Johnson and Nüsslein-Volhard, 1992). Mediation of the signal from activated Toll to the Cactus–Dorsal cytoskeletal complex is provided by the products of some other genes, such as Pelle and Tube. Pelle is a Ser/Thr protein kinase and Tube probably functions upstream of Pelle in the signaling pathway. Both proteins interact with each other.

The signaling pathway leading to Dorsal and the compounds involved in it seem to be very well conserved and resemble the signaling of NFκB (Grosshans *et al.,* 1994). A number of other cell–cell communication (and cell adhesion) events have been described in *Drosophila* and will certainly be further explored. Additional types of membrane proteins are identified in the transmembrane signaling processes of insect cells. Other components involved in these processes will have to be identified and their role will have to be elucidated. At this moment, however, it is already quite clear that many signaling compounds in (animal) cells have a conserved structure and function.

Many receptor types first discovered in vertebrates are now encountered in almost all animal phyla (Fig. 1). Transmembrane receptors containing ligand-induced enzymatic activities other than the ones described here have already been identified. These include protein phosphatases [protein tyrosine phosphatase (PTP) receptors have also been discovered in insects; Desai *et al.,* 1994] and guanylate cyclases (Garbers, 1991; Malva *et al.,* 1994). Although most signaling proteins appear to have a very ancient origin, it is not unlikely that certain cell surface receptors show a much less conserved structure and/or mode of action.

E. The Steroid and Orphan Receptor Family of Transcription Factors

Steroid hormone receptors and receptors for other small hydrophobic ligands (e.g., retinoic acid, thyroid hormone) form a family of transcription factors (Evans, 1988; O'Malley, 1990; Carson-Jurica *et al.,* 1990; Laudet *et al.,* 1994). A surprisingly large number of family members for which the ligand (if there is one) is unknown are present in insects. Many of these orphan receptors seem to be involved in embryonic and postembryonic development or in the ecdysteroid-triggered regulatory hierarchies which control molting and metamorphosis (Segraves, 1994). The first receptor

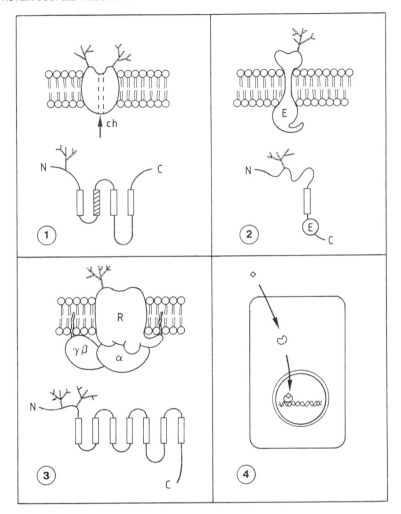

FIG. 1 Schematic overview of signal transducing receptor protein classes. (1) Ligand-gated ion channels consist of several subunits which contain four membrane-spanning regions. The channel is mainly formed by the second (M2) transmembrane domains. (2) Receptor proteins with an intrinsic enzymatic activity domain (E) at the cytoplasmic side. These proteins have a single membrane-spanning region. The extracellular portion of these receptors forms a binding site for the ligand, whereas the intracellular part contains an enzymatic activity (e.g., protein Tyr kinase, protein Ser/Thr kinase, protein phosphatase, guanylate cyclase activity) and regions for interaction with intracellular signaling factors. Some receptor molecules have a short cytoplasmic domain without an intrinsic enzymatic activity. They interact with additional intracellular proteins. (3) G-Protein-coupled receptors possess seven putative transmembrane domains (TMDs). The intracellular parts of the molecule are thought to interact with heterotrimeric G-proteins and with other regulatory molecules. (4) Nuclear receptors for steroid, ecdysteroid, thyroid, and retinoid ligands act as regulators of gene transcriptional activity.

in this hierarchy is the functional ecdysone receptor, which consists of a heterodimer of two steroid-receptor-like proteins: EcR and Usp, the product of the ultraspiracle gene (*usp*) (Yao *et al.*, 1993). The cloned *Drosophila* EcR (Koelle *et al.*, 1991) is by itself not capable of high-affinity DNA binding or transcriptional activation. These activities depend on heterodimer formation with Usp, the insect homolog of the vertebrate retinoid-X receptor (Thomas *et al.*, 1993). As outlined earlier, heterodimerization is a potential source of functional receptor diversification. The existence of isoforms of the combining partners might provide an additional mechanism for generating differential ecdysone responses. Downstream of the functioning of the ecdysone receptor EcR/Usp, three classes of genes appear to be regulated, generating a cascade of gene expression (Andres *et al.*, 1993; Meszaros and Morton, 1994): early or primary response genes, early/late or secondary response genes, and the midprepupal genes. Most of the insect receptor-like proteins appear to be more similar to their vertebrate counterparts than to their insect companions. This is probably an indication of their ancient role in cellular and developmental processes, although the ligands themselves might be much less conserved.

The weak structural similarity and pharmacological overlap between sesquiterpenoid juvenile hormones (JHs) and retinoids suggests that JHs might exert their functions through a member of the steroid–thyroid–retinoid receptor family. In fact, some of the orphan receptors might correspond to some JH ligands (Segraves, 1994). Many attempts have been made to identify JH-binding proteins from insect cells and tissues, but it is not yet entirely clear whether these are true receptors involved in JH signaling or just the carrier proteins (Wyatt, 1990; Riddiford, 1994). It is not unlikely that a number of different receptor types can be (in-)activated) by JHs. Premetamorphic actions of JHs are thought to be mediated by nuclear mechanisms at the level of the DNA. In adult insects, slow, nuclear, and fast JH effects have been reported. Several genes have been shown to be under the control of JHs in the locust fat body, e.g., two vitellogenin genes, 19K hemolymph protein, hexameric storage protein (persistent storage protein, PSP), and a female-specific 21K hemolymph protein (Braun and Wyatt, 1992; Zhang *et al.*, 1993; Wyatt *et al.*, 1994). These JH target genes contain putative JH-REs (responsive elements) as consensus sequences upstream of their transcriptional initiation site at which a "JH-responsive" transcription factor might bind (Wyatt, 1990). It is not clear whether this nuclear effect is directly or indirectly controlled by JHs. A nuclear protein factor of 29 kDa binds JH and might be linked to JH functioning. It is not a member of the steroid receptor superfamily of transcription factors and its exact role in JH signaling is not yet elucidated (Riddiford, 1993).

Non-nuclear mechanisms of action for steroids as well as for JHs have been described. The induction of oocyte maturation in amphibians and fish by progesterone-related steroids, some effects on neuronal activity by steroids, and the change in feeding behavior of crustaceans fed with ecdysteroids are examples of rapid effects induced by steroids. These effects probably originate from interactions exerted at the level of the plasma membrane and/or at the level of the cell matrix (De Loof et al., 1992; De Loof and Vanden Broeck, 1995a; Getzenberg et al., 1990). Ovarian follicle cells of Rhodnius and probably many other insect cells and tissues, such as the male accessory glands of the fruit fly, make use of another mechanism by which JHs produce fast cellular responses (Wyatt et al., 1994).

F. Properties of G-Protein-Coupled Receptors (GPCR)

A large number of transmembrane signaling receptors exerts their functions through the activation of membrane-associated G-proteins. These G-proteins are heterotrimers consisting of three subunits (α, β and γ). In the case of ligand-induced activation of a receptor, G-proteins can regulate the activity of certain target or "effector" proteins. This results in the generation of a cellular response which can be situated at the plasma membrane level (e.g., by regulating G-protein-gated ion channels) as well as at the level of nuclear, transcriptional activity. Many effector proteins are involved in second messenger production or regulation. Therefore, a lot of G-protein-mediated effects produce changes in secondary messengers (cAMP, cGMP, $InsP_3$, Ca^{2+}, DAG, arachidonic acid derivatives), which regulate the activity of other cellular components.

All "G-protein-coupled receptors" (GPCRs) have a similar topographical structure. Although the sequence similarity is not always obvious, they are thought to form a large receptor "superfamily" (Probst et al., 1992). The direct elucidation of the structure of a G-protein-coupled receptor by X-ray diffraction studies has not yet been achieved. All models are based on the structural data that were obtained with bacteriorhodopsin, which is thought to have a similar molecular topology. Computer modeling data are supported by results obtained through complementary approaches such as photoaffinity labeling, site-directed mutagenesis, deletion mutants, chimeric receptors, and the use of synthetic peptides. Based on hydropathy analysis, GPCRs possess seven (or eight) putative membrane-spanning regions (transmembrane domains, TMDs) of 20–30 hydrophobic amino acids forming an α-helical configuration (Fig. 2). The N-terminal polypeptide chain and three inter-TMD loops are situated at the extracellular side of the plasma membrane. This extracellular portion of the protein usually contains glycosylation sites. Some receptors (e.g., adenosine receptors; Olah and Stiles, 1992) have a very short N-terminal chain lacking glycosylation sites. The transmembrane regions are thought to be inserted into the plasma

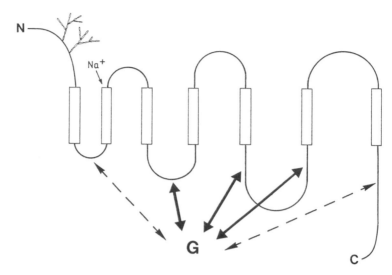

FIG. 2   General scheme of the topography of a G-protein-coupled receptor. The intracellular regions of the protein are thought to specifically interact with a given type of G-protein. The transmembrane regions and certain extracellular parts of the protein can be involved in providing epitopes for ligand-specific interactions.

membrane in such a way that they combine to form a ligand-binding pocket which is open toward the extracellular side. Hydrophilic residues included in the different α-helices are pointing toward each other, facing the pocket and forming the (more or less polar) ligand-binding site. Cationic amines probably interact with a negatively charged and evolutionarily conserved Asp residue in TMD3. The presence of some other residues might enhance the specificity of interaction with the agonist(s): Ser residues in TMD5 of adrenergic receptors probably form H-bonds with *meta*- and *para*-hydroxyl groups on the aromatic ring of the corresponding ligands; and some aromatic residues of TMD5 and 6 probably stabilize the interaction with the aryl ring of adrenergic ligands and serotonin (Hibert *et al.*, 1991). Binding experiments with a chimeric D1/D2 dopamine receptor suggest that the D2-selective agonist, quinpirole, gains much of its receptor subtype selectivity when TMD VI and VII of the D2 receptor are included (MacKenzie *et al.*, 1993).

As shown by mutant analysis, the seventh TMD of the human thromboxane A2 receptor is involved in the discrimination between agonist and antagonist binding sites. A clear distinction could be made between agonist and antagonist binding properties for a receptor variant in which a Trp residue of TMDVII was replaced by a Leu one. In this prostaglandin–eicosanoid receptor subfamily, TMDVII is highly conserved and may indeed form a critical portion of the ligand-binding pocket (Funk *et al.*, 1993). This was also suggested for the rhodopsins where the chromophore, retinal, is attached to a Lys residue of TMDVII. Receptor subclass-specific residues are thought to play a role in ligand binding selectivity or in subclass-specific activation or regulation processes. Similarly, less conserved residues at multiple discontinuous epitopes may contribute to subtype-specific binding selectivities for agonists and antagonists. Amino acid residues conserved in a large number of GPCRs probably have an essential role in protein folding, trafficking, or general receptor functioning. The highly conserved Pro residues in the TMDs might represent kinks or molecular switches involved in generating α-helix distortions and/or agonist-induced conformational changes.

Computer models for G-protein-coupled receptors are generally based on the high-resolution structural data of the 7TMD protein bacteriorhodopsin. Although no such data are available for the 7TMD receptors, these imperfect computer models appear to confirm the existence of a ligand-binding pocket, and several amino acid residues shown to be important for ligand binding by site-directed mutagenesis can also be predicted by the models (Venter *et al.*, 1989; Findlay and Eliopoulos, 1990; Hibert *et al.*, 1991; Brann *et al.*, 1993). Large ligands, such as glycoprotein hormones (FSH, LH, TSH), may interact with certain external parts of the receptor. The N-terminal region of the LH/FSH/TSH receptors is very large, is rich in Cys residues, and contains Leu-rich repeats. It therefore shows some similarities with members of the LRG, Leu-rich glycoprotein, family (Nagayama and Rapoport, 1992). Several Cys residues are probably involved in the formation of disulfide bonds and, by site-directed mutagenesis, some of the glycosylation sites have been shown to be important for the expression of functional receptors. A soluble binding protein formed by alternative RNA splicing was found to correspond to the N-terminal part of the LH/CG receptor. A "secondary" binding event with the membrane-associated regions of the receptor is thought to be essential for signal transduction. In this model, the N-terminal region is needed to bring part of the hormone/extracellular-receptor-site complex into the right conformation and position to interact with the membrane-associated receptor portion (and/or ligand binding pocket) (Quintana *et al.*, 1993).

A similar model has been proposed for the thrombin receptor. The large extracellular N-terminal region of this receptor contains the thrombin binding site as well as a putative thrombin cleavage site. A synthetic peptide, which consists of the N-terminal generated by cleavage, is a potent agonist for receptor (and platelet) activation. This indicates that the thrombin receptor is activated via a proteolytic mechanism. Thrombin binds to the uncleaved receptor and then cleaves the N-terminal part, generating a novel receptor N-terminal. This structure activates the receptor protein via an intramolecular secondary binding event. Indeed, uncleavable mutant receptors do not show any response to thrombin, but they are activated by the synthetic peptide agonist (Vu *et al.*, 1991). Receptors for smaller peptide ligands are probably activated by interaction of part of the peptide with a ligand-binding pocket, whereas other interactions occurring with extracellular parts of the receptors might be important for defining the selectivity or specificity of ligand binding. Studies with chimeric tachykinin receptors may confirm this hypothesis. Tachykinin peptides appear to bind to sites composed of multiple discontinuous regions on their receptors. The specificity of tachykinin variant binding to their respective receptors may reside in distinct receptor parts. By investigations with a number of chimeric NK1/NK3 receptors, Gether *et al.* (1993a) demonstrated that the binding selectivity for substance P and for neurokinin B appears to be present in distinct regions of the receptor. In contrast, the region from TMD V to VII is probably responsible for the fundamental interaction with all tachykinins (Yokota *et al.*, 1992; Nakanishi *et al.*, 1993). Nonpeptide antagonists did not necessarily interact with the same receptor parts and thus, some of these might act via allosteric inhibition mechanisms and not as true peptide mimetics.

Nonconserved residues in two epitopes at the top of TMD V and in one epitope at the top of TMD VI seem to be essential for specific binding of the nonpeptide, NK1 receptor antagonist, CP 96354, but they are not important for binding of substance P (Fong *et al.*, 1993; Gether *et al.*, 1993b). When His residues are introduced at this antagonist binding site, it can be converted into a high-affinity metal ion binding site. Zn^{2+} ions inhibit agonist binding to this mutagenized NK1 receptor and they efficiently block InsP3 production (Elling *et al.*, 1995). Differential binding domains for peptide and nonpeptide ligands were also observed in other peptide receptors. These findings may have important consequences for the development of nonpeptide antagonists or peptide mimetics (Xue *et al.*, 1994).

Many peptide families are characterized by the presence of conserved sequence parts usually located at their C-terminal. Therefore, the specificity of interaction of different members of such a peptide family with their corresponding receptor "subtype" is not provided by these conserved motifs, but by the less conserved peptide parts (the "address" sequences). The conserved motifs in peptide messengers, however, are generally indispensible for the preservation of their biological activity: they coincide with the minimal sequence which may still elicit a biological response (the "message" sequence). Binding of this "active core" region is responsible for receptor activation. This binding event may be situated in the more conserved receptor parts, especially in or next to the membrane-spanning domains, and may involve amino acid residues which are present in all peptide family member receptors. Insights on the binding and activation processes that occur at the level of receptor molecules coincide with the address-message concept that was worked out via structure and function studies on peptide messengers.

Ligand binding induces a conformational change of the receptor protein which is not restricted to the extracellular and intramembranic regions, but which is transduced toward the intracellular loops and C-terminal polypeptide chain. This results in an activation of the associated G-protein by enhancing the exchange rate of GTP for GDP: the GTP-bound G_α subunit releases its $G_{\beta\gamma}$ partner and both can then interact with their respective effector protein(s) to elicit a cellular response (Fig. 3). This implies that

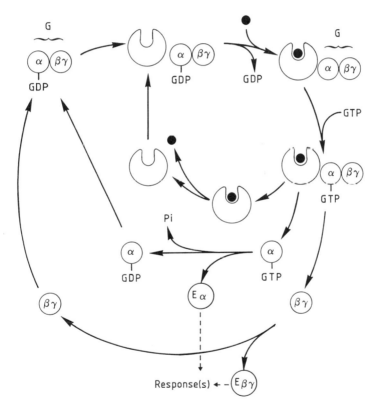

FIG. 3 Simplified scheme showing several steps involved in the production of ligand-induced responses by G-protein-coupled receptors. The ligand-bound receptor protein facilitates binding of GTP to the attached G_α subunit. The GTP-bound α subunit and the $\beta\gamma$ portion dissociate and can then interact with their respective effector proteins (E_α and $E_{\beta\gamma}$) to elicit further cellular responses. After GTP hydrolysis, the GDP-bound α subunit regains its affinity for $\beta\gamma$ and the GTP-ase cycle can start again.

G-protein-mediated effects are subjected to very subtle regulation mechanisms in which several components are involved. According to the intracellular situation and the type of G-protein involved, the G-$_\alpha$ and G-$_{\beta\gamma}$ subunits can influence similar (synergistically or antagonistically) and/or distinct effector systems. In addition, they can regulate each others activity. Furthermore, recent findings indicate that there are important routes that allow "cross-talk" between the major intracellular signaling pathways (Crespo et al., 1994; Chen et al., 1994; Norman et al., 1994; Tsukada et al., 1994).

Site-directed mutagenesis has revealed that certain conserved amino acid residues may play a crucial role in determining the signal-transducing properties of G-protein-coupled receptors. A reduction of signal transduction was obtained by mutation of some conserved residues, e.g., Pro in the second cytoplasmic loop (β-adrenergic receptor), Asp in the second loop adjacent to TMD III (-DRY-sequence in e.g., βAR, m1AChR, α2a-AR and the corresponding Glu in rhodopsin), Lys in the distal part of the third loop of rhodopsin (or the corresponding His in the βAR), the palmitoylated Cys in the C-terminal chain of the β2-AR. These residues might directly interfere with G-protein coupling. Some conserved residues in the TMDs also seem to be involved in determining the signal-transducing properties of receptor family members. A highly conserved Asp in TMDII is important for both agonist binding and signal transduction. This Asp residue is essential for the modulation of agonist binding affinity and signal transduction by monovalent cations (especially Na$^+$). This has been shown by site-directed mutagenesis in amine as well as in glycoprotein receptors (Quintana et al., 1993). Another conserved residue, Cys in TMDVI, is probably involved in signal transduction.

A number of highly conserved prolines and tryptophans were altered by mutational analysis of the m3 muscarinic receptor: site-directed mutagenesis (Pro to Ala and Trp to Phe) demonstrated remarkable changes in different aspects of receptor functioning. Mutation of Pro in TMDV, VI, or VII drastically decreased the B$_{max}$ values, indicating a defective protein traffic leading to fewer receptors being located in the plasma membrane. An m3 mAChR mutant (Pro in TMD IV) resulted in much lower binding affinities, whereas another one (Pro in TMD VII) yielded higher binding affinities but its biological activity was severely reduced. This highly conserved Pro in TMDVII is probably involved in agonist-induced changes in receptor conformation and, thus, in signal transduction. The effects of replacing Trp residues by Phe were less pronounced than the proline substitutions. Reduced binding affinities were observed for the Trp residues in TMDIV and VI (Wess et al., 1993). Site-directed mutagenesis of the conserved Tyr residue in TMD VII of the AT-1A angiotensin II receptor revealed that this residue is essential for coupling of the receptor to phospholipase C. The agonist binding properties of the mutant receptor were unaltered (Marie et al., 1994).

Other mutations have also been reported to have drastic effects on receptor functioning and G-protein activation. Some of these mutations might primarily decrease the efficiency of expression of the receptor protein at the plasma membrane. Others lead to an upregulation or even a constitutive activation that might result in a receptor protein with proto-oncogenic properties. In hyperfunctioning thyroid adenomas, two distinct somatic mutations were detected in the C-terminal portion of the third cytoplasmic loop of the TSH receptor. Expression of these mutants in COS7 cells resulted in higher constitutive activation of adenylate cyclase. In some pituitary adenomas and thyroid tumors, such activating mutations were found to be present in the $G_{S\alpha}$ protein itself. A similar constitutive activation was observed with mutagenesis in the terminal part of the third loop of the $\alpha1$ adrenergic receptor. Replacement of an Ala with Leu and/or of a Lys with His in the C-terminal region of the third cytoplasmic loop of the $\alpha1$-AR led to a remarkable increase in affinity and sensitivity of response. These substitutions might favor an active receptor conformation, resulting in more efficient (constitutive) coupling and G-protein activation even in the absence of agonists or in the presence of very low amounts of the endogenous agonist (Cotecchia *et al.*, 1990). Surprisingly, mutating the identical, conserved Ala residue in the TSHR and the $\alpha1$-AR resulted in a constitutive receptor activation. Mutation of the conserved Asp (or Glu) residue that is present in the same region of the TSHR also results in a higher agonist binding affinity and an increased sensitivity (Parma *et al.*, 1993). Other studies have shown that some GPCRs are potential oncoproteins. The *mas* oncogene was the first (orphan) GPCR with transforming properties that was discovered (Young *et al.*, 1986). It was detected by a tumorigenicity assay in nude mice and its expression in NIH 3T3 cells leads to the induction of foci. Later on, the *mas* oncogene was shown to correspond to an angiotensin receptor (Jackson *et al.*, 1988). In addition, the serotonin (5HT) 1c and muscarinic m1, m3, m5 receptors have been reported to possess a transforming potential under certain conditions (Gutkind *et al.*, 1991).

The specificity of the interaction of a given receptor with its corresponding G-protein(s) has been shown to reside in relatively short stretches of amino acids at membrane-proximal regions in the cytoplasmic portions of the receptor. This was shown by deletion studies and by the use of chimeric receptors. Replacement of the third cytoplasmic loop of the $\beta2$-AR with the corresponding region of an $\alpha2$-receptor gave rise to a chimeric receptor which was negatively coupled to adenylate cyclase. Substitution of the third loop of the $\beta2$-AR with that of the $\alpha1$-AR led to a receptor with $\beta2$ pharmacology that is coupled to an activation of phosphatidylinositol hydrolysis. Third-loop exchange was also sufficient to determine G-protein selectivity of m1/m2 and m2/m3 muscarinic receptor chimeras (Brann *et al.*,

1993). Deletion studies indicate that the cytoplasmic regions proximal to the TMDs are involved in G-protein coupling. The N-terminal of the third intracellular loop, next to TMD V, appeared to be most critical in defining the coupling specificity of muscarinic receptor subtypes.

Similar changes in coupling specificity were obtained with a number of other chimera experiments in which cytoplasmic receptor regions were replaced by others. Some of these regions are thought to contain peptide sequences that may form amphipathic α-helices. Small peptides with a similar amphipathic structure, such as the wasp venom mastoparan, have been shown to directly activate G-proteins and/or to compete with receptors for interacting with G-proteins (Taylor *et al.*, 1994). Point mutations and alanine insertions at the N-terminal of the third cytoplasmic loop of the muscarinic m3 receptor, which disrupt the putative amphipathic α-helical conformation, result in a decrease of PI turnover (Duerson *et al.*, 1993; Blüml *et al.*, 1994). On the other hand, the positive charges present in such a putative amphiphilic helical structure in the C-terminal part of the third cytoplasmic loop do not appear to be required for correct signaling of LH/CG receptors through Gs, but they do seem to be necessary for the expression of the receptor at the plasma membrane (Wang *et al.*, 1993).

The intracellular regions of the receptors are also subject to other interactions which may regulate receptor functioning. Protein kinases can phosphorylate many receptor proteins at certain amino acid residues. Specific G-protein-coupled receptor kinases (GRKs) and arrestins are involved in receptor desensitization mechanisms. Arrestin binding as a result of phosphorylation most probably inhibits coupling to a G-protein. Arrestin subtypes have been observed in insect tissues such as antennae (Raming *et al.*, 1993b) and photoreceptors (Komori *et al.*, 1994). Other receptor modifications have been reported. In some cases, palmitoylation of Cys residues in the C-terminal chain ensures the presence of a covalent attachment site for the plasma membrane and generates an additional intracellular loop which may be structurally important for interaction with the G-protein. The palmitoylated Cys in the cytoplasmic tail of the α2A-AR plays a role in agonist-dependent downregulation (Eason *et al.*, 1994). Other Cys residues can stabilize the three-dimensional structure of certain receptors by forming disulfide bridges.

As already mentioned, GPCRs all share a common general structure, but at the amino acid sequence level the similarities are not always obvious. Some receptors seem to belong to phylogenetically distinct protein families or subfamilies which adopted the same molecular model. Examples of GPCRs which do not show any significant sequence similarity to the classic GPCR group are members of the secretin receptor family (Ishihara *et al.*, 1991) or the metabotropic glutamate receptors. The properties of members of these (more recently discovered) subfamilies have not yet been investigated to the same extent as those of the classic GPCRs. The principles which were found to be important for understanding basic steps such as ligand binding, conformational changes, signal transduction, desensitization, and regulation may perhaps be extended toward these additional GPCRs.

III. G-Protein-Coupled Receptors in Insects

A. Muscarinic Acetylcholine Receptors

[^{3}H]Quinuclidinyl benzilate (QNB) has been shown to be a high-affinity ligand for the muscarinic receptor(s) in many insect species: the K_D of QNB binding is usually in the (sub)nanomolar range. For all insect species tested so far, the number of binding sites is a degree of magnitude lower than for α-bungarotoxin sites (nicotinic receptors). The opposite situation is found in the vertebrate brain. Muscarinic AChRs have been investigated by tritiated propylbenzylcholine mustard affinity labeling in *Drosophila* head preparations (Venter *et al.*, 1989). Cyclic GMP increase, PI breakdown, ion permeability changes, and cyclic AMP decrease have been described as possible responses elicited by muscarinic receptor agonists in insects.

A presynaptic role for mAChRs has been postulated: ACh release from synaptosomes is modulated by muscarinic ligands (Breer and Sattelle, 1987). High-affinity binding sites with muscarinic receptor properties have also been demonstrated in the locust (*Locusta migratoria*) CNS. Muscarinic binding sites in synaptosomal membranes display a low affinity for pirenzipine, an M1 antagonist, suggesting that in nerve terminals M2-like receptors might predominate. These locust synaptosomal mAChRs inhibit cAMP production (Knipper and Breer, 1988). Presynaptic as well as postsynaptic muscarinic receptors were detected in the nervous system of *Manduca sexta* (Trimmer and Weeks, 1989). Pharmacological identification of insect mAChRs using the same criteria as used in vertebrates suggests the presence of M1- as well as M2-type receptors, although the classification criteria used in vertebrates might be less relevant for insect mAChRs. Specific QNB binding revealed the presence of muscarinic-like receptors in cockroach brain membrane preparations. The receptor affinity for agonists was modulated by Mg^{2+} and guanine nucleotides, indicating the involvement of G-proteins. Pertussis toxin had no effect on muscarinic agonist binding and adenylate cyclase activity was not really inhibited by carbamylcholine. The localization of the muscarinic binding sites was detected in the brain by autoradiography. The receptors were present in the neuropil, especially in the mushroom body calyces, integrative brain structures which are known to receive sensory input from the antennal lobes (Orr *et al.,* 1991).

Muscarinic binding sites in the brains of honeybees, houseflies, and cockroaches were identified and compared by using different vertebrate mAChR-subtype-selective antagonists (Abdallah *et al.,* 1991). Although there are significant differences between insect and mammalian mAChRs, drug binding profiles suggest that these insect brain mAChRs might belong to a subtype which has pharmacological properties similar to the mammalian M1/M3 subtypes. A relatively large number of studies report electrophysiological effects mediated by insect muscarinic receptors (Davis and Pitman, 1991; Benson, 1992; Trimmer, 1992). Muscarinic receptors appear to modulate the excitability of the identified *Manduca sexta* proleg retractor motoneuron. Sensory neurons modulate this excitability by releasing ACh which acts through nAChRs and, via a separate pathway resulting in a slow excitatory postsynaptic potential, through mAChRs (Trimmer and Weeks, 1993). Muscarinic receptors may also be encountered in peripheral tissues (Wood *et al.,* 1992).

B. Biogenic Amine Receptors

1. Biogenic Amines in Insects

In vertebrates, classical amine transmitters and modulators have been shown to play a major role in the physiology of the brain and in behavioral responses. Similarly, biogenic amines have been shown to be synthesized in the insect nervous system. They function as neurotransmitters, neuromodulators, or sometimes as neurohormones, and are involved in many integrative processes of the nervous system including learning, memory, and behavior. Serotonin or 5-hydroxytryptamine, which is derived from the amino acid tryptophan and histamine, which is derived from histidine, have been shown to occur in both vertebrates and invertebrates. However, compared with the vertebrate aminergic systems, there are some remarkable differences too. In insects, the monophenolic amines, (D-)octopamine (OA) and (*p*-) tyramine (TA), are present in relatively large amounts, whereas in vertebrates these amines are thought to play only a minor role in neuronal signaling. Phenolamines as well as catecholamines are derivatives of the amino acid tyrosine (Tyr or Y). Catecholamines possess an additional hydroxyl group on the aromatic ring. The major signaling catecholamine present in the insect nervous system is dopamine (DA). Noradrenaline (norepinephrine) and adrenaline (epinephrine) play a major role in vertebrate aminergic systems. If they are present, these catecholamines reach only very low concentrations in the insect CNS. In insects, a large number of additional catecholamines has been shown to be involved in the processes of cuticle tanning, hardening, and sklerotization. These catecholamines are produced via synthesis pathways similar to the ones that are used in the CNS. They function as cross-linking reagents for cuticle proteins and chitin in the exoskeleton. Several *Drosophila* mutants which are defective in enzymes involved in amine synthesis or inactivation show behavioral and/or cuticular defects (Restifo and White, 1990; Homyk and McIvor, 1989; Piedrafita *et al.*, 1994).

Signaling biogenic amines are usually released from vesicular storage sites at nerve endings or varicosities (Fig. 4). As an extracellular signal, they selectively bind to specific receptors which are present at the plasma membrane of the target cell (e.g., the postsynaptic membrane). Most of these biogenic amine receptors belong to the GPCR superfamily, indicating

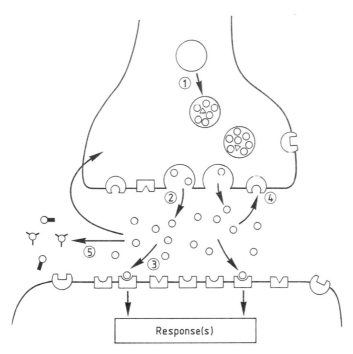

FIG. 4 Scheme illustrating the release, the action, and the inactivation of insect neuronal messenger molecules (e.g., biogenic amines). (1) The messenger molecules are synthesized and stored in vesicular storage sites. (2) As a result of neuronal stimulation, the vesicles fuse with the presynaptic membrane and the enclosed messengers are released. (3) The messengers specifically activate their receptor sites at the postsynaptic membrane, which results in a physiological response. (4) Presynaptic receptors may regulate (modulate) the release of the messengers. (5) Inactivation of the bioactive messengers by enzymatic modification or breakdown and/or by reuptake into nerve terminals.

that signaling events produced by biogenic amines are usually mediated by G-proteins. Inactivation of released amines is accomplished by re-uptake (e.g., into presynaptic nerve endings) and/or by acylation (e.g., *N*-acetylation by *N*-acetyltransferase, NAT, or *N*-beta-alanylation by *N*-beta-alanyltransferase) or sulfatation (by phenolsulfotransferase, PST). In contrast to the situation in vertebrates, monoamine oxidase (MAO), and catechol-*O*-methyltransferase (COMT) activities are very low in insect tissues.

Selective uptake systems for several biogenic amines have been described in insect tissues (Evans, 1978; Roeder and Gewecke, 1989). Specific reuptake of OA and DA in synaptosomal preparations was demonstrated by Scavone *et al.* (1994). Interestingly, the inhibitory effects of cocaine on these biogenic amine (especially OA) transporter systems might explain its insecticidal action (Nathanson *et al.*, 1993). In the locust CNS, Downer *et al.* (1993) have shown the existence of a specific uptake mechanism for tyramine (TA) which is distinct from DA and OA transport systems. TA is the direct precursor for the synthesis of OA and, until recently, it was not really considered a signaling amine. The hypothesis that this monoamine might indeed function as a neurotransmitter/modulator was supported by the observation that the distributions of TA and OA are not entirely linked and that TA can be released from locust cerebral ganglia under depolarizing conditions (high potassium ion concentrations) in a reserpine- and Ca^{2+}-dependent manner. These findings may be an encouragement for scientists in the field to reconsider the possible role of other amines and their metabolites as signal substances in the insect CNS.

2. Localization of Biogenic Amines in the CNS

The localization of several biogenic amines in insect tissues was analyzed by using histochemical and immunocytochemical methods. OA-immunoreactive cell populations have been localized in the nervous system. The presence of OA in dorsal and ventral unpaired median (DUM and VUM) neurons is well established (Eckert *et al.*, 1992). One of the DUM cells of the metathoracic ganglion, the DUMETi neuron, is well known to contain OA as a modulatory substance affecting the neuromuscular transmission process at the locust extensor tibiae muscle (Evans, 1981). Nearly all peripherally projecting DUM cells are probably octopaminergic (Stevenson *et al.*, 1992). OA-containing DUM cells also appear to contain taurine (2-aminoethanesulfonic acid, Bricker, 1992), as was shown in the cockroach *Periplaneta americana* by Nürnberger *et al.* (1993). Other OA-immunoreactive cells and fibers form a plurisegmental network, indicating that OA also fulfills an important role in the CNS itself. In the locust optic lobes, OA appears to be present in relatively high amounts (Downer *et al.*, 1993) and its presence was confirmed by the immunolocalization data (Konings *et al.*, 1988).

Tyrosine hydroxylase is the rate-limiting enzyme in catecholamine (dopamine, DA) synthesis. Its distribution in the ventral nerve cord of *Locusta migratoria* was mapped by immunocytochemistry (Orchard *et al.*, 1992). Most immunoreactive neurons appear to be interneurons showing many arborizations which are not restricted to a single segment. The neuronal processes are confined to the CNS. Only one pair of efferent neurons was found to be immunopositive: the so-called SN1 salivary neurons of the subesophageal ganglion. Indeed, it is known that the salivary nerve (7b) contains dopamine. Gifford *et al.* (1991) have demonstrated that DA is present in the subesophageal SN1 neurons of locusts and cockroaches, whereas the locust SN2 neurons contain 5HT. A comparable localization of DA was obtained by immunohistochemical studies performed in different insect species. In the insect CNS, the presence of DA seems to be restricted to a relatively small number of extensively branching interneurons. Major brain areas such as the central body and the mushroom bodies possess DA-like immunopositive material. Approximately 3000 cells in each optic lobe of the locust *Schistocerca gregaria* show DA immunoreactivity (Wendt and Homberg, 1992). The significant role of dopamine in CNS physiology was demonstrated by behavioral studies on honeybees: DA injection into the brain reduces conditioned responses to olfactory stimuli and thus influences the establishment of olfactory memory (Mercer and Menzel, 1982; Macmillan and Mercer, 1987).

The localization of serotonin (5HT) immunoreactivity in the CNS has been compared phylogenetically for distinct invertebrate phyla by Fujii and Takeda (1988). In insects, 5HT reactivity usually resides in a relatively small number of neurons. This number is a bit higher in adults (after metamorphosis) than in the larvae of holometabolic insects. Most neurons appear to be intrasegmental interneurons and 5HT-immunoreactivity patterns clearly reflect the original segmental plan of the CNS. Intersegmental interneurons and peripheral (afferent and efferent) fibers are also observed. Salivary and mandibular efferent neurons are known to contain 5HT. In fruit fly larvae, the ring gland and the pharyngeal muscle are targets for serotoninergic fibers. 5HT is also present in the optic lobes of fruitflies (Vallés and White, 1988). A superficial network of 5HT fibers has been described in several insects and might correspond to neurohemal release sites. Serotonin can thus act as a neurohormone (Lange *et al.*, 1988). 5HT synthesis, release, and reuptake by *Rhodnius prolixus* neurohemal tissue associated with abdominal nerves were examined by Orchard (1989). In the blowfly *Calliphora erythrocephala,* 5HT and 5HT-binding sites were colocalized by a combined immunohistochemical and autoradiographic approach. Both overlapping and mismatching regions of 5HT immunoreactivity and 5HT-binding sites were observed (Schmid *et al.*, 1993). The organization of 5HT immunoreactive cells and fibers in the central complex (protocerebral bridge and central body) of *Schistocerca* was revealed by Homberg (1991). Furthermore, 5HT was also localized in the stomatogastric nervous system of the stick insect *Carausius morosus* (Luffy and Dorn, 1992).

3. Some Functional Aspects of Insect Biogenic Amines

Signaling amines can be delivered via central or peripheral neuronal pathways as well as via the hemolymph. The action of biogenic amines have been reported in many research publications. As illustrated by Table I, biogenic amines are involved in many important physiological processes in insects. Nevertheless, novel functions may still be discovered and the detailed mechanisms of biogenic amine signaling are not yet entirely understood. Biogenic amines and their receptors have recently been reviewed by Roeder (1994) and by Downer and Hiripi (1994).

4. Dopamine

Both the immunocytochemical localization pattern and the abnormal levels of DA and related catecholamine compounds that are encountered in certain behavioral *Drosophila* mutants indicate that DA is involved in important integrative processes which take place in the CNS (Restifo and White, 1990). Nevertheless, DA appears to have a peripheral function via innervation of the salivary glands of a number of insects. In the locust glands, a DA-dependent stimulation of adenylate cyclase activity was observed. The pharmacological profile of this effect clearly differs from that of vertebrate D2 receptors, but may resemble that of the D1 receptor subtype (Lafon-Cazal and Bockaert, 1984). In many insect species, the CNS appears to contain a dopamine-sensitive adenylate cyclase activity (Orr *et al.*, 1988). [3H]Pifluthixol, a DA antagonist, was used as a ligand to characterize DA binding sites in the cockroach brain. A single class of high-affinity binding sites was obtained (Notman and Downer, 1987). In several electrophysiological studies, catecholamines have been shown to excite insect central neurons. In cockroaches, DA causes depolarization of an identified prothoracic common inhibitory motoneuron and excites it. This effect, however, is extremely voltage-dependent and Ca^{2+} (as well as some other ions) appears to play an important role in generating the DA response. In addition, cyclic AMP does not seem to be involved, suggesting that the DA receptor subtype that mediates this response might not be D1-like. In fact, the pharmacological profile does not correspond well to that of either mammalian D1 or D2 receptor subtypes. The study indicates that different

DA-dependent signal transduction pathways and receptor subtypes might occur in insect tissues (Davis and Pitman, 1991). Recently, putative DA receptors were localized in an organic culture of locust CNS explants by means of anti-idiotypic DA antibodies. Immunopositive structures were detected on membranes of outgrowing neurites as well as within these neurites (Vanhems *et al.*, 1994).

5. 5-Hydroxytryptamine

Several insect tissue preparations have been shown to produce physiological responses upon administration of 5HT (Table I). These preparations include the salivary glands of the blowfly (Berridge, 1970), the Malpighian tubules (Veenstra, 1988), the locust foregut (Banner *et al.*, 1987), and the cockroach CNS. The excellent pioneering work of Michael Berridge (1970, 1981, 1993, 1994) on 5HT responses of the blowfly salivary glands indicated the presence of two 5HT-induced signaling pathways, one via stimulation of adenylate cyclase, the other via InsP$_3$ production and Ca^{2+}.

5HT and OA also increase both the rate and the force of the heartbeat of *Manduca sexta* caterpillars. The pharmacology of the 5HT effect is similar to that of 5HT effects on Malpighian tubules in a variety of insects and is different from that of 5HT-induced effects on blowfly salivary glands (Platt and Reynolds, 1986). The mandibular closer muscles of the cricket, *Gryllus domesticus*, possess a 5HT-sensitive adenylate cyclase activity. Antagonists which are effective on this muscle preparation include spiperone, mianserin, ketanserin, and cyproheptadine, indicating a 5HT2-like pharmacology (R. A. Baines and Downer, 1991). [^{3}H] Serotonin binding sites have been identified in the locust brain by kinetic and pharmacological characterization (e.g., Hiripi and Downer, 1993). A single high-affinity binding site was observed. It shares some pharmacological characteristics with vertebrate 5HT1 receptors. No specific and saturable [^{3}H] ketanserin binding was obtained within the locust brain preparations, suggesting the absence (or low abundance) of a pharmacological 5HT2-like receptor type in this tissue. However, like the case with the insect DA receptors, the insect 5HT receptors are not easily classified by means of the pharmacological criteria used for mammalian 5HT receptors. Selective agonists/antagonists for mammalian 5HT receptors fail to classify satisfactorily the invertebrate 5HT receptors. Nevertheless, it is clear that insect tissues contain a number of distinct serotonin receptor subtypes and, therefore, there is a need for a classification system. The comparison of several 5HT receptor characterization studies indeed leads to the conclusion that the insect serotoninergic system is a quite complex one and that the multiplicity of insect 5HT receptors might be as high as in vertebrates (Roeder, 1994).

TABLE I
Effects of Biogenic Amines in Insect Tissues

Amine tissue	Effect(s)	Species	Reference
DA			
CNS	Olfactory memory, behavior	*Apis*	Macmillan and Mercer (1987)
Inhibitory motoneuron	Ca^{2+}, depolarization	Cockroach	Davis and Pitman (1991)
Salivary glands	cAMP elevation	Locust	Ali et al. (1993)
5HT			
CNS, muscles	Feeding behavior	*Rhodnius*	Lange et al. (1988)
Corpora allata	JH rise, cAMP rise	*Apis*	Rachinsky (1994)
Mandibular closer muscle	Modulation of contractions, cAMP rise	*Gryllus*	Baines and Downer (1991)
Visceral muscle	Modulation of contractility	Different insects	
Malpighian tubules	Fluid secretion	Different insects	
Malpighian tubules	Transepithelial voltage	*Aedes*	Veenstra (1988)
Salivary glands	Membrane responses	*Calliphora*	Berridge (1994)
Salivary glands	cAMP, $InsP_3$, and Ca^{2+} rise	*Calliphora*	Berridge (1970, 1994)
Salivary glands	cAMP elevation	Locust	Ali et al. (1993)
Haemocytes	Phagocytosis and nodulation	*Periplaneta*	Baines et al. (1992)
Haemocytes	cAMP rise	*Periplaneta*	Baines and Downer (1992)

OA

Organism	Effect	Insect	Reference
Organism	Arousal syndrome	Insects	Corbet (1991)
CNS	Protein phosphorylation	Locust	Rotondo et al. (1987)
NS	Neuronal activity, cAMP rise	Different insects	
Ventral nerve cord	cAMP rise	*Periplaneta*	Orr et al. (1992)
Ventral nerve cord	Na^+-dependent metabolism, Na^+-pump		Steele and Chan (1980)
Storage lobes CC	cAMP, cGMP rise	Locust	Pannabecker and Orchard (1988)
Corpora allata	JH rise, cAMP rise	*Apis*	Rachinsky (1994)
Extensor tibiae	Muscle contraction kinetics	Locust	Evans (1987)
Flight muscle	Contraction kinetics	Locust	Malamud et al. (1988)
Flight muscle	Glycolysis, contraction	Locust	Blau et al. (1994)
Oviducts	Contraction kinetics, cAMP	Locust	Lange and Tsang (1993)
Skeletal/visceral muscle	Modulation of contractility	Different insects	
Fat body	cAMP rise	Different insects	Fields and Woodring (1991)
Fat body/haemolymph	Sugar and lipid mobilization	*Acheta*	Orr et al. (1985)
Haemocytes	cAMP rise	*Periplaneta*	Jahagirdar et al. (1987)
Haemocytes	Ca^{2+} elevation	*Malacosoma*	Baines et al. (1992)
Haemocytes	Phagocytosis and nodulation	*Periplaneta*	Dunphy and Downer (1994)
Haemocytes	Phagocytosis and nodulation	*Galleria*	Baines and Downer (1994)
Haemocytes	Ca^{2+} and IP_3 rise	*Periplaneta*	
Firefly lantern	Light production, cAMP rise	*Photinus*	Nathanson (1979)
Sf9 cell line	cAMP rise	*Spodoptera*	Orr et al. (1992)
S2 cell line	cAMP rise	*Drosophila*	J. Vanden Broeck (our observation)

6. Phenolamines

OA receptors are considered potential targets for insecticidal or insectistatic agents. Formamidines (FAs) and other compounds that affect the octopaminergic system, e.g., cocaine (Nathanson *et al.,* 1993), have an influence on the viability of a number of insect pests. In addition, sublethal FA doses result in certain behavior effects such as hyperactivity, decreased feeding, increased locomotor activity leading to dispersal behavior and/or wing beating, detachment (e.g., of sucking ticks), uncoördinated movements, decreased oviposition and fecundity, scattered egg deposition, and decreased egg eclosion. FAs have been shown to act as OA receptor agonists, although FA-like compounds may also have some additional actions (Coats, 1982; Cascino *et al.,* 1989; M'Diaye and Bounias, 1991). OA can indeed be considered as the biogenic amine that regulates and coördinates insect responses to stress ("fight and flight") situations. The physiological state that corresponds to this "general arousal syndrome" prepares the organism for intense activity (Corbet, 1991).

Multiple effects of OA have been reported (see Table I) and the pharmacological characteristics of octopamine receptors have been well studied. Based on differences in pharmacological profiles and in second messenger production, several different OA receptors have been described in a large number of insect tissue preparations. The original classification which was used to define OA1, OA2A, and OA2B receptors is based on pharmacological differences observed when studying the modulatory effects of the octopaminergic DUMETi neuron on the locust extensor tibiae muscle (Evans, 1981). In fact, two OA effects are observed. First of all, OA modulates a myogenic rhythm of a specialized group of muscle fibers at the proximal end of the muscle. It is likely that this effect is mediated via an elevation of intracellular Ca^{2+} levels and the receptors responsible for this effect are usually classified as OA1. These OA1 receptors also display a different pharmacology than the ones that are involved in the second effect that was observed. This consists of a potentiation of both the amplitude and the relaxation rate of twitch tension generated by stimulation of a slow motoneuron. These receptors were defined as OA2 receptors. OA2ARs correspond to neuronal presynaptic receptors, which are present on the terminals of the slow motoneuron; OA2BRs are muscular postsynaptic receptors, mediating the direct OA responses on the muscle. Both OA2Rs are positively coupled to adenylate cyclase. Many other insect tissues have been shown to contain octopamine-sensitive adenylate cyclase activity with OA2A-like pharmacological properties. Observed heterogeneities in OAR pharmacology indicate that additional OAR subtypes might have to be defined (Evans and Robb, 1993; Roeder, 1991). It has therefore been proposed to extend the present classification scheme with an OA3R-type (or OA2CR) which was identified in the locust central nervous system by high-affinity binding of [^{3}H]octopamine (Roeder, 1991; Roeder and Nathanson, 1993). More selective agonists and antagonists may become useful tools for receptor discrimination and classification (Roeder, 1990a; Nathanson, 1993).

Another phenolamine which may function as a signaling substance is tyramine. TA fulfills important criteria for a transmitter candidate in the locust CNS (Downer *et al.*, 1993). Recently the characteristics of [^{3}H]tyramine and [^{3}H]octopamine binding sites were compared in a single study. Scatchard analysis and saturation curves revealed a high-affinity binding site for each of these phenolamines in locust brain membrane preparations. Both of them produced similar K_D (about 6 nM) and B_{max} (about 15–20 fmol/mg tissue) values, indicating a comparable affinity and concentration. Interestingly, these TA and OA binding sites possess distinct pharmacological profiles, suggesting that the locust CNS contains both specific TA and OA binding sites. The observed OA receptor pharmacology is similar to that of the above-mentioned OA3R. The TA binding site, however, displays a novel profile of agonist and antagonist binding and TA is indeed the ligand with the highest affinity of all substances which were tested (Table II). Guanine nucleotides clearly affect agonist binding affinities, suggesting that both phenolamine binding sites might be linked to G-proteins (Hiripi *et al.*, 1994). Further investigations will certainly lead to a better understanding of the molecular and physiological diversity of insect phenolamine receptors.

7. Histamine

Histamine (HA) is another biogenic amine which is present in the insect CNS. The retinae especially contain a lot of HA, which plays a physiological

TABLE II

Pharmacological Profiles for Phenolamine Receptor Subtypes in Locust Tissues

OA1	OA2A	OA2B	OA3	TA
Antagonists				
Phentolamine	Metoclopramide	Phentolamine	Mianserin	Chlorpromazine
Chlorpromazine	Mianserin	Metoclopramide	Phentolamine	Yohimbine
Cyproheptadine	Phentolamine	Mianserin	Chlorpromazine	Mianserin
Yohimbine	Cyproheptadine	Cyproheptadine	Cyproheptadine	Phentolamine
Mianserin	Chlorpromazine	Chlorpromazine	Metoclopramide	Cyproheptadine
Metoclopramide	Yohimbine	Yohimbine	Yohimbine	Metoclopramide
Agonists				
Tolazoline	Naphazoline	Naphazoline	Naphazoline	Tyramine
Naphazoline	Tolazoline	Tolazoline	Octopamine	Octopamine
Octopamine	Octopamine	Octopamine	Tolazoline	Dopamine
Tyramine	Tyramine	Tyramine	Tyramine	Naphazoline

role in the optic pathway. This HA-induced effect has a pharmacology which is different from that of vertebrate HA receptors and it is mediated via a HA-activated chloride channel (Hardie, 1988, 1989). In addition to its presence in the retina, some histaminergic neurons are found within the CNS (Nässel *et al.*, 1990). A high-affinity [^{3}H]mianserin binding site in the locust nervous tissue displays the pharmacological properties of a H1-like receptor type. Since mianserin is a high-affinity ligand for the neuronal locust OA3-receptor, these binding tests were performed with addition of 10^{-5} *M* of OA. All tested antihistaminic agents (e.g., promethazine, doxepin, amitriptyline, chlorpromazine) have a high affinity to this site. The affinity of the putative natural agonist, HA, appeared to be rather low (K_i in the millimolar range), but this might be an underestimation of the true value. Mg^{2+} ions affect HA binding affinity, suggesting that the binding site probably is a G-protein-coupled receptor (Roeder, 1990b; Roeder *et al.*, 1993).

C. Insect (Neuro-)Peptides

The improvement of preparative and analytical procedures has led to the identification of a rapidly growing number of bioactive peptides. Certain bioassays have been successfully employed to monitor the biological activity of tissue extracts and chromatographical fractions. Relatively low amounts of pure peptide material can now be sufficient for a structural characterization by high-resolution mass spectroscopy and Edman degradation. The results of this technological revolution are astonishing. In contrast to the vision of many classical endocrinologists that most physiological processes in insects are regulated by essentially two hormones, juvenile hormone and ecdysone, the total number of messenger substances involved in neuronal and endocrine signaling in insects might reach almost the same degree of diversity and complexity as in vertebrates. Large purification efforts in several insect endocrinology labs have yielded a multitude of novel peptide structures and the era of purifications and structural discoveries does not seem to be over yet. Thanks to this peptide purification work, enormous possibilities for further research have been created: these include the localization of peptides in insect tissues, the isolation and regulation of peptide precursor genes, the exploration of peptide-induced effects, the characterization and localization of binding sites, structure/activity studies, structural identification of peptide receptors, signal transduction research, molecular biological studies, and applications for insect pest control. Insect molecular neuroendocrinology has become a mature, but highly complicated research domain. Recently, some excellent reviews concerning insect neuropeptide research have been published (Schoofs *et al.*, 1993; Nagasawa, 1993; Nässel, 1993; Masler *et al.*, 1993; Kelly *et al.*, 1994).

Insect peptides have been shown to play a role in many physiological processes, but second messenger determinations and especially receptor characterization studies are still rare. Nevertheless, receptor characterization studies will become a hard necessity, if we want to obtain some insight into the increased complexity of insect physiology. Many insect peptides appear to elicit pleiotropic effects. Another important observation is the presence of a multitude of peptides belonging to the same peptide family and often derived from the same polypeptide precursor gene within a single insect species. How will we ever be able to explain why a single peptide can exert so many different biological (or pharmacological) effects? Why does such a tiny creature need so many variants of a single peptide? It is a major challenge for the (near) future to identify insect peptide receptors and thus to try to get an answer to these questions.

Compared with the high number of different peptides that are found in insects, very few studies on insect peptide pharmacology and receptor binding have been reported (e.g., opioid receptors, Stefano and Scharrer, 1981). Nevertheless, a large number of physiological effects elicited by peptides appear to be mediated via the action of G-proteins, indicating that, as in mammals, most small insect peptides act through GPCRs. To illustrate their complexity, we now look at a few examples of insect peptide effects.

1. Proctolin

The insect neuropeptide proctolin was originally isolated from the cockroach *Periplaneta americana* (Starrat and Brown, 1975). It is a pentapeptide (RYLPT) and its structure is most probably highly conserved among insect (and even other invertebrate) species. The physiological actions of proctolin have been studied mostly on a large number of insect muscle preparations where it induces (as a cotransmitter/modulator) slow tonic contracture or increases the frequency (accelerates) of myogenic contractions. These effects appear to be mediated via the formation of inositol triphosphate and diacylglycerol. This has been shown in locust oviducts, where the peptide-induced effects on muscular contraction were mimicked by phorbol esters and diacylglycerol analogs in a Ca^{2+}-dependent way (Lange, 1988). A similar observation was obtained by Baines *et al.* (1990) when studying the effect of proctolin on the locust mandibular closer muscle. The peptide is present in subesophageal cell bodies which send axons that innervate this muscle. It acts as a modulator by increasing the amplitude of neurally evoked contractions of this muscle. The effect was mimicked by $Ins(1,4,5)P_3$, and proctolin stimulated the production of $InsP_3$, while cAMP and cGMP levels were unaffected.

Proctolin analogs have been tested in the locust oviduct bioassay. Most analogs were at least three orders of magnitude less potent than the natural peptide, suggesting that all five amino acids are important for interaction with the receptor and for activation of this receptor. The supraanalog (the p-hydroxyl group of Tyr is replaced by p-OMe) and RYLAT displayed a relatively high activity, whereas RYSPT showed a high binding affinity but a low myotropic activity. Leuco-myosuppressin (-FLRFa as C-terminal) inhibited and enkephalins potentiated proctolin-induced contractions (Puiroux et al., 1993). A large number of proctolin analogs have now been tested. Some of these appear to be potent agonists: they have a polar group at the p-position of the aromatic side chain of Tyr. Others can be used as antagonists: they reduce the proctolin responsiveness. [α-methyl-L-tyrosine(2)]-proctolin, cyclic proctolin, and RYT have been identified as antagonists for proctolin-induced contractility of the locust foregut (Gray et al., 1994).

Proctolin is not only found at neuromuscular junctions, it is also present in the central nervous system and may act as a transmitter/modulator in the nervous system (Fitch and Djamgoz, 1988).

2. -FXPRLa Peptides

Many insect peptides display pleiotropic activities. Myotropic activity can be influenced by a very large number of insect peptides. Recently, this multitude of myotropic peptides was reviewed by Schoofs et al. (1993). The major question that arises from these findings is: "Are all these peptides physiologically involved in the in vivo modulation of muscular activity or might their primary function be a different one?" Unfortunately, the answer to this question cannot be given in a definitely "yes" or "no" form. Many signaling substances, including biogenic amines and peptides, seem to have a widespread distribution and they can elicit multiple responses, depending on the developmental stage, the type of tissue, etc. Therefore, pleiotropic activities are more the rule than the exception.

A family of peptides that clearly can produce multiple effects is the family of pyrokinin/PBAN peptides which share a common -FXPRLa C-terminal. Leucopyrokinin was isolated by using the Leucophaea hindgut assay. Fragments and analogs of this peptide have been tested in structural function studies and the C-terminal pentapeptide sequence -FTPRLa was identified as the active core of the peptide (Nachman et al., 1986). The conformationally restricted cyclic pyrokinin analog cyclo(-NTSFTPRL-) contains a β-turn structure, indicating that this is the conformational requirement for the activity retained within the active core region. This structure may provide us with a template to design and produce peptide mimetics for pyrokinin-mediated processes (Nachman et al., 1991). Other members of this peptide family were purified from lepidopteran species and display diapause-inducing (Bom-DH, *Bombyx mori*) and pheromone biosynthesis-stimulating (PBAN from different species) activities. Both substances have myotropic activities on the cockroach hindgut and the orthopteran pyrokinins stimulate silkworm diapause induction and pheromone biosynthesis (and with an even higher potency than the endogenous peptides). These cross-reactivities indicate that the receptors involved in mediating these distinct processes can recognize the conserved peptide structures. Moreover, the C-terminal sequence is essential for all -FXPRLa-induced effects. In addition, pyrokinins are myotropic for locust oviduct contraction and some pyrokinins also appear to induce cuticular melanization in the armyworm (Nachman *et al.*, 1993b). Recently, it was demonstrated that, in addition to cAMP, the stimulation of pheromone production by PBAN requires the involvement of Ca^{2+} (Ma and Roelofs, 1995).

3. Myosuppressins

Whereas many myotropic peptides have a stimulatory action on muscle contractility, some of them clearly have an inhibitory effect. The C-terminal pentapeptide VFLRFa has been identified as the active core (the shortest fragment of the peptide which elicits the inhibitory response) of the myosuppressin family of inhibitory myotropins. The C-terminal heptapeptide DHVFLRFa has a potency similar to the native myosuppressin. Several synthetic peptide analogs were screened for their myoinhibitory or myostimulatory activity and the structural requirements for these activities, as performed by myosuppressins and sulfakinins respectively, were determined (Nachman *et al.*, 1993a). Recently, high- and low-affinity binding sites for SchistoFLRFa were characterized in locust oviduct membrane preparations by using an *in vitro* binding assay with [125I]YDVDHVFLRFa. The binding was EDTA, Ca^{2+} and Mg^{2+} dependent and was competitively inhibited by adding nonlabeled myosuppressins, whereas proctolin did not affect binding of labeled [Y^1]-SchistoFLRFa. The two binding sites seem to be differentially located within the locust lateral oviducts: the high-affinity site is preferentially present in areas without innervation and the low-affinity site might be associated with the extensively innervated lower part of the oviducts (Wang *et al.*, 1994).

4. Allatostatins

The "allatostatins" form another insect peptide family. The initial purification was based on their inhibitory activity in a JH bioassay performed with corpora allata of cockroaches (*Diploptera punctata*). The C-terminal sequence part (-FGLa) is well conserved in all peptide family members and might thus be part of the active core. In the cockroach, multiple naturally occurring variants appear to be derived from a single large precursor polypeptide. Allatostatin-4 (AST4, DRLYSFGLa) is a small endogenous family member and was used as a reference for structure-activity studies. AST analogs were examined for their ability to reduce JH biosynthesis in corpora allata, *in vitro*. Ala or *D*-amino acid replacements were introduced in two series of AST4-analogs. The pharmacological profile obtained with the Ala-analogs indicated that the (side chains of) residues Y^4, F^6, G^7 and L^8 are extremely important for activity, whereas the profile of the *D*-amino acid analogs suggests that the three-dimensional conformation of the C-terminal of the peptide could be very important too (Hayes *et al.*, 1994).

In addition, putative allatostatin receptor proteins were studied by photoaffinity labeling in the cockroach *Diploptera punctata*. Therefore, Cusson *et al.* (1991) used a bioactive radioiodinated azidosalicylamide analog of allatostatin-1. By sodium dodecyl sulfate polyacrylamide gel electrophoresis, specific labeling of two bands (59 and 39 kDa) could be detected in the corpora allata preparation.

Membrane preparations derived from other tissues also resulted in specific binding of certain protein bands: a 41-kDa band in brain tissue and three bands of ca. 38, 41, and 60 kDa in the fat body. This study indicates that several allatostatin binding sites are present in different insect tissues and that these peptides might have additional functions besides the inhibitory role in cockroach JH biosynthesis. Furthermore, it would be interesting to know whether the different allatostatin binding sites display different affinities for the endogenous AST-variants.

5. Adipokinetic Hormones

Adipokinetic hormones (AKHs; pGlu-XXFXXXWa or pGlu-XXFXXXW-GXa) belong to a widespread arthropod peptide family (Orchard, 1987). In insects, they are mainly found in the glandular part of the corpora cardiaca and play an important neuroendocrine role as regulators of carbohydrate and lipid metabolism. They also can produce myotropic activities and accelerate the heart beat. Several AKH structure and activity studies have been performed in different insect species to identify active analogs which can be radiolabeled for receptor characterization studies. Ziegler *et al.* (1991) investigated the effects of several natural and synthetic analogs on the activity of glycogen phosphorylase in the fat body of *Manduca sexta*. They found that the N-terminal pyroglutamate (pGlu) structure is important for AKH activity. In this study, every amino acid position was shown to have some influence on receptor binding and/or activation. In another study (Gäde, 1993), the relative lipid mobilization potencies of a number of AKH analogs were determined by comparing complete dose-response curves. Three naturally occurring peptides with replacements at amino acid positions 2, 5, and/or 7 produced only slightly higher ED_{50} values than locust AKHI (pGlu-LNFTPNWGTa). Another group of analogs with changes at amino acid residues in positions 3, 6, 7, and/or 10 had about 50-fold higher ED_{50} values, whereas the third group showed very high ED_{50} values. Three other peptide analogs did not reach the maximum level of full activity even when added in high doses.

Two putative AKH receptor proteins (45 and 90 kDa) were solubilized from abdomens of the fly *Musca autumnalis* and partially purified by affinity chromatography. In addition, a ligand-binding assay was developed by using ^{125}I- and ^3H-labeled dipteran corpora cardiaca factor I (DCCI) (Minnifield and Hayes, 1992).

6. Eclosion Hormone

The insect eclosion hormone (EH) is a relatively large peptide which acts on the insect nervous system to elicit a coordinated behavioral response linked to the shedding of old cuticle at the time of eclosion or at the end of each molt (Kataoka *et al.*, 1987). The action of EH appears to involve an increase of cGMP levels in the CNS (Truman and Levine, 1982). This signaltransduction pathway has been further investigated. In *Manduca sexta,* EH does not seem to directly stimulate a membrane-bound form of guanylate cyclase (mammalian natriuretic peptides act through a membrane-bound guanylate cyclase type of receptor). Nitric oxide synthase inhibitors do not affect the EH stimulation of cGMP production, whereas inhibitors of arachidonic acid release and metabolism almost completely abolish the EH-induced cyclic GMP increase. Therefore, EH receptors are most probably coupled, directly or indirectly, to arachidonic acid release (e.g., via phospholipase A_2 stimulation) and/or metabolism (Morton and Giunta, 1992). In *Bombyx mori,* however, it has been shown that the synthesis of another second messenger, $InsP_3$, is also induced by EH. In addition, inhibitors of nitric oxide synthase, calmodulin, and phospholipase C inhibited the EH-induced increase in cGMP. Ins(1,4,5,)P_3 clearly stimulated the formation of cGMP in *Bombyx* abdominal ganglia. Therefore, the possible pathway of the EH signal transduction in *Bombyx* might start with the activation of phospholipase C, leading to an increase in $InsP_3$ followed by a rise in intracellular Ca^{2+}. Ca^{2+} calmodulin might then activate nitric oxide synthase, which in turn produces NO that diffuses and acts on soluble guanylate cyclase. The resulting rise in cGMP further elicits the ecdysis behavior (Schibanaka *et al.*, 1991).

7. Diuretic Hormone

Insect diuretic peptides show some limited sequence similarities with verte-brate CRFs. They have the ability to increase the cyclic AMP content in the Malpighian tubules of many different insect species and stimulate fluid secretion by these tubules (Kataoka *et al.*, 1989; Kay *et al.*, 1991). In *Man-duca*, the membrane-bound DH receptor from Malpighian tubules was characterized and showed high-affinity binding for insect DH peptides. Truncated DH analogs in which up to 20 N-terminal amino acids were removed maintained a high affinity for the receptor (Reagan *et al.*, 1993). This receptor was also solubilized by using the zwitterionic detergent CHAPS, 3-[(3-cholamidopropyl)dimethylammonio]-1-propanesulfonic acid. CHAPS-solubilized receptor showed time-dependent, saturable, and rever-sible specific binding with DH. In addition, biotinylated Mas-DH also had a high affinity for this receptor and therefore it can be useful for affinity chro-matographic purification of the solubilized receptor protein (Reagan *et al.*, 1994).

8. Trypsin-Modulating Oostatic Factors

In several insect species, females take a protein-rich meal before the onset of vitellogenesis. When female mosquitoes take a blood meal, the biosynthe-sis of trypsin in the gut is enhanced. It has been shown that the opposite process, which consists in downregulation of trypsin in the gut and leads to oöstasis, is regulated by a Pro-rich peptide hormone derived from the ovaries (Borovsky *et al.*, 1994). The exact mechanism of action of this hormone has not yet been elucidated, but the receptors for this peptide have recently been localized in the mosquito gut. This was achieved by using a complementary (or anti-) peptide approach. The *Aedes*-TMOF peptide has been shown to bind to its corresponding antipeptide and anti-bodies against this antipeptide (thus anti-antipeptide antibodies) were used in an immunohistochemical approach to localize the putative TMOF bind-ing sites.

Recently, peptides with similar functional activities were identified in the gray fleshfly, *Neobellieria bullata*. Neb-TMOF, a hexapeptide, has been shown to stimulate cAMP synthesis in the ring glands of *Calliphore* larvae, where it decreases the production of ecdysteroids (Bylemans, 1994; De Loof *et al.*, 1995).

D. Eicosanoids/Prostaglandins in Insects

Eicosanoids are biologically active metabolites of arachidonic acid (AA) and of a few other polyunsaturated fatty acids. They can mediate a broad array of biological functions, primarily as signaling and regulating agents. Usually, their synthesis is initiated by several agonists acting through phospholipase A_2 stimulation. PLA_2 is one of the cellular effector proteins that can be regulated via G-proteins. Several eicosanoid compounds produced via different metabolic pathways may act as secondary messengers and/or as paracrine signaling substances. In mammals, prostaglandins have been shown to mediate effects via GPCRs. Up to now, the study of these compounds has not been a high priority in insect physiology. The metabolism of arachidonic acid and the occurrence of prostaglandins and other eicosanoids in insect tissues has recently been reviewed by Stanley-Samuelson (1994).

E. Olfactory and Gustatory Signaling

In vertebrates, sensory processes such as olfaction, taste, and vision are mediated via GPCRs. Recently, a very large number of mammalian olfactory (and gustatory) receptors have been identified through molecular cloning techniques (Buck and Axel, 1992).

In insects, odor and taste perception are very important processes. Pheromones, such as sex pheromones and substances for communication by social insects, are produced by the insects themselves as chemical extraorganismal messengers. In addition, many other molecules in the environment can be detected by insects. As in mammals, insect olfactory signal transduction is thought to be regulated by G-proteins (Breer et al., 1988). Although several electrical and biochemical steps in the olfaction process have been described, the identification of the receptor molecules still presents a great challenge (Stengl, 1992).

Odor stimulation of antennal tissue appears to induce a rapid and transient increase in $InsP_3$ levels. This may lead to a rise of intracellular calcium levels. The electrical response might then be initiated through Ca^{2+}-sensitive ion channels (Wegener et al., 1993). Application of high pheromone doses elicited a delayed and sustained increase in cGMP concentration in *Heliothis virescens* antennal preparations. Cyclic GMP might therefore be involved in adaptation of the odor response, suggesting cross-talk between different second messenger ($InsP_3$ and cGMP) systems (Boekhoff et al., 1993). Recently, it has been demonstrated that $InsP_3$-dependent Ca^{2+} currents precede cation currents in insect olfactory receptor neurons. There appear to be no internal Ca^{2+} stores in the outer dendrites where the olfactory cascade is initiated, but another mechanism of Ca^{2+} recruitment seems to be present. Patch clamp measurements of cultured *Manduca sexta* olfactory neurons revealed that a transient pheromone-dependent inward current precedes the cation currents. This inward current most probably corresponds to an $InsP_3$-dependent Ca^{2+} current (Stengl, 1994).

F. Visual Signaling, Insect Opsins

As in vertebrates, light perception is mediated by a specialized type of chromoprotein (rhodopsin) composed of an apoprotein or opsin which is a GPCR and a chromophore molecule (generally 11-*cis*-retinal). This chromophore is covalently attached to the side chain of a lysine residue in the seventh transmembrane region of the opsin. Photoactivation of the rhodopsin initiates a process in which the energy of the absorbed photon is converted into a conformational change of the receptor molecule, leading to an activated GPCR form which is able to activate the G-protein transducin. Different light receptor molecules may differ in their absorption spectra. This may lead to "color" vision. The insect rhodopsin molecules will be discussed in a later section.

IV. Molecular Cloning of G-Protein-Coupled Receptors

A. Strategies for the Molecular Cloning of Receptor-Encoding cDNAs/Genes

1. Functional Approaches

a. The "Classic" Approach   Receptor proteins can be purified by "classic" preparative chromatographic and/or electrophoretic methods. Usually a radiolabeled ligand is needed to perform a binding assay. The ligand can also be covalently attached to its binding site by employing an affinity labelling procedure. Affinity chromatography might be used as an alternative method to obtain an enrichment of the binding protein(s). The isolation of the nicotinic acetylcholine receptor by α-bungarotoxin affinity chromatography is a famous example of this approach. Many receptors are glycoproteins and may therefore be partially purified by lectin-affinity chromatography.

Once a receptor is isolated, a partial amino acid sequence can be determined by using a combination of enzymatic digestion and microsequencing. Oligonucleotides based on this sequence can then be used as hybridization probes for screening the appropriate DNA library. Alternatively, they could serve as primers for PCR (polymerase chain reaction). Many different variations of this classic scheme might prove to be successful.

Other, more elegant cloning strategies based on functional and/or pharmacological criteria have been shown to be successful in receptor cloning experiments. In general, these methods make use of (well-defined) cellular systems for expression of cDNA or cRNA.

b. Xenopus *Oöcytes* *Xenopus* oöcytes are generally used as functional expression systems for mRNAs coding for receptors which can cause electrophysiological responses. The mRNA (or cRNA) is introduced by microinjection. Foreign receptor proteins produced by these oöcytes can therefore be characterized on a functional basis. The oöcyte system not only responds to ligand-gated ion channels, but has also the ability to couple several GPCRs to electrophysiological events through a G-protein-mediated intracellular Ca^{2+} rise. The assay can be sensitive enough to detect a particular receptor which is encoded by a relatively rare transcript. Therefore, this sensitive assay can be used to screen pools of cRNAs derived from a cDNA library. *Xenopus* oöcytes can translate these RNAs into functional receptor molecules that can be activated by administration of the ligand. Positive pools are further subdivided until the receptor cDNA is entirely purified. This approach was successful for vertebrate receptor cloning and has led to the identification of the receptors for a number of ligands such as substance K (Masu *et al.*, 1987) or serotonin (Julius *et al.*, 1988).

c. Expression Cloning in Transfected Cell Lines Another method which was successfully employed to clone vertebrate receptors might be adapted for insect receptor cloning. It is based on the expression of DNA in cultured cell lines. The screening method can be a very sensitive functional assay protocol or a radioligand-binding assay.

According to the size of the inserts, a cDNA library can be divided into several pools and these are amplified in *E. coli.* Cultured cells transfected with the different pools of expression constructs are then screened for high-affinity binding to the appropriate ligand. After binding, subsequent washing and fixation, these cells are exposed to a photographic emulsion (e.g., Kodak NBT2). Autoradiographic detection will then indicate which pool(s) contain the expected cDNA clone(s). The positive pools are further subdivided and the procedure can be repeated until the receptor clone is isolated. This approach is frequently used, and the cDNAs encoding the receptors for parathyroid hormone (Abou-Samra *et al.*, 1992) and gastrin (Kopin *et al.*, 1992) were isolated this way.

2. Comparative Approaches

Many receptor proteins contain regions which are well conserved during evolution. Therefore, the majority of cloned GPCRs have been detected by following a comparative approach based on existing sequence information. This approach is further combined with cDNA expression studies to analyze the functional and pharmacological properties of the receptor clones. The advent of insect cellular expression systems offers an opportunity to investigate cloned insect receptor proteins in a homologous cellular environment.

a. Hybridization under Reduced Stringency DNA fragments encoding known receptor proteins are frequently used as hybridization probes for screening libraries. Another possibility is the use of degenerate oligonucleotide probes derived from a conserved stretch of amino acids. Within most GPCR subfamilies, the transmembrane regions are quite well conserved and could be chosen for development of useful probes.

b. Cloning by PCR Consensus sequences may also be employed as primers for a polymerase chain reaction. PCR ensures the selective amplification of a DNA sequence between two oligonucleotides. The method proved to be extremely successful for cloning novel receptor sequences in mammals (Libert *et al.,* 1989; Buck and Axel, 1992). Moreover, it has also been employed to search for putative insect GPCRs (Vanden Broeck *et al.,* 1991).

B. Cloned Insect G-Protein-Coupled Receptors

1. Muscarinic Acetylcholine Receptor

The characterization of mAChRs in insect tissues indicates that there might be several muscarinic receptor subtypes present. In addition to these pharmacological indications, a muscarinic receptor cDNA was cloned from the nervous system of *Drosophila melanogaster* (Onai *et al.,* 1989; Shapiro *et al.,* 1989). Its sequence shows remarkable similarity ($>60\%$) to the mammalian mAChR subtypes. The pharmacology and signaling properties of this cloned receptor indeed confirmed this similarity: pirenzepine (M1 antagonist) binds to the expressed receptor and carbamylcholine stimulated PI turnover in mammalian cells expressing the insect receptor gene. In contrast to mammalian sequences, the insect mAChR contains a hydrophobic stretch of amino acids at its N-terminal which might correspond to a signal sequence. It also has a very large (putative) third cytoplasmic loop between TMD V and TMD VI.

The pharmacological properties of this receptor were further investigated through the expression of its cDNA in two independent transient expression systems, COS-7 cells and *Xenopus* oocytes. Binding of N-[^{3}H] methylscopolamine at COS-7 membranes was effectively displaced in the following order: atropine $>$ 4-diphenylacetoxy-N-methylpiperidine methiodide $>$ pirenzepine $>$ AFDX-116. Carbamylcholine-induced currents in the oöcyte system were also antagonized in the same order of effectiveness: 4-diphenylacetoxy-N-methylpiperidine methiodide (4-DAMP) $>$ pirenzepine $>$ AFDX-

116. This pharmacological profile is related to that of mammalian M1 and M3 mAChR subtypes. In the same study, the localization of the receptor protein in the nervous system of *Drosophila* was investigated by immunocytochemistry with an antiserum raised against a synthetic peptide that corresponds to the carboxy-terminal part of the molecule. The immunopositive reaction was found in the nervous system, especially in the neuropile, and very intense staining was observed in the glomeruli of the antennal lobes. Therefore, ACh may play an important role in the processing of olfactory signals (Blake *et al.*, 1993).

2. Dopamine Receptor

Recently, the first insect catecholamine receptor was cloned by exploiting the general similarity among biogenic amine GPCRs. A dopamine receptor homolog was identified in *Drosophila* by screening a library under reduced stringency conditions (the comparative approach). When the deduced amino acid sequence was compared with the vertebrate catecholamine receptor sequences, the conservation was very pronounced in the putative membrane-spanning domains (up to 78% for TMD III). The sequence identity with the mammalian D1 and D5 dopamine receptors is quite high. This D-Dop1 receptor is significantly larger than its mammalian counterparts, which is primarily due to its extended N-terminal portion. Again, this portion contains additional hydrophobic regions, one of which might function as a signal sequence. Consensus sites for N-linked glycosylation and phosphorylation are found in the extra- and intracellular portions respectively. Cys residues in the C-terminal part might be involved as putative palmitoylation sites.

The receptor cDNA was introduced in the pcDNA1 vector and the resulting expression construct was transfected into human embryonic kidney (HEK 293) cells. The cyclic AMP level of these cells was significantly increased by DA administration in a concentration-dependent way. SKF 38393, a D1 agonist, was also capable of stimulating cAMP production in these cells. The DA-induced cAMP stimulation was inhibited by known DA receptor antagonists such as butaclamol and flupentixol. The D1/5 antagonist SCH23390 was a less potent antagonist, indicating that pharmacological properties are not necessarily well conserved during evolution. [^{3}H]SCH 23390 did not detect a high-affinity binding site in transfected HEK 293 cells. The β2-AR antagonist pindolol did not substantially reduce the DA effect.

The tissue distribution of the D-Dop1 mRNA was investigated via Northern and *in situ* hybridization analysis. The positive hybridization signals were predominantly localized in perikarya of the optic lobes of adult fruit flies (Gotzes *et al.,* 1994). This observation is in agreement with the immuno-localization data showing that DA is also localized in the optic lobes.

3. Serotonin Receptors

Low stringency hybridization with oligonucleotides corresponding to the conserved TMD domains VI and VII of mammalian receptors resulted in the identification of three distinct *Drosophila* serotonin receptor cDNA and/or genomic clones. The sequences of these 5HT receptors—5HT-dro1, 5HT-dro2A, and 5HT-dro2B—exhibit clear sequence similarity with known mammalian serotonin receptors. The resemblance is highest within the transmembrane regions.

When stably expressed in NIH 3T3 cells, the 5HT-dro1 receptor proved to mediate an increase of cAMP. This effect was selectively induced by serotonin. The result suggests that coupling specificities for G-proteins might be well conserved during evolution. Mammalian as well as insect G-proteins are probably similar and may also act in a similar way. Nevertheless, it is not unlikely that *in vivo,* other signal transduction pathways can be initiated by this 5HT-dro1 receptor (Witz *et al.,* 1990).

The two other 5HT receptor clones are highly similar to one another and, in contrast to 5HT-dro1, contain a large third intracellular loop between TMD V and TMD VI. These 5HT-dro2 receptors both appear to be capable of inhibiting adenylate cyclase and stimulating PLC activity in stable mouse NIH 3T3 cell lines. Membrane preparations from COS-7 cells expressing these receptors display high-affinity binding to [^{125}I]LSD, and serotonin is the natural ligand with the highest affinity to these receptors.

In situ hybridization analysis of the expression of 5HT-receptor mRNAs in fruit fly (embryonic, larval, and adult) tissues showed differential expression patterns for all three receptors (Saudou *et al.,* 1992).

4. Phenolamine Receptors

A fruit fly phenolamine receptor was cloned independently by two separate research groups. They employed a homology-based cloning procedure and were able to demonstrate that their receptor is related to mammalian $\alpha 2$ adrenergic receptors. When expressed in mammalian cells, it displayed high-affinity binding to the $\alpha 2$-antagonist yohimbine. Octopamine proved to be a relatively good competitor for [^{3}H]yohimbine binding. In addition, OA was able to induce an attenuation of the activity of adenylate cyclase in stably transfected mammalian (CHO-K1) cells. Therefore, Arakawa *et al.* (1990) concluded that this receptor might be an OA1-type of receptor. The same receptor was also identified by Saudou *et al.* (1990). Although their results were similar to the findings of Arakawa *et al.* (1990), they came to the conclusion that the cloned receptor might not correspond to an OA receptor, but to a (putative) tyramine receptor. Tyramine is a naturally occurring insect amine and it is a transmitter/modulator candidate. It showed a higher affinity for the receptor and also proved to be more effective than OA in reducing cAMP levels in permanently transformed mammalian (NIH 3T3) cells. The observations were later confirmed with the stable CHO-K1 cells and some additional results obtained with this fruit fly phenolamine receptor were published by Evans and Robb (1993).

It appears now that the effect of adenylate cyclase attenuation is more efficiently mediated by TA, but that for the induction of an intracellular Ca^{2+} rise, OA might be a good agonist also. Therefore, the *in vivo* action of this phenolamine receptor is not yet entirely clarified. It is possible that differential activation could be achieved by a single receptor, depending on the relative concentrations of OA and/or TA which are present. Another interesting fact is that the optical isomers of OA are equally potent displacers of [^{3}H]yohimbine binding. Furthermore, the formamidines, demethylchlordimeform and amitraz, were shown to inhibit [^{3}H]yohimbine binding to the receptor expressed in CHO cells. Both show higher affinities than octopamine for this receptor, suggesting that this receptor is a potential target molecule for these insecticidal agents (Hall *et al.*, 1994).

OA effects and the pharmacology of its binding sites are well characterized in the locust *Locusta migratoria*. Recently, a locust homolog of the fruit fly tyramine–octopamine receptor was cloned by using a PCR-based cloning protocol with primers corresponding to the highly conserved regions in the TMDs III and VI of a number of mammalian biogenic amine receptors (Vanden Broeck *et al.*, 1994). This receptor indeed shows high sequence similarity with its fruit fly counterparts in these TMD regions (Fig. 5). Other parts of the molecule appear to be much less conserved and illustrate the long period of evolutionary divergence between dipteran and orthopteran species. The locust sequence is also similar to vertebrate $\alpha2$ adrenergic, 5HT1A, 5HT1E receptors and to the fruit fly 5HT receptors (30–40%). In order to investigate the receptor properties in an insect cellular environment, the cDNA was expressed in *Spodoptera* Sf9 and in *Drosophila* S2 cells. Membrane preparations of these cells exhibited high affinity binding to [^{3}H]yohimbine. The pharmacological profile of antagonist binding affinities was similar to that of the fruit fly phenolamine receptor and clearly differed from all known locust OA receptor subtypes. From all endogenous ligands tested, TA proved to be by far the best displacer of yohimbine binding. This means that the high-affinity binding properties to TA are very well conserved and may have been subject to selective pressure. In contrast, the dissociation constants for OA and for DA were at least two orders of magnitude lower than for TA. Moreover, cAMP production in forskolin-stimulated S2 cells, which permanently expressed receptor cDNA, was

```
                                        I
                        _____        __
TA-R-Loc     (xx)-PEWEVAVTAVSLSLIILITIVGNVLVVLSVFTYKPLRIVQNFFIV
TA-R-Dro     (xx)-PEWEALLTALVLSVIIVLTIIGNILVILSVFTYKPLRIVQNFFIV
5HT1-R-Dro   (xx)-PPLTSIFVSIVLLIVILGTVVGNVLVCIAVCMVRKLRRPCNYLLV
5HT2A-R-Dro  (xx)-QLLRMAVTSVLLGLMILVTIIGNVFVIAAIILERNLQNVANYLVA
5HT2B-R-Dro  (xx)-LLVKMIAMAVVLGLMILVTIIGNVFVIAAIILERNLQNVANYLVA
DA-R-Dro     (xx)-SLVSIVVVGIFLSVLIFLSVAGNILVCLAIYTDGSLPRIGNLFLA
M1ACh-R-Dro  (xx)-SLASMVVMGFVAAILSTVTVAGNVMVMISFKIDKQLQTISNYLF
                  .  .. ..  .. **..*  .       *      *  *  ..

                 II                            III
             _____            _____
             SLAVADLTVAVLVMPFNVAYSLIQRWVFGIVVCKMWLTCDVLCCTASILN
             SLAVADLTVALLVLPFNVAYSILGRWEFGIHLCKLWLTCDVLCCTSSILN
             SLALSDLCVALLVMPMALLYEVLEKWNFGPLLCDIWVSFDVLCCTASILN
             SLAVADLFVACLVMPLGAVYEISQGWILGPELCDIWTSCDVLCCTASILH
             SLAVADLFVACLVMPLGAVYEISNGWILGPELCDIWTSCDVLCCTASILH
             SLAIADLFVASLVMTFAGVNDLLGYWIFGAQFCDTWVAFDVMCSTASILN
             SLAIADFAIGTISMPLFAVTTILGYWPLGPIVCDTWLALDYLASNASVLN
             ***..*. .. . ...   ..   * .*   *. * .   . ...*.*.

                                            IV
                                  _____
             LCAIALDRYWAITDPINYAQKRTLRRVLAMIAGVWLLSGVISSPPLIGWN
             LCAIALDRYWAITDPINYAQKRTVGRVLLLISGVWLLSLLISSPPLIGWN
             LCAISVDRYLAITKPLEYGVKRTPRRMMLCVGIVWLAAACISLPPLLILG
             LVAIAVDRYWAVTN-IDYIHSRTSNRVFMMIFCVWTAAVIVSLAPQFGWK
             LVAIAADRYWTVTN-IDYNNLRTPRRVFLMIFCVWFAALIVSLAPQFGWK
             LCAISMDRYIHIKDPLRYGRWVTRRVAVITIAAIWLLAAFVSFVPISLGI
             LLIINFDRYFSVTRPLTYRAKGTTNRAAVMI-GAWGISLLLWPPWIYSW-
             *  *. ***  ..  . *       *     . *    . .

                                 V
                                 _____
             DWP-MEFNDT------TPCQLTEEQGYVIYSSLGSFFIPLFIMTIVYVEI
             DWP-DEFTSA------TPCELTSQRGYVIYSSLGSFFIPLAIMTIVYIEI
             NEH-EDEEGQ------PICTVCQNFAYQIYATLGSFYIPLSVMLFVYYQI
             DPDYLQRIEQ------QKCMVSQDVSYQVFATCCTFYVPLMVILALYWKI
             DPDYMKRIEE------QHCMVSQDVGYQIFATCCTFYVPLLVILFLYWKI
             HRPDQPLIFEDNGKKYPTCALDLTPTYAVVSSCISFYFPCVVMIGIYCRL
             --PYIEGKRTVPKDECYIQFIETNQYITFGTALAAFYFPVTIMCFLYWRI
                                  .      .. .*. *  .. .  .*   .

                                              VI
                                    _____
FIATK    -(xxx)- KQRISLSKERRAARTLGIIMGVFVVCWLPFFLMYV
FVATR    -(xxx)- KQKISLSKERRAARTLGIIMGVFVICWLPFFLMYV
FRAAR    -(xxx)- KLRFQLAKEKKASTTLGIIMSAFTVCWLPFFILAL
YQTAR    -(xxx)- KETLEAKRERKAAKTLAIITGAFVVCWLPFFVMAL
YIIAR    -(xxx)- RQLLEAKRERKAAQTLAIITGAFVICWLPFFVMAL
YCYAQ    -(xxx)- HSSPYHVSDHKAAVTGVIMGVFLICWVPFFCVNI
WRETK    -(xxx)- KKSQEKRQESKAAKTLSAILLSFIITWTPYNILVL
                 .  ..     .    . .*. *.. *   *  * *.  ..

                                    VII
                          _____
             IVPFCNPSCKPSPKLVNFITWLGYINSALNPIIYTIFNLDFRRAFKKLLH
             ILPFCQTCC-PTNKFKNFITWLGYINSGLNPVIYTIFNLDYRRAFKRLLG
             IRPF--ETMHVPASLSSLFLWLGYANSLLNPIIYATLNRDFRKPFQEILY
             TMPLCAA-CQISDSVASLFLWLGYFNSTLNPVIYTIFSPEFRQAFKRILF
             TMSLCKE-CEIHTAVASLFLWLGYFNSTLNPVIYTIFNPEFRRAFKRILF
             TAAFCKTC--IGGQTFKILTWLGYSNSAFNPIIYSIFNKEFRDAFKRILT
             IKPLTTCSDCIPTELWDFFYALCYINSTINPMSYALCNATFRRTYVRILT
             . ..          ...  * * ** .**. *.   . ..* .. .*

             FKT---------------------------------------------
             LN----------------------------------------------
             FRCS--------SLNTMMRENYYQDQYGEPPSQRVMLGDERHGARESFL
             GGHR--------PVH-------------------------YRSGKL
             GRKA--------AAR-------------------------ARSAKI
             MRNPWCCAQDVGNIHPRNSDRFITDYAAKNVVVMNSGRSSAELEQVSAI
             CKWH-----------------------------TRNREGMVRGVYN
```

FIG. 5 Sequence alignment of a number of insect amine receptors. The position of identical residues is indicated by "*". Highly conserved codon positions are marked with a ".". Stretches of amino acids, which are poorly conserved, are not shown in the alignment, but indicated as (xx) or (xxx). The seven putative transmembrane domains are indicated.

attenuated by addition of TA. TA did not stimulate cyclic AMP production in these cells. This is the first report of the expression and characterization of an insect GPCR in insect cell lines (Vanden Broeck *et al.*, 1995a). Northern blotting analysis and *in situ* hybridization studies showed a non-abundant but widespread distribution of the TA receptor mRNA within the locust CNS (Vanden Broeck *et al.*, 1995b).

As mentioned in another section, recent studies confirm the hypothesis that TA acts as a signaling substance in the locust CNS. In addition, a specific TA binding site has now been characterized by incubating locust CNS membrane preparations with [³H]tyramine (Hiripi *et al.*, 1994). Although this [³H]tyramine binding site exhibits only low affinity to many antagonistic ligands, the pharmacological profile of antagonist binding is remarkably similar to that of the cloned TA receptor. Nevertheless, it is not yet clear whether locusts might possess several distinct TA receptor subtypes.

5. Peptide Receptors

In mammals, receptors for most small peptide messengers belong to the GPCR superfamily. Moreover, during the past 5 years, the majority of these mammalian peptide receptors have been identified via molecular cloning studies. It is very reasonable to predict that this technical (r)evolution will be extended toward other phyla and that many peptide receptors encountered in insects will be cloned during the next 5 years. Although most insect peptides show a very limited degree of sequence similarity with mammalian peptides, certain peptide families have been shown to have members in vertebrates as well as in invertebrate species: e.g., vasotocin-like, CRF-like, gastrin-CCK-like, NPY/NPF-like, and tachykinin-like peptides are encountered in insects. Therefore, there is a good chance that the receptors, which mediate the effects exerted by those peptides, have been conserved by coevolutionary selective pressure, which tends to preserve conformational matching and basic functional mechanisms. In fact, the first insect peptide GPCRs have already been cloned and sequenced, whereas novel orphan receptors, which are probably signaling through putative insect peptides, are now being characterized via expression studies.

a. Orphan Peptide Receptors   Based on sequence conservation of the mammalian tachykinin peptide receptors, three novel fruit fly GPCR cDNAs were cloned and further identified. The sequences of these receptors indeed exhibit clear similarities with one another (Fig. 6) and with other mammalian peptide receptors.

Xenopus oöcytes expressing the DTKR (*Drosophila* tachykinin receptor) cRNA produce pertussis-toxin-sensitive electrophysiological responses to

```
                                      I
TLR1-Dro (xx)-RPYELPWEQKTIWAIIFGLMMFVAIAGNGIVLWIVTGHRSMRTVTNYFLL
TLR2-Dro (xx)-FAFVVPWWRQVLWSILFGGMVIVATGGNLIVVWIVMTTKRMRTVTNYFIV
NYR-Dro  (xx)-EDMWSSAYFKIIVYMLYIPIFIFALIGNGTVCYIVYSTPRMRTVTNYFIA
          .  ...   ...   . . *  ** .* .** . .*******.
              II                                III
NLSIADLLMSSLNCVFNFIFMLNSD-WPFGSIYCTINNFVANVTVSTSVF
NLSIADAMVSSLNVTFNYYYMLDSD-WPFGEFYCKLSQFIAMLSICASVF
SLAIGDILMSFFCEPSSFISLFILNYWPFGLALCHFVNYSQAVSVLVSAY
.*.*.*  ..*  .    .. ..   . ****   *  . ... . ... .*..
      ___                          IV
TLVAISFDRYIAIVDPLKRRTSRRKVRIILVLIWALSCVLSAPCLLYSSI
TLMAISIDRYVAIIRPLQPRMSKRCNLAIAAVIWLASTLISCPMMIIYRT
TLVAISIDRYIAIMWPLKPRITKRYATFIIAGVWFIALATALPIPIVS--
**.***.***.**. **..* ..*    *  . .*  .   . *   .
                                    V
MTKQYYNGKSRTVCFMMWPDGRYPTSMADYAYNLIILVLTTGIPMIVMLI
EEVPVRGLSNRTVCYPEWPDGPTNHSTMESLYNILIIILTYFLPIVSMTV
-GLDIPMSPWHTKCEKYICREMWPSRSQEYYYTLSLFALQFVVPLGVLIF
  .       .* *      .        *.... *   .*. .
                                      VI
CYSLMGRVPGGSRSIGEN-TDRQMESMKSKRKVVRMFIAIVSIFAICWLP
TYSRVGIELWGSKTIGEC-TPRQVENVRSKRRVVKMMIVVVLIFAICWLP
TYARITIRVWAKRPPGEAETNRDQRMARSKRKMVKMMLTVVIVFTCCWLP
*.  ..    .... **  * *.   .***..*.*....* .*. ****
                              VII
YHLFFIYAYHNNQVASTKYVQHMYLGFYWLAMSNAMVNPLIYYWMNKRFR
FHSYFIITSCYPAITEAPFIQELYLAIYWLAMSNSMYNPIIYCWMNSRFR
FNILQLLLN-DEEFAHWDPLPYVWFAFHWLAMSHCCYNPIIYCYMNARFR
 ..   .     . .     .. ......*****.   **.**..** ***

MYFQRIICCCC-(xx)
YGFKMVFRWCL-(xx)
SGFVQLMHRMP-(xx)
      *   ..
```

FIG. 6 Sequence alignment of three insect peptide receptors which show sequence similarity to mammalian tachykinin receptors. The position of identical residues is indicated by "*". Highly conserved codon positions are marked with a ".". TLR1 = NKD; TLR2 = DTKR; NYR = NPY-R. Stretches of amino acids, which are poorly conserved, are not shown in the alignment, but indicated as (xx). The seven putative transmembrane domains are indicated.

substance P and its agonists, but not to other vertebrate tachykinin-like peptides. This might suggest that the (mysterious) endogenous ligand for DTKR could be closely related to SP. In adult flies, the DTKR transcripts are preferentially localized in the head. *In situ* hybridization experiments indicate that the transcripts are predominantly present in neuronal cell bodies. Furthermore, the expression of the DTKR gene appears to be developmentally regulated, since in the later stages of embryogenesis, the highest levels of expression can be observed (Li *et al.,* 1991).

Another orphan receptor which shows sequence similarity with mammalian tachykinin receptors is NKD (*neurokinin receptor of Drosophila*).

Stable mammalian (NIH 3T3) cell lines expressing this receptor cDNA seem to be responsive to locustatachykinin-II (Schoofs *et al.*, 1993), but not to any other known tachykinin-like peptide. Locustatachykinin-II was shown to produce an elevation of the InsP3 content of these cells at low concentrations and in a very selective way. Therefore, it is possible that the natural ligand for this receptor is a close relative of the locust tachykinin-like peptide, locustatachykinin-II. The transcript of the NKD gene in the adult fruit fly is mainly localized in the head, although there is also some detection in the body. In addition, the mRNA is detected at very early stages in embryogenesis and reaches its highest amounts at the time of neuronal development (Monnier *et al.*, 1992).

A third GPCR which is similar to mammalian peptide receptors was cloned from *Drosophila*. The cDNA sequence was amplified and cloned by means of a PCR approach. Expressed in *Xenopus* oöcytes, this receptor mediated electrophysiological responses induced by mammalian members of the neuropeptide Y family in the following order of effectiveness: peptide YY > NPY ≫ PP. Interestingly, invertebrate and insect members of this peptide family have already been reported (NPF-like peptides). A number of other mammalian peptides, including tachykinins, were not effective. The receptor mRNA is present in the adult fly's head and body and, again, there is a developmental regulation of the transcript (Li *et al.*, 1992). More recently, the cDNA coding for a GPCR with significant sequence similarity to receptors for tachykinin-like peptides was cloned from the stable fly, *Stomoxys calcitrans* (Guerrero *et al.*, 1995).

b. The Diuretic Hormone Receptor   The receptor that mediates the stimulatory effect of the CRF-like diuretic hormone on fluid secretion by the Malpighian tubules has been cloned from *Manduca sexta*. This was achieved by using an expression cloning approach. RNA from Malpighian tubules was used as a template for the construction of a cDNA library. The size-fractionated cDNA was cloned in the mammalian expression vector pKS1. The plasmid-cDNA constructs were subdivided into 512 pools of about 2100 *E. coli* transformants. Plasmid DNA derived from these pools was isolated and used for transfection of COS-7 cells. After about 3 days, these COS-7 cell pools were screened for binding with iodinated [^{125}I]Mas-DH. Then the cells were washed, fixed, dried, and coated with photographic emulsion. Positive batches of pKS1-plasmid+cDNA (6/150) were further subdivided until the Mas-DH receptor cDNA was entirely purified.

The receptor was expressed in COS-7 cells and showed very high affinity binding to [^{3}H]Mas-DH ($K_D = 79$ pM). Furthermore, receptor activation by DH results in a stimulation of cAMP production within these COS-7 cells. The receptor protein contains seven putative transmembrane regions and shares sequence similarity (25–30%) with the mammalian CRF, secretin, calcitonin, vasoactive intestinal peptide (VIP), parathyroid hormone, glucagon-like peptide 1, growth hormone releasing hormone, PACAP, and glucagon receptors. The DH analogs display an order of affinity of Mas-DH = Pea-DH = Acd-DH > Lom-DH = Mas-DPII (Mas = *Manduca sexta*, Pea = *Periplaneta americana*, Acd = *Acheta domesticus*, Lom = *Locusta migratoria*). In the same order of potency, these peptides were also capable of stimulating adenylate cyclase activity in transfected COS-7 cells. The Mas-DHR transcript was shown to be present in Malpighian tubules, but could not be detected in brain and fat body of *Manduca* (Reagan, 1994).

Recently, recombinant baculovirus containing Mas-DHR was used to infect insect Sf9 cells. Membrane preparations of these infected cells displayed high-affinity binding to DH, but, owing to strong transcriptional activity of the baculovirus construct, the receptor expression levels were much higher in these cells than in COS-7 cells or than in the Malpighian tubules. The receptor protein was visualized by chemical cross-linking of [^{125}I]Mas-DH after SDS-PAGE as a broad band of 48–52 kDa. In addition, DH was able to stimulate cAMP production in the infected Sf9 cells. Interestingly, the N-terminal truncated analog of Mas-DH (amino acids 13–41), which binds to the receptor with high affinity, did not stimulate cAMP synthesis and was even able to act as a DH antagonist. Therefore, the N-terminal of DH might be necessary for functional receptor activation (Reagan, 1995). An additional DH receptor homolog was cloned from *Acheta domesticus* and shares about 53% of its sequence with the Mas-DHR.

The recent identification of insect GPCRs for peptide ligands indicates that many receptor subfamilies have an ancient origin which has to precede the divergence of proto- and deuterostomian animals.

6. Rhodopsins

Different insect opsin genes have been cloned and identified. In *Drosophila*, these visual pigments (rhodopsins, Rh 1-4) are preferentially expressed in distinct photoreceptor cells. The most abundant one, Rh1, is found in the outer photoreceptor cells R1-6, has an absorption maximum at 480 nm, and its gene corresponds to the ninaE locus (O'Tousa *et al.*, 1985; Zuker *et al.*, 1985). Two other opsins (Rh3, Rh4) have a different absorption spectrum and are expressed in nonoverlapping sets of the inner photoreceptor cells (R7) (Fryxell and Meyerowitz, 1987; Zuker *et al.*, 1987; Montell *et al.*, 1987). The Rh2 opsin has been shown to form the visual pigment which is predominantly encountered in the dorsal ocelli and has an absorption maximum at 420 nm (Cowman *et al.*, 1986; Feiler *et al.*, 1988; Pollock and Benzer, 1988). Additional opsins derived from other *Drosophila* species (*pseudo-obscura, viridis*) and from *Calliphora* have now been identified. All insect opsins appear to possess significant sequence similarities with the vertebrate ones. In the next section, some considerations concerning the molecular evolution of opsin genes are discussed.

C. Aspects of the Molecular Evolution of G-Protein-Coupled Receptors

GPCRs have a well-conserved general structure, although some of the amino acid or corresponding nucleotide sequences show a very low degree of similarity. It is not unlikely that the so-called superfamily consists of several monophyletic protein families which, by convergent evolution, have adopted a similar overall structure. The generally observed topography probably serves to fulfill the functional needs of a transmembrane signaling protein that acts through the activation of heterotrimeric G-proteins. A multitude of different ligands use this type of receptor to exert their biological functions: these range from environmental signals such as light, odor, and taste over many neurotransmitters, neuromodulators, paracrines, and pheromones to hormones and glycoproteins. Even if the GPCR receptor superfamily is not entirely monophyletic, this still implies that during evolution, many gene multiplications and duplications must have occurred to provide all these ligands with specific receptors. It is now clear that many of these events (but not all) took place before the great radiation of the present animal phyla. This is illustrated by the fact that a lot of receptor subfamilies are present in both proto- and deuterostomian species. The corresponding ligands themselves may have been conserved to a lesser extent. This probably explains why such a large diversity of different ligands for these GPCR receptors are encountered. Nevertheless, specific receptor–ligand interactions must have been subject to a coevolutionary selective pressure that has preserved the basic requirements for correct functioning and conformational matching. This is reflected in the evolutionary conservation of a number of extracellular signaling molecules, such as certain biogenic amines (e.g., acetylcholine, dopamine) or members of highly conserved peptide messenger families (e.g., vasopressin-like peptides, tachykinins, CRF-like peptides, NPY/NPF-like peptides).

The functional and structural constraints of GPCR-mediated signal transduction mechanisms include other interactions, especially those with G-proteins, which have been well preserved. Owing to their relative importance as key regulators of cellular physiology, G-proteins and their intracellular signaling properties, including the structural motif responsible for protein–protein interactions, may have been very well conserved. This explains why mammalian GPCRs can interact with and activate G-proteins which are endogenously present in insect cellular expression systems and vice versa. Therefore, the basic signaling properties of evolutionarily conserved receptor subtypes are probably well conserved too. At first sight, this does not seem to be the case with mammalian and insect visual pigments. The different types of visual pigments combine distinct absorption spectra with different opsin proteins, which may be differentially expressed. Opsin sequence alignments have shown that insect and mammalian opsin genes and proteins are conserved. The sequence alignment dendrograms, however, put the insect (*Drosophila*) and mammalian (human) opsins in separate clusters (Fig. 7). This might indicate that color vision in humans and in fruit flies has been established by using the same type of receptor proteins, but that the different visual pigments (and their opsins) may have arisen independently after the evolutionary divergence point between proto- and deuterostomians.

Within each of the (insect or mammalian) clusters, the third cytoplasmic domain is well conserved, whereas between insect and human sequences,

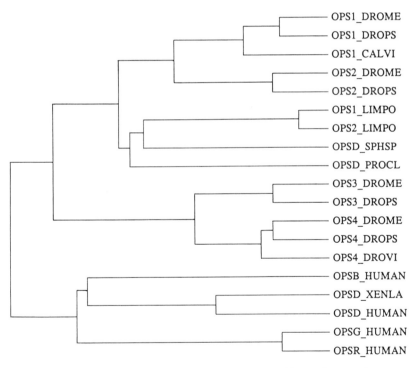

FIG. 7   Dendrogram of an alignment of different vertebrate and invertebrate opsin sequences present in the "SWISSPROT" protein sequence database (performed by the CLUSTAL program of PCGENE, Intelligenetics). All arthropod opsins are found in the large upper cluster of sequences, whereas the vertebrate (and human) opsins form a separate cluster. The functional and evolutionary significance of this observation is discussed in the text.

this part shows the lowest degree of sequence similarity (Fryxell and Myerowitz, 1991). Many experiments with GPCRs indicate that certain cytoplasmic regions and especially the third intracellular loop, are very important for determining the specificity of the interaction with the corresponding G-protein(s). In vertebrates, a specialized G-protein, transducin, mediates the light-induced responses via cGMP-phosphodiesterase regulation. In insects, however, the light-induced signal-transduction pathway is distinct from the vertebrate one: InsP$_3$ is the second messenger which mediates visual signaling in the insect photoreceptors. Therefore, the insect phototransducing G$_\alpha$ subunit belongs to a different class of G-proteins and is similar to vertebrate G$_{q\text{-}\alpha}$ (Lee *et al.*, 1994). The existence of distinct coupling mechanisms for insect and mammalian visual signal transduction may explain why the respective receptors have such a low degree of sequence identity in their third cytoplasmic loops. The visual pigments of insects and mammals might indeed be the result of a separate but convergent evolution leading to "color" vision by using distinct (but distantly related) signaling pathways.

Receptors coupled to G-proteins have an ancient origin: GPCRs have even been found in yeast (*Saccharomyces cerevisiae*) and in the slime mold (*Dictyostelium discoideum*). In yeast, the products of the STE2 and STE3 genes have been demonstrated to be putative 7TMD membrane proteins which act as receptors for the pheromone α and a mating factor respectively. These pheromones are extracellular peptide messengers that are involved in generating the sexual mating response between a and α-type cells (Hagen *et al.*, 1986). G-Protein subunits are responsible for further intracellular signal transduction (Nomoto *et al.*, 1990). In yeast, the G-protein β and γ subunits (gene STE4 and STE18 products) probably act as positive regulators that propagate the mating signal. The G-α subunit (GPA1 gene product) probably functions as a negative regulator which, in the absence of mating pheromone, in its GDP-bound form associates with β and γ to inhibit their activity. In *Dictyostelium,* chemoattraction regulates a developmental program which leads to the aggregation of ameba-like cells into a multicellular organism in which certain differentiation processes take place. Cyclic AMP serves as the extracellular chemoattractant signal which initiates and regulates the chemotactical aggregation. The chemoattractant receptor is a 7TMD protein that shares some very limited sequence similarity with animal GPCRs. The receptor is thought to act through a G-protein. Cells transformed to express antisense receptor mRNA fail to produce the cAMP receptor and do not enter the aggregation stage (Klein *et al.*, 1988). More recently, a cytosolic protein, CRAC, containing a PH (pleckstrin homology) domain was found to be required for receptor and G-protein-mediated adenylyl cyclase activation. This protein is probably involved as a signal transducer or regulator acting between the G-protein and the effector enzyme, adenylyl cyclase (Insall *et al.,* 1994).

V. Identification of G-Protein-Coupled Receptors in Cellular Expression Systems

A. The Use of Cultured Insect Cells to Study G-Protein-Coupled Receptors

In vertebrates as well as in invertebrates, cloning techniques have allowed the identification of many novel receptor genes and cDNAs. Receptor proteins and mRNAs can be localized *in situ* by employing immunocyto-chemistry and hybridization technology. Physiological and pharmacological properties of receptors can be studied *in vivo.* A single organ, however, may contain a multitude of different receptor (sub)types and some of these may be rarely expressed. Therefore, the characterization of certain receptors and the elucidation of their functional significance is a difficult task: *in vivo* situations are complicated. I already mentioned that *Xenopus* oöcytes and cultured mammalian cells are extensively used as expression systems for cloned receptor cDNAs. They allow production of a receptor protein in a less complex environment. Thus, the affinities of a number of receptor ligands can be determined and agonist-induced responses which are generated in these cells can be measured. Proteins produced in these expression systems can be purified and antibodies can be raised against them. Furthermore, modifications such as glycosylation or palmitoylation can be identified. In addition, the expression systems allow investigation of structure–function relationships by expression and characterization of mutagenized receptors (deletion, insertion, chimeric or site-directed muta-genesis). Mutated genes can be introduced into mouse embryonic stem cells. These provide a means to generate transgenic mice with altered physiological properties. Animals with a defective receptor gene may be valuable tools for studying the *in vivo* role of a receptor.

Physiological effects can be elicited by cloned receptors in cultured cell lines. Although the true *in vivo* effects might be different, the coupling specificities of GPCRs to their respective G-proteins appear to be well preserved. Therefore, coupling events occur within intracellular environments that differ from the natural ones. This is illustrated by the fact that certain activated insect receptors induce similar responses in both the original insect tissue and the mammalian expression system. This is an indication that some basic coupling mechanisms of GPCRs with G-proteins are similar in insect and in mammalian cells. The indication is further supported by results obtained with mammalian receptors expressed in insect cells. Nevertheless, since several distinct types and subtypes of heterotrimeric G-proteins are encountered in mammals as well as in insects, some (evolutionary) differences may not be excluded. Therefore it is important to verify in detail the coupling specificities within homologous cellular expression systems.

Insect cells are very popular for large-scale expression of cloned genes. Recombinant baculovirus production has become a standard technique in university labs and biotech companies all over the world. Unlike mammalian cell lines, insect cells can be easily grown without the need for expensive incubators. Extremely high expression levels are reached in lepidopteran cell lines, such as the Sf9 one, when these are infected with recombinant virus containing the foreign gene under the control of a "late" viral promoter. The strong polyhedrin and P10 promoters become active in the later stages of the baculoviral infection cycle (for transfer vectors and practical protocols, see O'Reilly *et al.*, 1992; King and Possee, 1992). Moreover, to get even higher yields, the infected cells can be grown at high densities in spinner bottles or in bioreactors. As shown in Table III, many mammalian GPCR-cDNAs have been successfully expressed by employing the baculovirus system. Research labs and companies are using the system to make (and even sell) membrane preparations containing high levels of a given receptor type as a source for receptor-specific binding assays.

Although the baculoviral expression system is a very successful one, it also has some disadvantages. First of all, the virus is lytic and it will kill the infected cells. Therefore, the expression is transient and changes according to the time after infection. In addition, the normal physiological state of the virally infected cells is altered. The production of cellular proteins is inhibited and viral protein synthesis becomes predominant, especially in the later stage after infection. At these late stages of the infection cycle, the foreign gene is overexpressed and this might be less beneficial for obtaining a correct protein modification (e.g., glycosylation) or localization (e.g., by intracellular trafficking toward the plasma membrane). Overexpression and altered cellular physiology are less favorable properties of an expression system that has to be employed for functional characterization studies. It is clear that nondisruptive expression systems are needed in order to eliminate most of these disadvantages.

Methods for the production of stably transformed insect cells have now been established by several research groups. These methods appear to be extremely well suited for expression of membrane glycoproteins. Although the expression levels are often lower than those obtained with the baculovirus-based systems, protein modifications and trafficking seem to be performed in a more efficient way. The fraction of correctly expressed receptor proteins obtained in these stably transfected cells can therefore be larger

TABLE III

Expression and Characterization of Some Mammalian G-Protein-coupled Receptors in Insect Cells

Receptor	Effect(s) studied	Reference
Human D4 dopamine receptor	Radioligand binding assays	Mills *et al.* (1993)
Human D2 dopamine receptor	Radioligand binding/photolabeling/glycosylation	Javitch *et al.* (1994)
Human D1 dopamine receptor	cAMP rise/photolabeling/palmitoylation/phosphorylation/desensitization	Ng *et al.* (1994)
Human β2-adrenergic receptor	cAMP rise/binding/photolabeling/palmitoylation/desensitization	Mouillac *et al.* (1992)
Human β2-adrenergic receptor	cAMP rise/GTP-ase activity/binding/glycosylation/desensitization	Kleymann *et al.* (1993)
Turkey β, human M1/M2 rec.	Binding/glycosylation/reconstitution in phospholipid vesicles	Parker *et al.* (1991)
Human β2-adrenergic receptor	Screening of fluorescent β-adrenoceptor ligands	Heithier *et al.* (1994)
Human 5HT1A receptor	Binding/coupling to a Go-like protein/pertussis toxin ADP ribosylation	Mulheron *et al.* (1994)
Human 5HT1 receptors	Inhibition of forskolin-stimulated cAMP formation/binding	Parker *et al.* (1994)
Human 5HT1B receptor	Attenuation of adenylyl cyclase/phosphorylation/palmitoylation	Ng *et al.* (1993)
Human and rat mACh rec.	Binding/production/solubilization	Rinken *et al.* (1994)
Rat M3 mACh receptor	Potassium currents/pertussis toxin	Vasudevan *et al.* (1992)
Rat M5 mACh receptor	Ca^{2+} rise	Hu *et al.* (1994)
Human M2 mACh receptor	Phosphorylation and desensitization by receptor kinases	Richardson *et al.* (1993)
Rat cannabinoid receptor	Binding/immunoblotting	Petit *et al.* (1994)
Human TSH receptor	Post-translational modification	Misrahi *et al.* (1994)
Human *N*-formyl peptide rec.	Binding/lack of G-protein coupling	Quehenberger *et al.* (1992)
Rat odorant receptors	InsP$_3$ rise	Raming *et al.* (1993a)

than in virus-infected cells (Joyce *et al.,* 1993). Relatively high levels of expression can be reached by using a novel method to produce permanently transformed *Drosophila* S2-cells (Vulsteke *et al.,* 1993). The expression constructs are based on the presence of a baculoviral immediate early gene promoter which seems to be active in almost every insect cell line without requiring additional elements. This procedure is based on cotransfection of the expression construct with a hygromycin B resistance construct. Selection of the cells with hygromycin B leads to the genomic integration of both constructs at a relatively high copy number. When comparing this method with the baculovirus system, similar amounts of receptor protein could be detected in the membrane fraction of the S2 cells and of infected Sf9 cells (Vanden Broeck *et al.,* 1995a). The major advantage of the use of stable S2 cells is the nondisruptive and permanent character of the expression. There is no infection-dependent expression and the properties of the cells are not changed to the same extent as by a viral infection. The method might prove to be particularly well suited for the introduction of hetero-oligomeric ligand-gated ion channels composed of different combinations of subunits. The interactions between GPCRs, G-protein subunits, and effector protein isoforms could also be investigated in this way. Moreover, stably transformed insect cells are useful tools for functional receptor characterization studies and they might be adapted for expression cloning purposes. There is a need to investigate the signal-transduction mechanisms which are present in the insect cellular expression systems.

B. Effects Measured on Cultured Insect Cells Generated by Coupling to G-Proteins

Several mammalian and a few insect G-protein-coupled receptors have been successfully expressed by employing the baculoviral expression system. The system is suitable for large-scale production of receptor proteins. In addition, mammalian receptors can elicit agonist-induced responses in insect cells. cAMP rise, cAMP inhibition, InsP$_3$ rise, Ca^{2+} rise, electrophysiological effects, phosphorylation, palmitoylation, and desensitization are some of the insect cellular responses that were observed after agonist administration. The various results of a number of receptor expression studies are summarized in Table III.

C. Future Prospects

The interaction of GPCRs with their ligands and with the corresponding G-proteins can be studied in insect cell lines. As in mammalian systems, site-directed mutagenesis, deletion–insertion mutagenesis, and chimeric proteins could be used to define important regions or amino acid residues for ligand binding or functional activation (Buckley *et al.,* 1990). The cellular expression systems can be employed to monitor biochemical changes (e.g., protein phosphorylation), second messenger changes (cAMP, InsP$_3$, DAG, Ca^{2+}, arachidonic acid metabolites, cGMP, etc.), or electrophysiological effects. Confocal laser and microphysiometer technology might be used to analyze cellular responses to administration of agonists and antagonists. The cell lines are useful for characterization of novel cloned (orphan) GPCRs. The expression systems might also be adapted for functional receptor cloning purposes.

Cultured cells can become useful tools for studying G-protein signaling in insects. Although the basic mechanisms of G-protein coupling are probably well conserved during evolution, insect-specific features of G-protein-mediated effects are still poorly understood: the exact mechanisms by which insect G-proteins regulate the activity of cellular effector systems and the specificities of the molecular interactions involved deserve our attention. As in mammals and yeast, G-protein subunits (α and β-γ) are key regulatory factors taking part in intracellular signaling. They may interact with other major signaling pathways described earlier in this chapter (e.g., signaling by tyrosine kinases, phosphatases, Ser/Thr kinases, small G-proteins, ion channels, and transcription factors) to integrate incoming information formed by environmental or intercellular messages and to allow the cell to produce an appropriate response. Regulation and desensitization of activated receptors are also part of such a response. Specific receptor kinases can phosphorylate amino acid residues in intracellular regions of GPCRs to alter their conformation and/or to allow binding of arrestin-like molecules. Insect cells are systems in which these processes could be studied in more detail.

Moreover, insect cell lines are efficient expression systems for economically important proteins. Receptors are possible target molecules for medical or agricultural applications. Insect cells producing large quantities of a given receptor type can be used as a stable receptor source for compound screening. Real-time measurements obtained with biosensor or microphysiometer technology can become interesting tools for this type of screening test. In combination with improving protein modeling abilities, possible receptor ligands such as peptide mimetics could then be designed, synthesized, and screened on permanently transformed insect cells. Insect receptors are possible target molecules for insecticidal agents. These might be administered to insects by direct feeding, by genetic engineering of plants, or by manipulated (baculo)viruses. Another possible source of useful compounds are the ligands for insect pheromone receptors. In fact, the study of evolutionary aspects of GPCRs and of their signaling pathways may reveal some insect-specific features of these processes. More insight into these features might result in novel, less disruptive methods for insect pest control (Keeley and Hayes, 1987; Spindler-Barth, 1992; Masler *et al.*, 1993; Nachman *et al.*, 1993c; Kelly *et al.*, 1994).

VI. Concluding Remarks

Insect cells appear to employ the same main classes of signal-transducing molecules as mammalian cells. This indicates that the molecular principles of physiological functioning are most probably similar in all metazoans. Although the major classes and several subclasses of receptor proteins may have had their evolutionary origin before the great radiation of the distinct animal phyla, a lot of gene duplications must have occurred later in evolution in order to give rise to a multitude of ligand-specific receptors. During evolution, the nature and especially the numbers of the molecular constituents involved in environmental and cell–cell communication processes have been changing. In fact, biological evolution might basically be an evolution of communication systems (De Loof and Vanden Broeck, 1995b). Important intracellular pathways have been well conserved due to a "(co-) evolutionary" selective pressure that has preserved functional molecular mechanisms and conformational matching between interacting partners (Stefano, 1988). Specific intermolecular interactions appear to be extremely important for generating a large diversification of the communication systems. This diversification is enlarged by the existence of several variants or isoforms of extracellular messengers, receptors, intracellular signal transducers, and effector proteins. These can have distinct specificities and/ or activities.

In mammals, many extracellular messenger molecules exert their functions through the activation of G-proteins. This process is mediated by a binding event at the level of the plasma membrane where a specific interaction of the messenger molecule with an integral membrane protein, a G-protein-coupled receptor (GPCR), takes place. All GPCRs appear to have a similar three-dimensional structure and activation mechanism.

In insects, the number of extracellular messenger molecules that are identified is rapidly growing. It seems that the intercellular communication events that occur within these animals reach a degree of complexity which was not suspected before. In addition, insects possess many distinct receptor proteins, and a relatively small number of these have been identified by molecular cloning. The characterized proteins fulfill physiological functions which are similar to those of their mammalian counterparts. Insect-specific features of such receptor proteins might lead to the development of novel methods and agents for insect pest control.

Insect cell lines prove to be well suited for high-efficiency expression and characterization of cloned mammalian and insect GPCRs (Vanden Broeck *et al.,* 1995a). Stably transformed insect cell lines can be employed as a permanent source of receptor proteins for radioligand binding assays and for screening procedures. In addition, they could become interesting and versatile tools to be used in signal-transduction research. This, in combination with technological improvements of the scientific instrumentarium, may lead to novel or more detailed insights into complicated biological signaling processes.

Acknowledgments

All those who contributed to the making of this review are gratefully acknowledged. The author especially thanks M. Van der Eeken and J. Puttemans who helped with typesetting and figure design. In addition, the author is very grateful to the K. U. Leuven Research Council, the E.C., the N.F.W.O., and the I.W.T. for financial support of research projects.

References

Abdallah, E. A. M., Eldefrawi, M. E., and Eldefrawi, A. T. (1991). Pharmacologic characterization of muscarinic receptors of insect brains. *Arch. Insect Biochem. Physiol.* **17,** 107–118.

Abou-Samra, A. B., Jüppner, H., Force, T., Freeman, M. W., Kong, X.-F., Schipani, E., Urena, P., Richards, J., Bonventre, J. V., Potts, J. T., Kronenberg, H. M., and Segré, G. V. (1992). Expression cloning of a common receptor for parathyroid hormone and parathyroid hormone-related peptide from rat osteoblast-like cells. *Proc. Natl. Acad. Sci. U.S.A.* **89,** 2732–2736.

Albert, J. L., and Lingle, C. J. (1993). Activation of nicotinic acetylcholine receptors on cultured *Drosophila* and other insect neurones. *J. Physiol. (London)* **463,** 605–630.

Ali, D. W., Orchard, I., and Lange, A. B. (1993). The aminergic control of locust (*Locusta migratoria*) salivary glands: Evidence for dopaminergic and serotoninergic innervation. *J. Insect Physiol.* **39,** 623–632.

Andres, A. J., Fletcher, J. C., Karim, F. D., and Thummel, C. S. (1993). Molecular analysis of the initiation of insect metamorphosis: A comparative study of *Drosophila* ecdysteroid-regulated transcription. *Dev. Biol.* **160,** 388–404.

Arakawa, S., Gocayne, J. D., McCombie, W. R., Urquhart, D. A., Hall, L. M., Fraser, C. M., and Venter, J. C. (1990). Cloning, localization, and permanent expression of a *Drosophila* octopamine receptor. *Neuron* **4,** pp. 343–354.

Baines, D., and Downer, R. G. H. (1992). 5-Hydroxytryptamine-sensitive adenylate cyclase affects phagocytosis in cockroach haemocytes. *Arch. Insect Biochem. Physiol.* **21,** 303–316.

Baines, D., and Downer, R. G. H. (1994). Octopamine enhances phagocytosis in cockroach haemocytes: Involvement of inositol triphosphate. *Arch. Insect Biochem. Physiol.* **26,** 249–261.

Baines, D., DeSantis, T., and Downer, R. G. H. (1992). Octopamine and 5-hydroxytryptamine enhance the phagocytic and nodule formation activities of cockroach (*Periplaneta americana*) haemocytes. *J. Insect. Physiol.* **38,** 905–914.

Baines, R. A., and Downer, R. G. H. (1991). Pharmacological characterization of a 5-hydroxy-tryptamine-sensitive receptor/adenylate cyclase complex in the mandibular closer muscles of the cricket, *Gryllus domestica. Arch. Insect Biochem. Physiol.* **16,** 153–163.

Baines, R. A., Lange, A. B., and Downer, R. G. H. (1990). Proctolin in the innervation of the locust mandibular closer muscle modulates contractions through the elevation of inositol triphosphate. *J. Comp. Neurol.* **297,** 479–486.

Banner, S. E., Osborne, R. H., and Cattel, K. J. (1987). The pharmacology of the isolated foregut of the locust *Schistocerca gregaria:* II. Characterization of a 5HT2-like receptor. *Comp. Biochem. Physiol. C* **88C,** 139–144.

Basler, K., Christen, B., and Hafen, E. (1991). Ligand-independent activation of the sevenless receptor tyrosine kinase changes the fate of cells in the developing *Drosophila* eye. *Cell (Cambridge, Mass.)* **64,** 1069–1081.

Benson, J. A. (1992). Electrophysiological pharmacology of the nicotinic and muscarinic cholinergic responses of isolated neuronal somata from locust thoracic ganglia. *J. Exp. Biol.* **170,** 203–233.

Berridge, M. J. (1970). The role of 5-hydroxytryptamine and cyclic AMP in the control of fluid secretion by isolated salivary glands. *J. Exp. Biol.* **53,** 171–186.

Berridge, M. J. (1981). Electrophysiological evidence for the existence of separate receptor mechanisms mediating the action of 5-hydroxytryptamine. *Mol. Cell. Endocrinol.* **23,** 91–104.

Berridge, M. J. (1993). Inositol triphosphate and calcium signaling. *Nature (London)* **361,** 315–325.

Berridge, M. J. (1994). Relationship between latency and period for 5-hydroxytryptamine-induced membrane responses in the *Calliphora* salivary gland. *Biochem. J.* **302,** 545–550.

Bicker, G. (1992). Taurine in the insect central nervous system. *Comp. Biochem. Physiol. C* **103C,** 423–428.

Blake, A. D., Anthony, N. M., Chen, H. H., Harrison, J. B., Nathanson, N. M., and Satelle, D. B. (1993). *Drosophila* nervous system muscarinic acetylcholine receptor: Transient functional expression and localization by immunocytochemistry. *Mol. Pharmacol.* **44,** 716–724.

Blau, C., Wegener, G., and Candy, D. J. (1994). The effect of octopamine on the glycolytic activator fructose 2,6-bisphosphate in perfused locust flight muscle. *Insect Biochem. Mol. Biol.* **24,** 677–683.

Bloomquist, J. R., Roush, R. T., and ffrench-Constant, R. H. (1992). Reduced neuronal sensitivity to dieldrin and picrotoxinin in a cyclodiene-resistant strain of *Drosophila melanogaster. Arch. Insect Biochem. Physiol.* **19,** 17–25.

Blüml, K., Mutschler, E., and Wess, J. (1994). Insertion mutagenesis as a tool to predict the secondary structure of a muscarinic receptor domain determining specificity of G-protein coupling. *Proc. Natl. Acad. Sci. U.S.A.* **91,** 7980–7984.

Boekhoff, I., Seifert, E., Göggerle, S., Lindemann, M., Krüger, B. W., and Breer, H. (1993). Pheromone-induced second-messenger signaling in insect antennae. *Insect Biochem. Mol. Biol.* **23,** 757–762.

Boguski, M. S., and McCormick, F. (1993). Proteins regulating Ras and its relatives. *Nature (London)* **366,** 643–653.

Borovsky, D., Powell, C. A., Nayar, J. K., Blalock, J. E., and Hayes, T. K. (1994). Characterization and localization of mosquito-gut receptors for trypsin modulating oostatic factor using a complementary peptide and immunocytochemistry. *FASEB J.* **8,** 350–355.

Bossy, B., Ballivet, M., and Spierer, P. (1988). Conservation of neural nicotine acetylcholine receptors from *Drosophila* to vertebrate central nervous systems. *EMBO J.* **7,** 611–618.

Brann, M. R., Klimkowski, V. J., and Ellis, J. (1993). Structure/function relationships of muscarinic acetylcholine receptors. *Life Sci.* **52,** 405–412.

Braun, R. P., and Wyatt, G. R. (1992). Modulation of DNA-binding proteins in *Locusta migratoria* in relation to juvenile hormone action. *Insect Mol. Biol.* **1**, 99–107.

Breer, H., and Sattelle, D. B. (1987). Molecular properties and functions of insect acetylcholine receptors. *J. Insect Physiol.* **33**, 771–790.

Breer, H., Raming, K., and Boekhoff, I. (1988). G-proteins in the antennae of insects. *Naturwissenschaften* **75**, 627.

Brummel, T. J., Twombly, V., Marqués, G., Wrana, J. L., Newfeld, S. J., Attisano, L., Massagué, J., O'Connor, M. B., and Gelbart, W. M. (1994). Characterization and relationship of Dpp receptors encoded by the saxophone and thick veins genes in *Drosophila*. *Cell* (*Cambridge, Mass.*) **78**, 251–261.

Brunner, D., Dücker, K., Oellers, N., Hafen, E., Scholz, H., and Klämbt, C. (1994). The ETS domain protein Pointed-P2 is a target of MAP kinase in the Sevenless signal transduction pathway. *Nature* (*London*) **370**, 386–389.

Buck, L., and Axel, R. (1992). A novel multigene family may encode odorant receptors: A molecular basis for odor recognition. *Cell* (*Cambridge, Mass.*) **65**, 175–187.

Buckley, N. J., Hulme, E. C., and Birdsall, N. J. M. (1990). Use of clonal cell lines in the analysis of neurotransmitter receptor mechanisms and function. *Biochim. Biophys. Acta* **1055**, 43–53.

Bylemans, D. (1994). Isolation, characterization and mode of action of two novel folliculostatins of the grey fleshfly, *Neobellieria bullata*. Ph.D. Thesis, K. U. Leuven, Belgium.

Carson-Jurica, M., Schrader, W. T., and O'Malley, B. (1990). Steroid receptor family: Structure and functions. *Endocr. Rev.* **11**, 201–220.

Casanova, J., and Struhl, G. (1993). The torso receptor localizes as well as transduces the spatial specifying terminal body pattern in *Drosophila*. *Nature* (*London*) **362**, 152–155.

Casida, J. E. (1993). Insecticide action at the GABA-gated chloride channel: Recognition, progress, and prospects. *Arch. Insect Biochem. Physiol.* **22**, 13–23.

Cascino, P., Nectoux, M., Guiraud, G., and Bounias, M. (1989). The formamidine amitraz as a hyperglycemic alpha-agonist in worker honeybees (*Apis mellifera mellifera* L.) *in vivo*. *Biomed. Environ.* **2**, 106–114.

Chao, M. V. (1992). Neurotrophin receptors: A window into neuronal differentiation. *Neuron* **9**, 583–593.

Chen, Y-H., Pouysségur, J., Courtneidge, S. A., and Van Obberghen-Schilling, E. (1994). Activation of Src family kinase activity by the G protein-coupled thrombin receptor in growth-responsive fibroblasts. *J. Biol. Chem.* **269**, 27372–27377.

Childs, S. R., Wrana, J. L., Arora, K., Attisano, L., O'Connor, M. B., and Massagué, J. (1993). Identification of a *Drosophila* activin receptor. *Proc. Natl. Acad. Sci. U.S.A.* **90**, 9475–9479.

Clifford, R., and Schüpbach, T. (1994). Molecular analysis of *Drosophila* EGF receptor homolog reveals that several genetically defined classes of alleles cluster in subdomains of the receptor protein. *Genetics* **137**, 531–550.

Coats, J. R. (1982). "Insecticide Mode of Action." Academic Press, New York and London.

Corbet, S. A. (1991). A fresh look at the arousal syndrome of insects. *Adv. Insect Physiol.* **23**, 81–116.

Cotecchia, S., Exum, S., Caron, M. G., and Lefkowitz, R. J. (1990). Regions of the alpha1-adrenergic receptor involved in coupling to phosphatidylinositol hydrolysis and enhanced sensitivity of biological function. *Proc. Natl. Acad. Sci. U.S.A.* **87**, 2896–2900.

Cowman, A. F., Zuker, C. S., and Rubin, G. M. (1986). An opsin gene expressed in only one photoreceptor cell type of *Drosophila* eye. *Cell* (*Cambridge, Mass.*) **44**, 705–710.

Crespo, T., Xu, N., Daniotti, J. L., Troppmair, J., Rapp, U. R., and Gutkind, J. S. (1994). Signaling through transforming G protein-coupled receptors to NIH 3T3 cells involves c-Raf activation. *J. Biol. Chem.* **269**, 21103–21109.

Cusson, M., Prestwich, G. D., Stay, B., and Tobe, S. S. (1991). Photoaffinity labeling of allostatin receptor proteins in the corpora allata of the cockroach, *Diploptera punctata*. *Biochem. Biophys. Res. Commun.* **181**, 736–742.

Cutforth, T., and Rubin, G. M. (1994). Mutations in Hsp83 and cdc37 impair signaling by the sevenless receptor kinase in *Drosophila. Cell (Cambridge, Mass.)* **77,** 1027–1036.

Davis, J. P. L., and Pitman, R. M. (1991). Characterization of receptors mediating the actions of dopamine on an identified inhibitory motoneurone of the cockroach. *J. Exp. Biol.* **155,** 203–217.

De Loof, A., and Vanden Broeck, J. (1995a). Hormones and the cytoskeleton: An overview. *Neth. J. Zool.* (in press).

De Loof, A., and Vanden Broeck, J. (1995b). Communication, the key to defining "Life," "Death" and the force driving Evolution. *Belg. J. Zool.* **125,** 5–28.

De Loof, A., Callaerts, P., and Vanden Broeck, J. (1992). The pivotal role of the plasmamembrane-cytoskeletal complex and of epithelium formation in differentiation in animals. *Comp. Biochem. Physiol. A* **101A,** 639–651.

De Loof, A., Bylemans, D., Schoofs, L., Janssen, I., Spittaels, K., Vanden Broeck, J., Huybrechts, R., Borovsky, D., Hua, Y.-J., Koolman, J., and Sower, S. (1995). Folliculostatins, gonadotropins and a model for control of growth in the grey fleshfly, *Neobellieria bullata. Insect Biochem. Mol. Biol.* **25,** 661–667.

Desai, C. J., Popova, E., and Zinn, K. (1994). A *Drosophila* receptor tyrosine phosphatase expressed in the embryonic CNS and larval optic lobes is a member of the set of proteins bearing the "HRP" carbohydrate epitope. *J. Neurosci.* **14,** 7272–7283.

Donaldson, C. J., Mathews, L. S., and Vale, W. W. (1992). Molecular cloning and binding properties of the human type II activin receptor. *Biochem. Biophys. Res. Commun.* **184,** 310–316.

Downer, R. G. H., and Hiripi, L. (1994). Biogenic amines in insects. *In* "Insect Neurochemistry and Neurophysiology 1993 (A. B. Borkovec and M. J. Loeb, eds.), pp. 23–38. CRC Press, Boca Raton, FL.

Downer, R. G. H., Hiripi, L., and Juhos, S. (1993). Characterization of the tyraminergic system in the central nervous system of the locust, *Locusta migratoria migratoides. Neurochem. Res.* **18,** 1245–1248.

Ducrson, K., Carroll, R., and Clapham, D. (1993). α-helical distorting substitutions disrupt coupling between m3 muscarinic receptor and G proteins. *FEBS Lett.* **324,** 103–108.

Dunphy, G. B., and Downer, R. G. H. (1994). Octopamine, a modulator of the haemocytic nodulation response of non-immune *Galleria mellonella* larvae. *J. Insect Physiol.* **40,** 267–272.

Eason, M. G., Jacinto, M. T., Theiss, C. T., and Liggett, S. B. (1994). The palmitoylated cysteine of the cytoplasmic tail of α2A-adrenergic receptors confers subtype-specific agonist-promoted downregulation. *Proc. Natl. Acad. Sci. U.S.A.* **91,** 11178–11182.

Eckeret, M., Rapus, J., Nürnberger, A., and Penzlin, H. (1992). A new specific antibody reveals octopamine-like immunoreactivity in cockroach ventral nerve cord. *J. Comp. Neurol.* **322,** 1–15.

Elling, C. E., Moller-Nielsen, S., and Schwartz, T. W. (1995). Conversion of antagonist-binding site to metal-ion site in the tachykinin NK-1 receptor. *Nature (London)* **374,** 74–77.

Evans, P. D. (1978). Octopamine: A high affinity uptake mechanism in the nervous system of the cockroach. *J. Neurochem.* **30,** 1015–1022.

Evans, P. D. (1981). Multiple receptor types for octopamine in the locust. *J. Physiol. (London)* **318,** 99–122.

Evans, P. D. (1987). Phenyliminoimidazolidine derivatives activate both OA1 and OA2 receptor subtypes in locust skeletal. *J. Exp. Biol.* **129,** 239–250.

Evans, P. D., and Robb, S. (1993). Octopamine receptor subtypes and their modes of action. *Neurochem. Res.* **18,** 869–874.

Evans, R. M. (1988). The steroid and thyroid hormone receptor superfamily. *Science* **240,** 889–895.

Feiler, R., Harris, W. A., Kirschfeld, K., Wehrhahn, C., and Zuker, C. S. (1988). Targetted misexpression of a *Drosophila* opsin gene leads to altered visual function. *Nature (London)* **333,** 737–741.

ffrench-Constant, R. H. (1994). The molecular and population genetics of cyclodiene insecticide resistance. *Insect Biochem. Mol. Biol.* **24**, 335–345.

Fields, P. E., and Woodring, J. P. (1991). Octopamine mobilization of lipids and carbohydrates in the house cricket, *Acheta domesticus. J. Insect Physiol.* **37**, 193–199.

Findlay, J., and Eliopoulos, E. (1990). Three-dimensional modelling of G protein-linked receptors. *Trends Pharmacol. Sci.* **11**, 492–499.

Fitch, J., and Djamgoz, M. B. A. (1988). Proctolin potentiates synaptic transmission in the central nervous system of an insect. *Comp. Biochem. Physiol. C* **89C**, 109–112.

Fong, T. M., Cascieri, M. A., Yu, H., Bansal, A., Swain, C., and Strader, C. D. (1993). Amino-aromatic interaction between histidine 197 of the neurokinin-1 receptor and CP 96345. *Nature (London)* **362**, 350–353.

Fraser, S. P., Djamgoz, M. B. A., Usherwood, P. N. R., O'Brien, J., Darlison, M. G., and Barnard, E. A. (1990). Amino acid receptors from insect muscle: Electrophysiological characterization in *Xenopus* oocytes following expression by injection of mRNA. *Mol. Brain Res.* **8**, 331–341.

Freeman, R. S., and Donoghue, D. J. (1991). Protein kinases and protooncogenes: Biochemical regulators of the eukaryotic cell cycle. *Biochemistry* **30**, 2293–2302.

Fryxell, K. J., and Meyerowitz, E. M. (1987). An opsin gene that is expressed only in the R7 photoreceptor cell of *Drosophila. EMBO J.* **6**, 443–451.

Fryxell, K. J., and Meyerowitz, E. M. (1991). The evolution of rhodopsins and neurotransmitter receptors. *J. Mol. Evol.* **33**, 367–378.

Fujii, K., and Takeda, N. (1988). Phylogenetic detection of serotonin immunoreactive cells in the central nervous system of invertebrates. *Comp. Biochem. Physiol. C* **89C**, 233–239.

Funk, C. D., Furci, L., Moran, N., and Fitzgerald, G. A. (1993). Point mutation in the seventh hydrophobic domain of the human thromboxane A2 receptor allows discrimination between agonist and antagonist binding sites. *Mol. Pharmacol.* **44**, 934–939.

Gäde, G. (1993). Structure-activity relationships for the lipid-mobilizing action of further bioanalogues of the adipokinetic hormone/red pigment concentrating hormone family of peptides. *J. Insect Physiol.* **39**, 375–383.

Garbers, D. L. (1991). Identification of a cell surface receptor common to germ and somatic cells. *Biol. Reprod.* **44**, 225–230.

Gether, U., Johansen, T. E., Snider, R. M., Lowe, J. A., Emonds-Alt, X., Yokota, Y., Nakanishi, S., and Schwartz, T. W. (1993a). Binding epitopes for peptide and non-peptide ligands on the NK1 (substance P) receptor. *Regul. Pept.* **46**, 49–56.

Gether, U., Johansen, T. E., Snider, R. M., Lowe, J. A., Nakanishi, S., and Schwartz, T. W. (1993b). Different binding epitopes on the NK-1 receptor for substance P and a non-peptide antagonist. *Nature (London)* **362**, 345–348.

Getzenberg, R. H., Pienta, K. J., and Coffey, D. S. (1990). The tissue matrix: Cell dynamics and hormone action. *Endocr. Rev.* **11**, 399–417.

Gifford, A. N., Nicholson, R. A., and Pitman, R. M. (1991). The dopamine and 5-hydroxytryptamine content of locust and cockroach salivary neurones. *J. Exp. Biol.* **161**, 405–414.

Gotzes, F., Balfanz, S., and Bauman, A. (1994). Primary structure and functional characterization of a *Drosophila* dopamine receptor with high homology to human D1/5 receptors. *Recept. Channels* **2**, 131–141.

Gray, A. S., Osborne, R. H., and Jewess, P. J. (1994). Pharmacology of proctolin receptors in the isolated foregut of the locust *Schistocerca gregaria*- Identification of α-methyl-L-tyrosine²-proctolin as a potent receptor antagonist. *J. Insect Physiol.* **40**, 595–600.

Grosshans, J., Bergmann, A., Haffter, P., and Nüsslein-Volhard, C. (1994). Activation of the kinase Pelle by Tube in the dorsoventral signal transduction pathway of *Drosophila* embryo. *Nature (London)* **372**, 563–566.

Guerrero, F. D., Clottens, F., and Holman, M. (1995). Purification of a cDNA encoding a tachykinin-like neuropeptide hormone receptor. *J. Cell. Biochem.* **21A**, 216.

Gutkind, J. S., Novotny, E. A., Brann, M. R., and Robbins, K. C. (1991). Muscarinic acetylcholine receptor subtypes as agonist-dependent oncogenes. *Proc. Natl. Acad. Sci. U.S.A.* **88,** 4703–4707.

Hafen, E., Basler, K., Edstroem, J.-E., and Rubin, G. M. (1987). Sevenless, a cell-specific homeotic gene of *Drosophila,* encodes a putative transmembrane receptor with a tyrosine kinase domain. *Science* **236,** 55–63.

Hagen, D. C., McCaffrey, G., and Sprague, G. F., Jr. (1986). Evidence that the yeast STE3 gene encodes a receptor for the peptide pheromone a factor: Gene sequence and implications for the structure of the presumed receptor. *Proc. Natl. Acad. Sci. U.S.A.* **83,** 1418–1422.

Hall, L. M., Hannan, F., Feng, G., Eberl, D. F., Hon, Y. Y., and Kousky, C. T. (1994). Cloning and molecular analysis of G-protein coupled receptors in insects. *In* "Insect Neurochemistry and Neurophysiology 1993" (A. B. Borkovec and M. J. Loeb, eds.), pp. 53–72. CRC Press, Boca Raton, FL.

Hardie, R. C. (1988). Effects of antagonists on putative histamine receptors in the first visual neuropil of the housefly (*Musca domestica*). *J. Exp. Biol.* **138,** 221–241.

Hardie, R. C. (1989). A histamine-activated chloride channel involved in neurotransmission at a photoreceptor synapse. *Nature (London)* **339,** 704–706.

Hart, A. C., Krämer, H., Van Vector, D. L., Paidhungat, M., and Zipursky, S. L. (1990). Induction of cell fate in the *Drosophila* retina: The bride of sevenless protein is predicted to contain a large extracellular domain and seven transmembrane segments. *Genes Dev.* **4,** 1835–1847.

Hayes, T. K., Guan, X-C., Johnson, V., Strey, A., and Tobe, S. S. (1994). Structure-activity studies of allostatin 4 on the inhibition of juvenile hormone biosynthesis by corpora allata: The importance of individual side chains and stereochemistry. *Peptides (N.Y.)* **15,** 1165–1171.

Heithier, H., Hallmann, D., Bocge, F., Reiländer, H., Dees, C., Jaeggi, K. A., Arndt-Jovin, D., Jovin, T. M., and Helmreich, E. J. M. (1994). Synthesis and properties of fluorescent β-adrenoceptor ligands. *Biochemistry* **33,** 9126–9134.

Henderson, J. E., Knipple, D. C., and Soderlund, D. M. (1994). PCR-based homology probing reveals a family of GABA receptor-like genes in *Drosophila melanogaster. Insect Biochem. Mol. Biol.* **24,** 363–371.

Hermans-Borgmeyer, I., Zopf, D., Ryseck, R.-P., Hovemann, B., Betz, H., and Gundelfinger, E. D. (1986). Primary structure of a developmentally regulated nicotinic acetylcholine receptor protein from *Drosophila. EMBO J.* **5,** 1503–1508.

Hibert, M. F., Trumpp-Kallmeyer, S., Bruinvels, A., and Hoflack, J. (1991). Three-dimensional models of neurotransmitter G-binding-protein-coupled receptors. *Mol. Pharmacol.* **40,** 8–15.

Hiripi, L., and Downer, R. G. H. (1993). Characterization of serotonin binding sites in insect (*Locusta migratoria*) brain. *Insect Biochem. Mol. Biol.* **23,** 303–307.

Hiripi, L., Juhos, S., and Downer, R. G. H. (1994). Characterization of tyramine and octopamine receptors in the insect (*Locusta migratoria migratorioides*) brain. *Brain Res.* **633,** 119–126.

Homberg, U. (1991). Neuroarchitecture of the central complex in the brain of the locust *Schistocerca gregaria* and *S. americana* as revealed by serotonin immunocytochemistry. *J. Comp. Neurol.* **303,** 245–254.

Homyk, T., Jr., and McIvor, W. (1989). A mutation that causes muscle defects also affects catecholamine metabolism in *Drosophila. J. Neurogenet.* **6,** 57–73.

Hu, Y., Rajan, L., and Schilling, W. P. (1994). Ca^{2+} signaling in Sf9 insect cells and the functional expression of a rat M5 muscarinic receptor. *Am. J. Physiol.* **266,** C1736–C1743.

Hunter, T., and Karin, M. (1992). The regulation of transcription by phosphorylation. *Cell (Cambridge, Mass.)* **70,** 375–387.

Insall, R., Kuspa, A., Lilly, P. J., Shaulsky, G., Levin, L. R., Loomis, W. F., and Devreotes, P. (1994). CRAC, a cytosolic protein containing a pleckstrin homology domain, is required for receptor and G-protein-mediated activation of adenylyl cyclase in *Dictyostelium. J. Cell Biol.* **126,** 1537–1545.

Ishihara, T., Nakamura, S., Kaziro, Y., Takahashi, T., Takahashi, K., and Nagata, S. (1991). Molecular cloning and expression of a cDNA encoding the secretin receptor. *EMBO J.* **10,** 1635–1641.

Jackson, T. R., Blair, L. A. C., Marshall, J., Goedert, M., and Hanley, M. R. (1988). The mas oncogene encodes an angiotensin. *Nature (London)* **335,** 437–440.

Jaffe, K., and Blanco, M. E. (1994). Involvement of amino acids, opioids, nitric oxide, and NMDA receptors in learning and memory consolidation in crickets. *Pharmacol. Biochem. Behav.* **47,** 493–496.

Jahagirdar, A. P., Milton, G., Viswanatha, T., and Downer, R. G. H. (1987). Calcium involvement in mediating the action of octopamine and hypertrehalosemic peptides on insect haemocytes. *FEBS Lett.* **219,** 83–87.

Javitch, J. A., Kaback, J., Li, X., and Karlin, A. (1994). Expression and characterization of human dopamine D2 receptor in baculovirus-infected insect cells. *J. Recept. Res.* **14,** 99–117.

Johansen, K. M., Fehon, R. G., and Artavanis-Tsakonas, S. (1989). The Notch gene product is a glycoprotein expressed on the cell surface of both epidermal and neuronal precursor cells during *Drosophila* development. *J. Cell Biol.* **109,** 2427–2440.

Joyce, K. A., Atkinson, A. E., Bermudez, I., Beadle, D. J., and King, L. A. (1993). Synthesis of functional GABAA receptors in stable insect cell lines. *FEBS Lett.* **335,** 61–64.

Julius, D., MacDermott, A. B., Axel, R., and Jessell, T. M. (1988). Molecular characterization of a functional cDNA encoding the serotonin 1c receptor. *Science* **241,** 558–564.

Kammermeyer, K. L., and Wadsworth, S. C. (1987). Expression of *Drosophila* epidermal growth factor receptor homologue in mitotic cell populations. *Development (Cambridge, UK)* **100,** 201–210.

Kataoka, H., Troetschler, R. G., Kramer, S. J., Cesarin, B. J., and Schooley, D. A. (1987). Isolation and primary structure of the eclosion hormone of the tobacco hornworm. *Manduca sexta. Biochem. Biophys. Res. Commun.* **146,** 746–750.

Kataoka, H., Troetschler, R. G., Li, J. P., Kramer, S. J., Carney, R. L., and Schooley, D. A. (1989). Isolation and identification of a diuretic hormone from the tobacco hornworm *Manduca sexta. Proc. Natl. Acad. Sci. U.S.A.* **86,** 2976–2980.

Kay, I., Wheeler, C. H., Coast, G. M., Totty, N. F., Cusinato, O., Patel, M., and Goldsworthy, G. J. (1991). Characterization of a diuretic peptide from *Locusta migratoria. Biol. Chem. Hoppe-Seyler* **372,** 929–934.

Keeley, L. L., and Hayes, T. K. (1987). Speculations on biotechnology applications for insect neuroendocrine research. *Insect Biochem.* **17,** 639–651.

Kelly, P. A., Djiane, J., Postel-Vinay, M.-C., and Edery, M. (1991). The prolactin/growth hormone receptor family. *Endocr. Rev.* **12,** 235–251.

Kelly, T. J., Masler, E. P., and Menn, J. J. (1994). Insect neuropeptides: Current status and avenues for pest control. *In* "Natural and Derived Pest Management Agents" (P. A. Hedin, J. J. Menn, and R. M. Hollingworth, eds.). Vol. 551, 292–318. Am. Chem. Soc. Symp. Ser. Washington, DC.

King, L. A., and Possee, R. D. (1992). "The Baculovirus Expression System, a Laboratory Guide." Chapman & Hall, London.

Klein, P. S., Sun, T. J., Saxe, C. L., Kimmel, A. R., Johnson, R. L., and Devreotes, P. N. (1988). A chemoattractant receptor controls development in *Dictyostelium discoideum. Science* **241,** 1467–1472.

Kleymann, G., Boege, F., Hahn, M., Hampe, W., Vasudevan, S., and Reiländer, H. (1993). Human β2-adrenergic receptor produced in stably transformed insect cells is functionally coupled via endogenous GTP-binding protein to adenylyl cyclase. *Eur. J. Biochem.* **213,** 797–804.

Knipper, M., and Breer, H. (1988). Subtypes of muscarinic receptors in insect nervous system. *Comp. Biochem. Physiol. C* **90C,** 275–280.

Knust, E., Dietrich, U., Bremer, K. A., Weigel, D., Vässin, H., and Campos-Ortega, J. A. (1987). EGF homologous sequences encoded in the genome of *Drosophila melanogaster,* and their relation to neurogenic genes. *EMBO J.* **6,** 761–766.

Koelle, M. R., Talbot, W. S., Segraves, W. A., Bender, M. T., Cherbas, P., and Hogness, D. S. (1991). The *Drosophila* EcR gene encodes an ecdysone receptor, a new member of the steroid receptor superfamily. *Cell (Cambridge, Mass.)* **67,** 59–77.

Komori, N., Usukura, J., Kurien, B., Shichi, H., and Matsumoto, H. (1994). Phosrestin I, an arrestin homolog that undergoes light-induced phosphorylation in dipteran photoreceptors. *Insect Biochem. Mol. Biol.* **24,** 607–617.

Konings, P. N. M., Vullings, H. G. B., Geffard, M., Buijs, R. M., Diederen, J. H. B., and Jansen, W. F. (1988). Immunocytochemical demonstration of octopamine-immunoreactive cells in the nervous system of *Locusta migratoria* and *Schistocerca gregaria. Cell Tissue Res.* **251,** 371–379.

Kopin, A. S., Lee, Y.-M., McBride, E. W., Miller, L. J., Lu, M., Lin, H. Y., Kolakowski, L. F., and Beinborn, M. (1992). Expression cloning and characterization of the canine parietal cell gastrin receptor. *Proc. Natl. Acad. Sci. U.S.A.* **89,** 3605–3609.

Lafon-Cazal, M., and Bockaert, J. (1984). Pharmacological characterization of dopamine-sensitive adenylate cyclase in the salivary glands of *Locusta migratoria* L. *Insect Biochem.* **14,** 541–545.

Lange, A. B. (1988). Inositol phospholipid hydrolysis may mediate the action of proctolin on insect visceral muscle. *Arch. Insect Biochem. Physiol.* **10,** 201–209.

Lange, A. B., and Tsang, P. K. C. (1993). Biochemical and physiological effects of octopamine and selected octopamine agonists on the oviducts of *Locusta migratoria. J. Insect Physiol.* **39,** 393–400.

Lange, A. B., Orchard, I., and Lloyd, R. J. (1988). Immunohistochemical and electrochemical detection of serotonin in the nervous system of the blood-feeding bug, *Rhodnius prolixus. Arch. Insect Biochem. Physiol.* **8,** 187–201.

Laudet, V., Hänni, C., Coll, J., Catzeflis, F., and Stehelin, D. (1994). Evolution of the nuclear receptor gene superfamily. *EMBO J.* **11,** 1003–1013.

Lee, H.-J., Rocheleau, T., Zhang, H.-G., Jackson, M. B., and ffrench-Constant, R. H. (1993). Expression of a *Drosophila* GABA receptor in a baculovirus insect cell system. Functional expression of insecticide susceptible and resistant GABA receptors from the cyclodiene resistance gene Rdl. *FEBS Lett.* **335,** 315–318.

Lee, Y-J., Shah, S., Suzuki, E., Zars, T., O'Day, P. M., and Hyde, D. R. (1994). The *Drosophila* dgq gene encodes a Gα protein that mediates phototransduction. *Neuron* **13,** 1143–1157.

Leitch, B., Watkins, B. L., and Burrows, M. (1993). Distribution of acetylcholine receptors in the central nervous system of adult locusts. *J. Comp. Neurol.* **334,** 47–58.

Li, X.-J., Wolfgang, W., Wu, Y.-N., North, R. A., and Forte, M. (1991). Cloning, heterologous expression and developmental regulation of a *Drosophila* receptor for tachykinin-like peptides. *EMBO J.* **10,** 3221–3229.

Li, X.-J., Wu, Y.-N., North, R. A., and Forte, M. (1992). Cloning, functional expression and developmental regulation of a neuropeptide Y receptor from *Drosophila melanogaster. J. Biol. Chem.* **267,** 9–12.

Libert, F., Parmentier, M., Lefort, A., Dinsart, C., Van Sande, J., Maenhaut, C., Simons, M.-J., Dumont, J. E., and Vassart, G. (1989). Selective amplification and cloning of four new members of the G protein-coupled receptor family. *Science* **244,** 569–572.

Livneh, E., Glazer, L., Segal, D., Schlessinger, J., and Shilo, B-Z. (1985). The *Drosophila* EGF receptor gene homolog: Conservation of both hormone binding and kinase domains. *Cell (Cambridge, Mass.)* **40,** 599–607.

Luffy, D., and Dorn, A. (1992). Immunohistochemical demonstration in the stomatogastric nervous system and effects of putative neurotransmitters on the motility of the isolated midgut of the stick insect, *Carausius morosus. J. Insect Physiol.* **38,** 287–299.

Lummis, S. C. R. (1990). GABA receptors in insects. *Comp. Biochem. Physiol. C* **95C,** 1–8.

Ma, P. W. K., and Roelofs, W. L. (1995). Calcium involvement in the stimulation of sex pheromone production by PBAN in the european corn borer, *Ostrinia nubilalis* (Lepidoptera: Pyralidae). *Insect Biochem. Mol. Biol.* **25,** 467–473.

MacKenzie, R. G., Steffey, M. E., Manelli, A. M., Pollock, N. J., and Frail, D. E. (1993). A D1/D2 chimeric dopamine receptor mediates a D1 response to a D2-selective agonist. *FEBS Lett.* **323,** 59–62.

Macmillan, C. S., and Mercer, A. R. (1987). An investigation of the role of dopamine in the antennal lobes of the honeybee, *Apis mellifera. J. Comp. Physiol. A* **160A,** 359–366.

Malamud, J. G., Mizisin, A. P., and Josephson, R. K. (1988). The effects of octopamine on contraction kinetics and power output of a locust flight muscle. *J. Comp. Physiol. A* **162A,** 827–835.

Malva, C., Gigliotti, S., Cavaliere, V., Manzi, A., and Graziani, F. (1994). Characterization and expression of a membrane guanylate cyclase *Drosophila* gene. *In* "Insect Neurochemistry and Neurophysiology 1993" (A. B. Borkovec and M. J. Loeb, eds.), pp. 363–366. CRC Press, Boca Raton, FL.

Marie, J., Maigret, B., Joseph, M.-P., Larguier, R., Nouet, S., Lombard, C., and Bonnafous, J.-C. (1994). Tyr-292 in the seventh transmembrane domain of the AT-1A angiotensin II receptor in essential for its coupling to phospholipase C. *J. Biol. Chem.* **269,** 20815–20818.

Marshall, J., Buckingham, S. D., Shingai, R., Lunt, G. G., Goosey, M. W., Darlison, M. G., Sattelle, D. B., and Barnard, E. A. (1990). Sequence and functional expression of a single α-subunit of an insect nicotinic acetylcholine receptor. *EMBO J.* **9,** 4391–4398.

Martin, J.-R., Raibaud, A., and Ollo, R. (1994). Terminal pattern elements in *Drosophila* embryo induced by the torso-like protein. *Nature (London)* **367,** 741–745.

Masler, E. P., Kelly, T. J., and Menn, J. J. (1993). Insect neuropeptides: Discovery and application in insect management. *Arch. Insect Biochem. Physiol.* **22,** 87–111.

Masu, Y., Nakayama, K., Tamaki, H., Harada, Y., Kuno, M., and Nakanishi, S. (1987). cDNA cloning of bovine substance-K receptor through oöcyte expression system. *Nature (London)* **329,** 836–838.

Mathews, L. S., and Vale, W. W. (1991). Expression cloning of an activin receptor, a predicted transmembrane serine kinase. *Cell (Cambridge, Mass.)* **65,** 973–982.

M'Diaye, K., and Bounias, M. (1991). Sublethal effects of the formamidine amitraz on honeybees gut lipids, following *in vivo* injections. *Biomed. Environ. Sci.* **4,** 376–383.

Mercer, A. R., and Menzel, R. (1982). The effects of biogenic amines on conditioned and unconditioned responses to olfactory stimuli in the honey bee *Apis mellifera. J. Comp. Physiol.* **145,** 363–368.

Meszaros, M., and Morton, D. B. (1994). Isolation and partial characterization of a gene from trachea of *Manduca sexta* that requires and is negatively regulated by ecdysteroids. *Dev. Biol.* **162,** 618–630.

Mills, A., Allet, B., Bernard, A., Chabert, C., Brandt, E., Cavegn, C., Chollet, A., and Kawashima, E. (1993). Expression and characterization of human D4 dopamine receptors in baculovirus-infected insect cells. *FEBS Lett.* **320,** 130–134.

Minnifield, N. M., and Hayes, D. K. (1992). Partial purification of a receptor for the adipokinetic hormones from *Musca autumnalis* face flies. *Prep. Biochem.* **22,** 215–228.

Misrahi, M., Ghinea, N., Sar, S., Saunier, B., Jolivet, A., Loosfelt, H., Cerutti, M., Devauchelle, G., and Milgröm, E. (1994). Processing of the precursors of the human thyroid-stimulating hormone receptor in various eukaryotic cells (human thyrocytes, transfected L cells and baculovirus-infected insect cells). *Eur. J. Biochem.* **222,** 711–719.

Monnier, D., Colas, J.-F., Rosay, P., Hen, R., Borrelli, E., and Maroteaux, L. (1992). NKD, a developmentally regulated tachykinin receptor in *Drosophila. J. Biol. Chem.* **267,** 1298–1302.

Montell, C., Jones, K., Zuker, C., and Rubin, G. (1987). A second opsin gene expressed in the ultraviolet-sensitive R7 photoreceptor cells of *Drosophila melanogaster. J. Neurosci.* **7,** 1558–1566.

Morton, D. B., and Giunta, M. A. (1992). Eclosion hormone stimulates cyclic GMP levels in *Manduca sexta* nervous tissue via arachidonic acid metabolism with little or no contribution from the production of nitric oxide. *J. Neurochem.* **59,** 1522–1530.

Mouillac, B., Caron, M., Bonin, H., Dennis, M., and Bouvier, M. (1992). Agonist-modulated palmitoylation of β2-adrenergic receptor in Sf9-cells. *J. Biol. Chem.* **267,** 21733–21737.

Mulheron, J. G., Casanas, S. J., Arthur, J. M., Garnovskaya, M. N., Gettys, T. W., and Raymond, J. R. (1994). Human 5HT1A receptor expressed in insect cells activates endogenous G_0-like G protein(s). *J. Biol. Chem.* **269,** 12954–12962.

Nachman, R. J., Holman, G. M., and Cook, B. J. (1986). Active fragments and analogs of the insect neuropeptide leucopyrokinin: Structure-function studies. *Biochem. Biophys. Res. Commun.* **137,** 936–942.

Nachman, R. J., Roberts, V. A., Dyson, H. J., Holman, G. M., and Tainer, J. A. (1991). Active conformation of an insect neuropeptide family. *Proc. Natl. Acad. Sci. U.S.A.* **88,** 4518–4522.

Nachman, R. J., Holman, G. M., Hayes, T. K., and Beier, R. C. (1993a). Structure-activity relationships for inhibitory insect myosuppressins: Contrast with the stimulatory sulfakinins. *Peptides (N.Y.)* **14,** 665–670.

Nachman, R. J., Holman, G. M., Schoofs, L., and Yamashita, O. (1993b). Silkworm diapause induction activity of myotropic pyrokinin (FXPRLa) insect neuropeptides. *Peptides (N.Y.)* **14,** 1043–1048.

Nachman, R. J., Holman, G. M., and Haddon, W. F. (1993c). Leads for insect neuropeptide mimetic development. *Arch. Insect Biochem. Physiol.* **22,** 181–197.

Nagasawa, H. (1993). Recent advances in insect neuropeptides. *Comp. Biochem. Physiol. C* **106C,** 295–300.

Nagayama, Y., and Rapoport, B. (1992). The throtropin receptor 25 years after its discovery: New insight after its molecular cloning. *Mol. Endocrinol.* **6,** 145–156.

Nakanishi, S., Nakajima, Y., and Yokota, Y. (1993). Signal transduction and ligand-binding domains of the tachykinin receptors. *Regul. Pept.* **46,** 37–42.

Nässel, D. R. (1993). Neuropeptides in the insect brain: A review. *Cell Tissue Res.* **273,** 1–29.

Nässel, D. R., Pirvola, U., and Panula, P. (1990). Histamine-like immunoreactive neurons innervating putative neurohaemal areas and central neurophil in the thoracic-abdominal ganglia of the flies *Drosophila* and *Calliphora. J. Comp. Neurol.* **297,** 525–536.

Nathanson, J. A. (1979). Octopamine receptors, adenosine 3′,5′-monophosphate, and neuronal control of firefly flashing. *Science* **203,** 165–168.

Nathanson, J. A. (1993). Identification of octopaminergic agonists with selectivity for octopamine receptor subtypes. *J. Pharmacol. Exp. Ther.* **265,** 509–515.

Nathanson, J. A., Hunnicut, E. J., Kantham, L., and Scavone, C. (1993). Cocaine as a naturally occurring insecticide. *Proc. Natl. Acad. Sci. U.S.A.* **90,** 9645–9648.

Nellen, D., Affolter, M., and Basler, K. (1994). Receptor Serine/Threonine kinases implicated in the control of *Drosophila* body pattern by decapentaplegic. *Cell (Cambridge, Mass.)* **78,** 225–237.

Ng, G. Y. K., George, S. R., Zastawny, R. L., Caron, M., Bouvier, M., Dennis, M., and O'Dowd, B. F. (1993). Human serotonin-1B receptor expression in Sf9 cells: Phosphorylation, palmitoylation and adenylyl cyclase inhibition. *Biochemistry* **32,** 11727–11733.

Ng, G. Y. K., Mouillac, B., George, S. R., Caron, M., Dennis, M., Bouvier, M., and O'Dowd, B. F. (1994). Desensitization, phosphorylation and palmitoylation of the human dopamine D1 receptor. *Eur. J. Pharmacol. Mol. Pharmacol. Sect.* **267,** 7–19.

Nomoto, S., Nakayama, N., Arai, K.-I., and Matsumoto, K. (1990). Regulation of the yeast pheromone response pathway by G protein subunits. *EMBO J.* **9**, 691–696.

Norman, J. C., Price, L. S., Ridley, A. J., Hall, A., and Koffer, A. (1994). Actin filament organization in activated Mast cells in regulated by heterotrimeric and small GTP-binding proteins. *J. Cell Biol.* **126**, 1005–1015.

Notman, H. J., and Downer, R. G. H. (1987). Binding of [^{3}H]-pifluthixol, a dopamine antagonist in the brain of the American cockroach *Periplaneta americana. Insect Biochem.* **17**, 587–590.

Nürnberger, A., Rapus, J., Eckert, M., and Penzlin, H. (1993). Taurine-like immunoreactivity in octopaminergic neurones of the cockroach, *Periplaneta americana* (L.). *Histochemistry* **100**, 285–292.

Olah, M. E., and Stiles, G. L. (1992). Adenosine receptors. *Annu. Rev. Physiol.* **54**, 211–225.

O'Malley, B. (1990). The steroid receptor superfamily: More excitement predicted for the future. *Mol. Endocrinol.* **4**, 363–369.

Onai, T., FitzGerald, M. G., Arakawa, S., Gocayne, J. D., Urquhart, D. A., Hall, L. M., Fraser, C. M., McCombie, W. R., and Venter, J. C. (1989). Cloning, sequence analysis and chromosome localization of a *Drosophila* muscarinic acetylcholine receptor. *FEBS Lett.* **255**, 219–225.

Orchard, I. (1987). Adipokinetic hormones-an update. *J. Insect Physiol.* **33**, 451–463.

Orchard, I. (1989). Serotoninergic neurohaemal tissue in *Rhodnius prolixus:* Synthesis, release and uptake of serotonin. *J. Insect Physiol.* **35**, 943–947.

Orchard, I., Lange, A. B., and Brown, B. B. (1992). Tyrosine-hydroxylase-like immunoreactivity in the ventral nerve cord of the locust, *Locusta migratoria,* including neurones innervating the salivary glands. *J. Insect Physiol.* **38**, 19–27.

O'Reilly, D. R., Miller, L. K., and Luckow, V. A. (1992). "Baculovirus Expression Vectors, a Laboratory Manual." Freeman, New York.

Orr, G. L., Gole, J. W. D., and Downer, R. G. H. (1985). Characterization of an octopamine-sensitive adenylate cyclase in haemocyte membrane fragments of the american cockroach *Periplaneta americana* L. *Insect Biochem.* **15**, 695–701.

Orr, G. L., Gole, J. W. D., Notman, H. J., and Downer, R. G. H. (1988). The regulation of basal and dopamine-sensitive adenylate cyclase by salts of monovalent cations in brain of the american cockroach, *Periplaneta americana* L. *Insect Biochem.* **18**, 79–86.

Orr, G. L., Orr, N., and Hollingworth, R. M. (1991). Distribution and pharmacological characterization of muscarinic-cholinergic receptors in the cockroach brain. *Arch. Insect Biochem. Physiol.* **16**, 107–122.

Orr, N., Orr, G. L., and Hollingworth, R. M. (1992). The Sf9 cell line as a model for studying insect octopamine-receptors. *Insect Biochem. Mol. Biol.* **22**, 591–597.

O'Tousa, J. E., Baehr, W., Martin, R. L., Hirsh, J., Pak, W. L., and Applebury, M. L. (1985). The *Drosophila* ninaE gene encodes an opsin. *Cell (Cambridge, Mass.)* **40**, 839–850.

Padgett, R. W., St. Johnston, R. D., and Gelbart, W. M. (1987). A transcript from a *Drosophila* pattern gene predicts a protein homologous to the transforming growth factor-beta family. *Nature (London)* **325**, 81–84.

Padgett, R. W., Wozney, J. M., and Gelbart, W. M. (1993). Human BMP sequences can confer normal dorsal-ventral patterning in the *Drosophila* embryo. *Proc. Natl. Acad. Sci. U.S.A.* **90**, 2905–2909.

Pannabecker, T., and Orchard, I. (1988). Receptors for octopamine in the storage lobe of the locust (*Locusta migratoria*) corpus cardiacum: Evidence from studies on cyclic nucleotides. *Insect Physiol.* **34**, 815–820.

Parker, E. M., Kameyama, K., Higashijima, T., and Ross, E. M. (1991). Reconstitutively active G protein-coupled receptors purified from baculovirus-infected insect cells. *J. Biol. Chem.* **266**, 519–527.

Parker, E. M., Grisel, D. A., Iben, L. G., Nowak, H. P., Mahle, C. D., Yocca, F. D., and Gaughan, G. T. (1994). Characterization of human 5HT1 receptors expressed in Sf9 insect cells. *Eur. J. Pharmacol. Mol. Pharmacol. Sect.* **268**, 43–53.

Parma, J., Duprez, L., Van Sande, J., Cochaux, P., Gervy, C., Mockel, J., Dumont, J., and Vassart, G. (1993). Somatic mutations in the thyrotropin receptor gene cause hyperfunctioning thyroid adenomas. *Nature (London)* **365**, 649–651.

Pawson, T. (1995). Protein modules and signalling networks. *Nature (London)* **373**, 573–580.

Pazin, M. J., and Williams, L. T. (1992). Triggering signaling cascades by receptor tyrosine kinases. *Trends Biochem. Sci.* **17**, 374–378.

Penton, A., Chen, Y., Staehling-Hampton, K., Wrana, J. L., Attisano, L., Szidonya, J., Cassill, J. A., Massagué, J., and Hoffman, F. M. (1994). Identification of two bone morphogenetic protein type I receptors in *Drosophila* and evidence that Brk25D is a decapentaplegic receptor. *Cell (Cambridge, Mass.)* **78**, 239–250.

Petit, D. A. D., Showalter, V. M., Abood, M. E., and Cabral, G. A. (1994). Expression of a cannabinoid receptor in baculovirus-infected insect cells. *Biochem. Pharmacol.* **48**, 1231–1243.

Petruzzelli, L., Herrera, R., Arenas-Garcia, R., Fernandez, R., Birnbaum, M. J., and Rosen, O. M. (1986). Isolation of a *Drosophila* genomic sequence homologous to the kinase domain of the human insulin receptor and detection of the phosphorylated *Drosophila* receptor with an anti-peptide antibody. *Proc. Natl. Acad. Sci. U.S.A.* **83**, 4710–4714.

Piedrafita, A. C., Martinez-Ramirez, A. C., and Silva, F. J. (1994). A genetic analysis of the aromatic amino acid hydroxylases involvement in DOPA synthesis during *Drosophila* adult development. *Insect Biochem. Mol. Biol.* **24**, 581–588.

Platt, N., and Reynolds, S. E. (1986). The pharmacology of the heart of a caterpillar, the tobacco hornworm, *Manduca sexta*. *J. Insect Physiol.* **32**, 221–230.

Pollock, J. A., and Benzer, S. (1988). Transcript localization of the four opsin genes in the three visual organs of *Drosophila;* RH2 is ocellus specific. *Nature (London)* **333**, 779–782.

Pouysségur, J., and Seuwen, K. (1992). Transmembrane receptors and intracellular pathways that control cell proliferation. *Annu. Rev. Physiol.* **54**, 195–210.

Probst, W. C., Snyder, L. A., Schuster, D. I., Brosius, J., and Sealfon, S. C. (1992). Sequence alignment of the G-protein coupled receptor superfamily. *DNA Cell Biol.* **11**, 1–20.

Puiroux, J., Pedelaborde, A., and Loughton, B. G. (1993). The effect of proctolin analogues and other peptides on locust oviduct muscle contractions. *Peptides (N.Y.)* **14**, 1103–1109.

Quehenberger, O., Prossnitz, E. R., Cochrane, C. G., and Ye, R. D. (1992). Absence of Gi proteins in the Sf9 insect cell. Characterization of the uncoupled recombinant N-formyl peptide receptor. *J. Biol. Chem.* **267**, 19757–19760.

Quintana, J., Wang, H., and Ascoli, M. (1993). The regulation of the binding affinity of the luteinizing hormone/choriogonadotropin receptor by sodium ions is mediated by a highly conserved aspartate located in the second transmembrane domain of G protein-coupled receptors. *Mol. Endocrinol.* **7**, 767–775.

Rachinsky, A. (1994). Octopamine and serotonin influence on corpora allata activity in honey bee (*Apis mellifera*) larvae. *J. Insect Physiol.* **40**, 549–554.

Raming, K., Krieger, J., Strotmann, J., Boekhoff, I., Kubick, S., Baumstark, C., and Breer, H. (1993a). Cloning and expression of odorant receptors. *Nature (London)* **361**, 353–356.

Raming, K., Freitag, J., Krieger, J., and Breer, H. (1993b). Arrestin-subtypes in insect antennae. *Cell. Signal.* **5**, 69–80.

Reagan, J. D. (1994). Expression cloning of an insect diuretic hormone receptor. *J. Biol. Chem.* **269**, 9–12.

Reagan, J. D. (1995). Functional expression of a diuretic hormone receptor in baculovirus-infected insect cells: Evidence suggesting that the N-terminal region of diuretic hormone is associated with receptor activation. *Insect Biochem. Mol. Biol.* **25**, 535–539.

Reagan, J. D., Li, J. P., Carney, R. L., and Kramer, S. J. (1993). Characterization of a diuretic hormone receptor from the tobacco hornworm, *Manduca sexta*. *Arch. Insect Biochem. Physiol.* **23**, 135–146.

Reagan, J. D., Patel, B. C., Li, J. P., and Miller, W. H. (1994). Characterization of a solubilized diuretic hormone receptor from the tobacco hornworm, *Manduca sexta. Insect Biochem. Mol. Biol.* **24,** 569–572.

Restifo, L. L., and White, K. (1990). Molecular and genetic approaches to neurotransmitter and neuromodulator systems in *Drosophila. Adv. Insect Physiol.* **22,** 115–219.

Richardson, R. M., Kim, C., Benovic, J. L., and Hosey, M. M. (1993). Phosphorylation and desensitization of human M2 muscarinic cholinergic receptors by two isoforms of the β-adrenergic receptor kinase. *J. Biol. Chem.* **268,** 13650–13656.

Riddiford, L. M. (1993). Hormone receptors and the regulation of insect metamorphosis. *Receptor* **3,** 203–209.

Riddiford, L. M. (1994). Cellular and molecular actions of juvenile hormone. I. General considerations and premetamorphic actions. *Adv. Insect Physiol.* **24,** 213–274.

Rinken, A., Kameyama, K., Haga, T., and Engström, L. (1994). Solubilization of muscarinic receptor subtypes from baculovirus-infected Sf9 insect cells. *Biochem. Pharmacol.* **48,** 1245–1251.

Roeder, T. (1990a). High-affinity antagonists of the locust neuronal octopamine receptor. *Eur. J. Pharmacol.* **191,** 221–224.

Roeder, T. (1990b). Histamine H1-receptor-like binding sites in the locust nervous tissue. *Neurosci. Lett.* **116,** 331–335.

Roeder, T. (1991). A new octopamine receptor class in locust nervous tissue, the octopamine 3 (OA_3) receptor. *Life Sci.* **50,** 21–28.

Roeder, T. (1994). Biogenic amines and their receptors in insects. *Comp. Biochem. Physiol. C* **107C,** 1–12.

Roeder, T., and Gewecke, M. (1989). Octopamine uptake systems in thoracic ganglia and leg muscles of *Locusta migratoria. Comp. Biochem. Physiol. C* **94C,** 143–147.

Roeder, T., and Nathanson, J. A. (1993). Characterization of insect neuronal octopamine receptors (OA_3 receptors). *Neurochem. Res.* **18,** 921–925.

Roeder, T., Vossfeldt, R., and Gewecke, M. (1993). Pharmacological characterization of the locust neuronal ^{3}H-mianserin binding site, a putative histamine receptor. *Comp. Biochem. Physiol. C* **106C,** 503–507.

Rothberg, J. M., Jacobs, J. R., Goodman, C. S., and Artavanis-Tsakonas, S. (1990). Slit: An extracellular protein necessary for development of midline glia and commissural axon pathways contains both EGF and LRR domains. *Genes Dev.* **4,** 2169–2187.

Rotondo, D., Vaughan, P. F. T., and Donnellan, J. F. (1987). Octopamine and cyclic AMP stimulate protein phosphorylation in the central nervous system of *Schistocerca gregaria. Insect Biochem.* **17,** 283–290.

Sattelle, D. B. (1990). GABA receptors of insects. *Adv. Insect Physiol.* **22,** 1–113.

Sattelle, D. B. (1992). Receptors for L-glutamate and GABA in the nervous system of an insect (*Periplaneta americana*). *Comp. Biochem. Physiol. C* **103C,** 429–438.

Saudou, F., Amlaiky, N., Plassat, J.-L., Borrelli, E., and Hen, R. (1990). Cloning and characterization of a *Drosophila* tyramine receptor. *EMBO J.* **9,** 3611–3617.

Saudou, F., Boschert, U., Amlaiky, N., Plassat, J.-L., and Hen, R. (1992). A family of *Drosophila* serotonin receptors with distinct intracellular signalling properties and expression patterns. *EMBO J.* **11,** 7–17.

Sawruk, E., Schloss, P., Betz, H., and Schmitt, B. (1990). Heterogeneity of *Drosophila* nicotinic acetylcholine receptors: SAD, a novel developmentally regulated alpha-subunit. *EMBO J.* **9,** 2671–2677.

Scavone, C., McKee, M., and Nathanson, J. A. (1994). Monoamine uptake in insect synaptosomal preparations. *Insect Biochem. Mol. Biol.* **24,** 589–597.

Schejter, E. D., Segal, D., Glazer, L., and Shilo, B.-Z. (1986). Alternative 5′ exons and tissue-specific expression of the *Drosophila* EGF receptor homolog transcripts. *Cell (Cambridge, Mass.)* **46,** 1091–1101.

Schlessinger, J., and Ullrich, A. (1992). Growth factor signaling by receptor tyrosine kinases. *Neuron* **9**, 383–391.

Schloss, P., Hermans-Borgmeyer, I., Betz, H., and Gundelfinger, E. D. (1988). Neuronal acetylcholine receptors in *Drosophila:* The ARD protein is a component of a high-affinity α-bungarotoxin binding complex. *EMBO J.* **7**, 2889–2894.

Schmid, A., Scheidler, A., Kaulen, P., and Brüning, G. (1993). Serotonin-immunoreactivity and serotonin binding sites in the brain of the blowfly *Calliphora erythrocephala:* A combined immunohistochemical and autoradiographical study. *Comp. Biochem. Physiol. C* **104C**, 193–197.

Schneider, D. S., Jin, Y., Morisato, D., and Anderson, K. V. (1994). A processed form of the Spätzle protein defines dorsal-ventral polarity in the *Drosophila* embryo. *Development (Cambridge, UK)* **120**, 1243–1250.

Schoofs, L., Vanden Broeck, J., and De Loof, A. (1993). The myotropic peptides of *Locusta migratoria:* Structures, distribution, functions and receptors. *Insect Biochem. Mol. Biol.* **23**, 859–881.

Segraves, W. A. (1994). Steroid receptors and orphan receptors in *Drosophila* development. *Semin. Cell Biol.* **5**, 105–113.

Shapiro, R. A., Wakimoto, B. T., Subers, E. B., and Nathanson, N. M. (1989). Characterization and functional expression in mammalian cells of cDNA and genomic clones encoding a *Drosophila* muscarinic acetylcholine receptor. *Proc. Natl. Acad. Sci. U.S.A.* **86**, 9039–9043.

Shibanaka, Y., Hayashi, H., Umemura, I., Fujisawa, Y., Okamoto, M., Takai, M., and Fujita, N. (1994). Eclosion hormone-mediated signal transduction in the silkworm abdominal ganglia: Involvement of a cascade from inositol(1,4,5)triphosphate to cyclic GMP. *Biochem. Biophys. Res. Commun.* **198**, 613–618.

Smith, C. L., and DeLotto, R. (1994). Ventralizing signal determined by protease activation in *Drosophila* embryogenesis. *Nature (London)* **368**, 548–551.

Smyth, K.-A., Parker, A. G., Yen, J. L., and McKenzie, J. A. (1992). Selection of dieldrin-resistant strains of *Lucilia cuprina* (Diptera : Calliphoridae) after ethyl methanesulfonate mutagenesis of a susceptible strain. *J. Econ. Entomol.* **85**, 352–358.

Spindler-Barth, M. (1992). Endocrine strategies for the control of ectoparasites and insect pests. *Parasitol. Res.* **78**, 89–95.

Staehling-Hampton, K., Hoffmann, F. M., Baylies, M. K., Rushton, E., and Bate, M. (1994). dpp induces mesodermal gene expression in *Drosophila*. *Nature (London)* **372**, 783–786.

Stanley-Samuelson, D. W. (1994). Postaglandins and related eicosanoids in insects. *Adv. Insect Physiol.* **24**, 115–212.

Starrat, A. N., and Brown, B. E. (1975). Structure of the pentapeptide proctolin, a proposed neurotransmitter in insects. *Life Sci.* **17**, 1253–1256.

Steele, J. E., and Chan, F. (1980). Na$^+$-dependent respiration in the insect nerve cord and its control by octopamine. *Proc. Soc. Chem. Ind.*, pp. 347–350.

Stefano, G. B. (1988). The evolvement of signal systems: Conformational matching, a determining force stabilizing families of signal molecules. *Comp. Biochem. Physiol. C* **90C**, 287–294.

Stefano, G. B., and Scharrer, B. (1981). High affinity binding of an enkephalin analogue in the cerebral ganglion of the insect *Leucophaea maderae* (Blattaria). *Brain Res.* **225**, 107–114.

Stein, D., Roth, S., Vogelsang, E., and Nüsslein-Volhard, C. (1991). The polarity of the dorsoventral axis in the *Drosophila* embryo is defined by an extracellular signal. *Cell (Cambridge, Mass.)* **65**, 725–735.

Stengl, M. (1992). Peripheral processes in insect olfaction. *Annu. Rev. Physiol.* **54**, 665–681.

Stengl, M. (1994). Inositol-triphosphate-dependent calcium currents precede cation currents in insect olfactory receptor neurons *in vitro*. *J. Comp. Physiol. A* **174A**, 187–194.

Stevenson, P. A., Pflüger, H. J., Eckert, M., and Rapus, J. (1992). Octopamine immunoreactive cell populations in the locust thoracic-abdominal nervous system. *J. Comp. Neurol.* **315**, 382–397.

St. Johnston, D., and Nüsslein-Volhard, C. (1992). The origin of pattern and polarity in the *Drosophila* embryo. *Cell (Cambridge, Mass.)* **68,** 201–219.

Taylor, J. M., Jacob-Mosier, G. G., and Lawton, R. G., Remmers, A. E., and Neubig, R. R. (1994). Binding of an $\alpha 2$ adrenergic receptor third intracellular loop peptide to $G\beta$ and the amino terminus of $G\alpha$. *J. Biol. Chem.* **269,** 27618–27624.

Thomas, H. E., Stunnenberg, H. G., and Stewart, A. F. (1993). Heterodimerization of the *Drosophila* ecdysone receptor with retinoid X receptor and ultraspiracle. *Nature (London)* **362,** 471–475.

Thompson, K., Decker, S. J., and Rosner, M. R. (1985). Identification of a novel receptor in *Drosophila* for both epidermal growth factor and insulin. *Proc. Natl. Acad. Sci. U.S.A.* **82,** 8443–8447.

Thompson, M., Shotkoski, F., and ffrench-Constant, R. (1993). Cloning and sequencing of the cyclodiene insecticide resistance gene from the yellow fever mosquito *Aedes aegypti.* Conservation of the gene and resistance associated mutation with *Drosophila*. *FEBS Lett.* **3,** 187–190.

Trimmer, B. A. (1992). The regulation of a muscarinic current in an identified insect motoneuron. *Soc. Neurosci. Abstr.* **18,** 472.

Trimmer, B. A., and Weeks, J. C. (1989). Effects of nicotinic and muscarinic agents on an identified motoneurone and its direct inputs in larval *Manduca sexta. J. Exp. Biol.* **144,** 303–337.

Trimmer, B. A., and Weeks, J. C. (1993). Muscarinic acetylcholine receptors modulate the excitability of an identified insect motoneuron. *J. Neurophysiol.* **69,** 1821–1836.

Truman, J. W., and Levine, R. B. (1982). The action of peptides and cyclic nucleotides on the nervous system of an insect. *Fed. Proc. Fed. Am. Soc. Exp. Biol.* **41,** 2929–2932.

Tsukada, S., Simon, M. I., Witte, O. N., and Katz, A. (1994). Binding of $\beta\gamma$ subunits of heterotrimeric G proteins to the PH domain of Bruton tyrosine kinase. *Proc. Natl. Acad. Sci. U.S.A.* **91,** 11256–11260.

Ullrich, A., and Schlessinger, J. (1990). Signal transduction by receptors with tyrosine kinase activity. *Cell (Cambridge, Mass.)* **61,** 203–212.

Ultsch, A., Schuster, C. M., Laube, B., Betz, H., and Schmitt, B. (1993). Glutamate receptors of *Drosophila melanogaster.* Primary structure of a putative NMDA receptor protein expressed in the head of the adult fly. *FEBS Lett.* **324,** 171–177.

Usherwood, P. N. R. (1994). Insect glutamate receptors. *Adv. Insect Physiol.* **24,** 309–341.

Vallés, A. M., and White, K. (1988). Serotonin-containing neurons in *Drosophila melanogaster:* Development and distribution. *J. Comp. Neurol.* **268,** 414–428.

Vanden Broeck, J., Hoste, J., and De Loof, A. (1991). Cloning of insect G-protein coupled receptor cDNA fragments by means of PCR. *In* "Molecular Mechanisms of Signal Transduction," EMBO-EMBL 1991 Symp. Abstr., p. 198.

Vanden Broeck, J., Verhaert, P., and De Loof, A. (1994). Cloning, sequencing and tissue distribution of a putative locust G protein-coupled receptor. *In* "Insect Neurochemistry and Neurophysiology 1993" (A. B. Borkovec and M. J. Loeb, eds.), pp. 367–370. CRC Press, Boca Raton, FL.

Vanden Broeck, J., Vulsteke, V., Huybrechts, R., and De Loof, A. (1995a). Characterization of a cloned locust tyramine receptor cDNA by functional expression in permanently transformed Drosophila S2 cells. *J. Neurochem.* **6,** 2387–2395 .

Vanden Broeck, J., Schoofs, L., Vulsteke, V., Verhees, M., Vandersmissen, V., Huybrechts, R., and De Loof, A. (1995b). Expression of a locust tyramine receptor cDNA and localization of the mRNA in the central nervous system of *Locusta migratoria. In* "Proceedings of the 1st International Conference of Insects: Chemical, Physiological and Environmental Aspects" (in press).

Vanhems, E., Delbos, M., Geffard, M., and Viellemaringe, J. (1994). Detection of putative dopamine receptors in neurites outgrowing from locust central nervous system explants using anti-idiotypic dopamine antibodies. *Neuroscience* **58,** 649–655.

Vässin, H., Bremer, K. A., Knust, E., and Campos-Ortega, J. A. (1987). The neurogenic gene delta of *Drosophila melanogaster* is expressed in neurogenic territories and encodes a putative transmembrane protein with EGF-like repeats. *EMBO J.* **6**, 3431–3440.

Vasudevan, S., Premkumar, L., Stowe, S., Gage, P. W., Reiländer, H., and Chung, S.-H. (1992). Muscarinic acetylcholine receptor produced in recombinant baculovirus infected Sf9 insect cells couples with endogenous G-proteins to activate ion channels. *FEBS Lett.* **311**, 7–11.

Veenstra, J. A. (1988). Effects of 5-hydroxytryptamine on the Malpighian tubules of *Aedes aegypti. J. Insect Physiol.* **34**, 299–304.

Venter, J. C., Fraser, C., Kerlavage, A., and Buck, M. (1989). Molecular biology of adrenergic and muscarinic cholinergic receptors. *Biochem. Pharmacol.* **38**, 1197–1208.

Vu, T.-K. H., Hung, D. T., Wheaton, V. I., and Coughlin, S. R. (1991). Molecular cloning of a functional thrombin receptor reveals a novel proteolytic mechanism of receptor activation. *Cell (Cambridge, Mass.)* **64**, 1057–1068.

Vulsteke, V., Janssen, I., Vanden Broeck, J., De Loof, A., and Huybrechts, R. (1993). Baculovirus immediate early gene promoter based expression vectors for transient and stable transformation of insect cells. *Insect Mol. Biol.* **2**, 195–204.

Wadsworth, S. C., Vincent, W. S., and Bilodeau-Wentworth, D. (1986). A *Drosophila* genomic sequence with homology to human epidermal growth factor receptor. *Nature (London)* **314**, 178–180.

Wadsworth, S. C., Rosenthal, L. S., Kammermeyer, K. L., Potter, M. B., and Nelson, D. J. (1988). Expression of a *Drosophila melanogaster* acetylcholine receptor-related gene in the central nervous system. *Mol. Cell. Biol.* **8**, 778–785.

Wang, H., Jaquette, J., Collison, K., and Segaloff, D. L. (1993). Positive charges in a putative amphiphillic helix in the carboxyl-terminal region of the third intracellular loop of the luteinizing hormone/chorionic gonadotropin receptor are not required for hormone-stimulated cAMP production but are necessary for expression of the receptor at the plasma membrane. *Mol. Endocrinol.* **7**, 1437–1444.

Wang, Z., Orchard, I., and Lange, A. B. (1994). Identification and characterization of two receptors for SchistoFLRFamide on locust oviduct. *Peptides (N.Y.)* **15**, 875–882.

Washio, H., and Tanaka, Y. (1992). Some effects of octopamine, proctolin and serotonin on dorsal unpaired median neurons of cockroach (*Periplaneta americana*) thoracic ganglia. *J. Insect Physiol.* **38**, 511–517.

Wegener, J. W., Boekhoff, I., Tareilus, E., and Breer, H. (1993). Olfactory signalling in antennal receptor neurons of the locust (*Locusta migratoria*). *J. Insect Physiol.* **39**, 153–163.

Wendt, B., and Homberg, U. (1992). Immunocytochemistry of dopamine in the brain of the locust *Schistocerca gregaria. J. Comp. Neurol.* **321**, 387–403.

Wess, J., Nanavati, S., Vogel, Z., and Maggio, R. (1993). Functional role of proline and tryptophan residues highly conserved among G protein-coupled receptors studied by mutational analysis of the m3 muscarinic receptor. *EMBO J.* **12**, 331–338.

Wharton, K. A., Johansen, K. M., Xu, T., and Artavanis-Tsakonas, S. (1985). Nucleotide sequence from the neurogenic locus notch implies a gene product that shares homology with proteins containing EGF-like repeats. *Cell (Cambridge, Mass.)* **43**, 567–581.

Wilson, C., Goberdhan, D. C. I., and Steller, H. (1993). Dror, a potential neurotrophic receptor gene, encodes a *Drosophila* homolog of the vertebrate Ror family of Trk-related receptor tyrosine kinases. *Proc. Natl. Acad. Sci. U.S.A.* **90**, 7109–7113.

Witz, P., Amlaiky, N., Plassat, J.-L., Maroteaux, L., Borrelli, E., and Hen, R. (1990). Cloning and characterization of a *Drosophila* serotonin receptor that activates adenylate cyclase. *Proc. Natl. Acad. Sci. U.S.A.* **87**, 8940–8944.

Wood, S. J., Osborne, R. H., Gomez, S., and Jewess, P. J. (1992). Pharmacology of the isolated foregut of the locust *Schistocerca gregaria*. IV. Characterization of a muscarinic acetylcholine receptor. *Comp. Biochem. Physiol. C* **103C**, 527–534.

Wrana, J. L., Attisano, L., Wieser, R., Ventura, F., and Massagué, J. (1994). Mechanism of activation of the TGF-β receptor. *Nature (London)* **370**, 341–347.

Wyatt, G. R. (1990). Developmental and juvenile hormone control of gene expression in locust fat body. *In* "Molecular Insect Science" (H. H. Hagedorn *et al.,* eds.), pp. 163–172. Plenum, New York.

Wyatt, G. R., Braun, R. P., Edwards, G. C., Glinka, A. V., and Zhang, J. (1994). Juvenile hormone in adult insects: Two modes of action. *Perspect. Comp. Endocrinol.* pp. 202–208.

Xue, J.-C., Chen, C., Zhu, J., Kunapuli, S., DeRiel, J. K., Yu, L., and Liu-Chen, L.-Y. (1994). Differential binding domains of peptide and non-peptide ligands in the cloned rat k opioid receptor. *J. Biol. Chem.* **269**, 30195–30199.

Yao, T.-P., Forman, B. M., Jiang, Z., Cherbas, L., Chen, J.-D., McKeown, M., Cherbas, P., and Evans, R. M. (1993). Functional ecdysone receptor is the product of EcR and Ultraspiracle genes. *Nature (London)* **366**, 476–479.

Yokota, Y., Akazawa, C., Ohkubo, H., and Nakanishi, S. (1992). Delineation of structural domains involved in the subtype specificity of tachykinin receptors through chimaeric formation of substance P/substance K receptors. *EMBO J.* **11**, 3585–3591.

Young, D., Waitches, G., Birchmeier, C., Fasano, O., and Wigler, M. (1986). Isolation and characterization of a new cellular oncogene encoding a protein with multiple potential transmembrane domains. *Cell (Cambridge, Mass.)* **45**, 711–719.

Zhang, J., McCracken, A., and Wyatt, G. R. (1993). Properties and sequence of a female-specific, juvenile hormone-induced protein from locust hemolymph. *J. Biol. Chem.* **268**, 3282–3288.

Ziegler, R., Eckart, K., Jasensky, R. D., and Law, J. H. (1991). Structure-activity studies on adipokinetic hormones in *Manduca sexta. Arch. Insect Biochem. Physiol.* **18**, 229–237.

Zuker, C. S., Cowman, A. F., and Rubin, G. M. (1985). Isolation and structure of a rhodopsin gene from *D. melanogaster. Cell (Cambridge, Mass.)* **40**, 851–858.

Zuker, C. S., Montell, C., Jones, K., Laverty, T., and Rubin, G. M. (1987). A rhodopsin gene expressed in photoreceptor cell R7 of the *Drosophila* eye: Homologies with other signal-transducing molecules. *J. Neurosci.* **7**, 1550–1557.

Microtubules and Microtubule Motors: Mechanisms of Regulation

Catherine D. Thaler and Leah T. Haimo

Department of Biology, University of California, Riverside, California

Microtubule-based motility is precisely regulated, and the targets of regulation may be the motor proteins, the microtubules, or both components of this intricately controlled system. Regulation of microtubule behavior can be mediated by cell cycle-dependent changes in centrosomal microtubule nucleating ability and by cell-specific, microtubule-associated proteins (MAPs). Changes in microtubule organization and dynamics have been correlated with changes in phosphorylation. Regulation of motor proteins may be required both to initiate movement and to dictate its direction. Axonemal and cytoplasmic dyneins as well as kinesin can be phosphorylated and this modification may affect the motor activities of these enzymes or their ability to interact with organelles. A more complete understanding of how motors can be modulated by phosphorylation, either of the motor proteins or of other associated substrates, will be necessary in order to understand how bidirectional transport is regulated.

KEY WORDS: Dynein, Kinesin, Microtubules, Tubulin, Cytoskeleton, MAPS, Motility, Phosphorylation.

I. Introduction

Directed movements are a hallmark of living organisms, and eukaryotic cells exhibit a wide range of motile phenomena which occur with precise spatial and temporal specificity. For example, the Golgi apparatus in many eukaryotic cells is localized perinuclearly, at the centrosomes, the minus ends of the interphase microtubule network. Chromosome separation during mitosis occurs only after control checkpoints in the cell cycle have been traversed. In order for directed movements to occur, eukaryotic cells must have a system not only to produce motive force but also to regulate it by turning motile systems on and off or by controlling the direction in which

International Review of Cytology, Vol. 164

269

movements occur. Motive force is generated by molecular motors, members of either the myosin, kinesin, or dynein gene families, which use energy from ATP hydrolysis to drive transport relative to an appropriate filamentous structure—either actin filaments or microtubules. These filaments have been extensively studied for more than two decades, and their biochemistry and assembly properties have been well characterized. The past several years have yielded an enormous amount of information on the motor molecules themselves. The multigene families encoding these proteins are being characterized and the individual properties and sites of action of many of the various members of these families are being elucidated (Cheney *et al.*, 1993; Endow and Titus, 1992; Goldstein, 1993; Mooseker, 1993; Titus, 1993; Walker and Sheetz, 1993; Holzbaur and Vallee, 1994; Hoyt, 1994; Mitchell, 1994; Schroer, 1994; Weiss *et al.*, 1994; Ruppel and Spudich, 1995). An understanding of the regulatory processes controlling movements is now beginning to emerge, and this area promises to yield exciting information in coming years.

Despite the diversity of movements characterizing eukaryotic cells, two common themes underlie the regulation of motility: regulation can occur through control of the assembly or structure of cytoskeletal components or through control of the activity of the motor molecule itself. Whether it is the motor or the track that is affected by the regulatory process, the mechanism through which regulation is mediated is more often than not a change in the state of protein phosphorylation of an essential component of the system. Changes in phosphorylation occur as a consequence of a cell cycle-dependent pathway or as a function of a signal transduction event that occurs following stimulation of cell surface receptors. In many motile systems, the target proteins of the phosphorylation pathway are not yet known. Thus, the next few years will likely greatly enhance the depth of our understanding of regulation of motility as the target proteins become identified and their specific roles in motility are understood.

This review focuses on microtubules and microtubule motors. In-depth discussions of regulation of actin assembly and myosin activity appear in excellent recent reviews (Tan *et al.*, 1992; Condeelis, 1993). Our current understanding of the mechanisms by which the cell regulates microtubule assembly and motor function to control its architecture and motile events is presented here.

II. Regulation of Microtubule Behavior

The microtubule array in cells is not static but is remodeled as the cell moves through its cycle. As the cell progresses from interphase into mitosis, the polarized array of microtubules emanating from the centrosomes is disassembled and then reassembled to form the more extensive bipolar array of the mitotic spindle. The centrosome and the changes that it undergoes during the cell cycle may account in large part for the restructuring of the microtubule cytoskeleton. Furthermore, the behavior of the microtubules themselves changes dramatically from interphase to mitosis; the long, persistent microtubules of interphase are converted into the shorter, more dynamic microtubules of the M-phase. These changes are under the control of cell cycle-dependent protein kinases. In addition, motile cells reorient their centrosomes and their microtubule array as they change direction; signal transduction, resulting in protein phosphorylation, might provide the basis for this cytoskeletal restructuring, as well. These changes in microtubule behavior are most likely an essential component of motility. Microtubule behavior can be controlled by a number of proteins; these induce microtubule assembly, or confer stability upon the microtubules, or induce their severing and depolymerization.

A. Centrosomal Control of Microtubule Behavior

1. Function of γ-Tubulin

The spatial distribution of microtubules in most eukaryotic cells is controlled by restricting microtubule assembly to that nucleated from microtubule organizing centers (MTOCs). Although centrioles are often found as components of MTOCs, they are not required for MTOC function and indeed, are absent from many organisms, including higher plants. While MTOCs may have diverse structures, they most likely contain common proteins, thus permitting them to function similarly. One protein of the MTOC has been identified and shown to be an essential component of all MTOCs studied to date. The discovery of this protein, γ-tubulin, stemmed from the observation that mutations in the *Aspergillus nidulans* gene, *mipA*, were capable of partly suppressing mutations in the β-tubulin gene, suggesting that β-tubulin and the product of the *mipA* gene might interact. Quite surprisingly, cloning and sequencing of the *mipA* gene revealed its product to be closely related to α- and β-tubulin. Indeed, it was as closely related to these tubulins (35% identical) as they are to each other and thus was recognized as a distinct member of the tubulin superfamily and named γ-tubulin (Oakley and Oakley, 1989). Subsequent studies have shown γ-tubulin to be an essential, ubiquitous, and highly conserved protein (Stearns *et al.*, 1991; Zheng *et al.*, 1991; Joshi *et al.*, 1992). Disruption of the γ-tubulin gene in *A. nidulans* (Oakley *et al.*, 1990), or in the fission yeast *Schizosaccharomyces pombe* (Stearns *et al.*, 1991) is lethal. Expression of the human γ-tubulin gene in *S. pombe* can rescue normal levels of growth in cells in which the endogenous γ-tubulin gene has been disrupted (Horio and Oakley, 1994), indicating that the important elements of the molecule remain conserved between highly divergent species.

γ-Tubulin appears to be essential for microtubule assembly from spindle pole bodies in lower eukaryotes and from the centrosomes in higher eukaryotes. Antibodies against γ-tubulin have been used to localize it to the spindle pole bodies of *A. nidulans* (Oakley *et al.*, 1990) and to the centrosomes of *Drosophila* and mammalian cells (Zheng *et al.*, 1991; Stearns *et al.*, 1991; Joshi *et al.*, 1992). Interestingly, microtubules do not stain along their length with antibodies against γ-tubulin, and thus this tubulin appears not to be a component of the wall of the microtubule (Stearns *et al.*, 1991). However, γ-tubulin can associate with the minus ends of microtubules which exhibit γ-tubulin staining when asters are induced with dimethylsulfoxide (DMSO) to form in the absence of centrosomes (Stearns and Kirschner, 1994). Microinjection of an antibody directed against a highly conserved domain of γ-tubulin into a variety of cell types stains the centrosomes but does not interfere with the assembled microtubule array. In contrast, when the microtubules in the injected cells are depolymerized at 4°C, they do not repolymerize upon rewarming. Moreover, when cells are injected with γ-tubulin antibody immediately prior to mitosis, they fail to form a mitotic spindle (Joshi *et al.*, 1992). These studies indicate that γ-tubulin is required for the *in vivo* nucleation of microtubules.

γ-Tubulin is also required for microtubule assembly from centrosomes *in vitro*. When lysed cells whose endogenous microtubules have been depolymerized are incubated in antibody against γ-tubulin, microtubule regrowth is completely inhibited (Joshi *et al.*, 1992). Centrioles introduced into *Xenopus* egg extracts are able to nucleate microtubules only after acquisition by these centrioles of γ-tubulin from the extract (Stearns and Kirschner, 1994; Félix *et al.*, 1994). That γ-tubulin is required for microtubule assembly from centrosomes (Joshi *et al.*, 1992), that it interacts with β-tubulin (Oakley and Oakley, 1989) and with the minus ends of microtubules (Stearns and Kirschner, 1994), and that it is present in centrosomes (Stearns *et al.*, 1991; Zheng *et al.*, 1991; Joshi *et al.*, 1992) suggest that γ-tubulin serves as the microtubule nucleator, inducing microtubules to assemble and linking them to the MTOC. If so, then γ-tubulin is responsible, at least in part, for organizing the microtubule array in cells. While these studies demonstrate that γ-tubulin is necessary for microtubule assembly from centrosomes, γ-tubulin may not be sufficient for this assembly.

2. Phosphorylation of Centrosomal Components

The enhanced nucleating ability of mitotic relative to interphase centrosomes (Kuriyama and Borisy, 1981; Bré *et al.*, 1990) may be correlated with a phosphorylation event (Centonze and Borisy, 1990; Buendia *et al.*, 1992) as well as with an increase in γ-tubulin at the centrosomes during mitosis (Stearns *et al.*, 1991). In one study (Centonze and Borisy, 1990), centrosomes in mitotic CHO cell extracts nucleated more microtubules than those in interphase extracts or in mitotic extracts containing an antibody that recognizes phosphorylated mitotic epitopes (the MPM-2 antibody; Davis *et al.*, 1983). Further, preabsorption of the MPM-2 antibody with mitotic, but not interphase extracts resulted in subsequent levels of assembly in mitotic extracts containing the mitotic-preabsorbed antibody that were equivalent to those in non-MPM-2 treated extracts. These observations suggest that there is a phosphorylated molecule, recognized by the MPM-2 antibody, that appears during mitosis and which is required for the higher levels of assembly from mitotic centrosomes. Indeed, treatment of the mitotic extracts with alkaline phosphatase to induce dephosphorylation resulted in decreased microtubule assembly (Centonze and Borisy, 1990). Similarly, centrosomes isolated from mammalian cells and treated with urea were unable to nucleate microtubules unless these centrosomes were first incubated in *Xenopus* egg extracts. Centrosomes incubated in interphase extracts nucleated fewer microtubules than those incubated in prophase extracts. Treatment of the prophase extracts with protein kinase inhibitors reduced the ability of these extracts to enhance microtubule assembly from centrosomes. Conversely, treatment of interphase extracts with cyclin A to activate p34^{cdc2} kinase or a *cdc2*-like kinase increased the ability of these extracts to promote centrosomal microtubule nucleating activity (Buendia *et al.*, 1992).

Additional evidence that mitotic centrosomes acquire an MPM-2 phosphorylated epitope comes from immunofluorescent studies demonstrating that, in addition to acquiring γ-tubulin when centrioles are incubated in egg extracts, as discussed earlier, these centrioles also exhibit no MPM-2 staining until after they have been incubated in egg extracts (Félix *et al.*, 1994; Stearns and Kirschner, 1994). Together, these studies suggest that acquisition of both γ-tubulin and the MPM-2 epitope appears to be required for the increased nucleating activity of mitotic centrosomes. It will be of considerable interest to determine if it is γ-tubulin itself or other centrosomal components that become phosphorylated during mitosis and thereby contribute to the cell cycle-dependent behavior of the centrosomes.

B. MAP Control of Microtubule Behavior

1. Properties of MAPs

Microtubule-associated proteins (MAPs) are a group of proteins which copurify with tubulin through cycles of assembly and disassembly. Some of these MAPs appear to be specifically localized to neuronal cells, while others are present in non-neuronal cells. In neurons, the high-molecular-weight MAPs, MAP1A and B, are present in axons and dendrites (Bloom *et al.*, 1984) and MAP2 is localized to dendrites (Bernhardt and Matus, 1984; Burgoyne and Cumming, 1984; Caceres *et al.*, 1984; DeCamilli *et al.*, 1984), while tau, a lower molecular weight MAP, is restricted to axons (Binder *et al.*, 1985). MAP4 is the major MAP present in mammalian, non-neuronal cells (Bulinski and Borisy, 1980). These proteins profoundly influence microtubule behavior *in vitro*. In particular, they lower the critical tubulin concentration required for assembly, stimulate microtubule assembly and stabilize microtubules by inhibiting disassembly, and cross-bridge microtubules (Gelfand and Bershadsky, 1991; Hirokawa, 1994). MAP2 and tau exhibit 67% amino acid similarity within their carboxy-terminal 185 amino acids. Within this domain are three to four imperfect 18-mer repeats that are responsible for microtubule binding (Lewis *et al.*, 1988). MAP4 also contains three repeats of this same 18-mer microtubule binding domain (Chapin and Bulinski, 1991) while MAP1A and B contain a distinct microtubule binding site (Noble *et al.*, 1989). Thus MAP2, MAP4, and tau are related proteins that most likely bind to the same site on the microtubule lattice, while MAP1A and B are distinct MAPs.

2. Effect of MAPs on Cell Architecture

The expression of MAP2 or tau in cells that normally do not express these proteins, or the inhibition of their expression in cells that do, can have profound effects on the architecture and behavior of the cells. Inhibition of tau (Caceres and Kosik, 1990) or MAP2 (Dinsmore and Solomon, 1991; Caceres *et al.*, 1992) synthesis using antisense RNAs results in an inhibition of neurite outgrowth. Interestingly, the effect of inhibition of expression of the two proteins is not identical. In MAP2 antisense studies, embryonic carcinoma cells, which normally extend neurite processes in culture when stimulated by retinoic acid, do not do so if MAP2 antisense RNA sequences are expressed in these cells before retinoic acid induction. Moreover, those few neurites that are extended are much shorter than in control cells (Dinsmore and Solomon, 1991). Similar results are obtained using cultured cerebellar macroneurons (Caceres *et al.*, 1992). Inhibition of tau synthesis, on the other hand, blocks primary cultures of neurons from rat embryonic cerebellum from extending a single elongated neurite, but the cells do elaborate numerous short neurites (Caceres and Kosik, 1990). These studies suggest that MAP2 may be required for initial neurite extension, and that tau may then stabilize a single extension so that it develops into the elongated axon.

The role of MAPs, in general, and tau, in particular, in neurite formation and extension has been brought into question by a recent study. Despite the seeming importance of tau for axon formation, discussed earlier, disruption of the tau gene in mice produces animals that appear to be neurologically normal. In addition, primary cultures of small-caliber neurons from these tau-deficient mice elongate axon-like processes *in vitro* (Harada *et al.*, 1994). Thus, tau may be not be essential for neurite elongation or for neuronal function. However, it is also possible that tau's function is redundantly encoded into the genome since these mice, lacking tau, upregulate MAP1A in proportion to the extent of their tau loss (Harada *et al.*, 1994).

Both tau and MAP2 are capable of inducing neurite-like processes in cells that are normally round and without extensions. Transfection of moth ovary cells with baculovirus containing a tau or MAP2 cDNA insert and expression of these genes result in the elaboration of very long neurite-like processes which contain extensive microtubule bundles (Baas *et al.*, 1991; Knops *et al.*, 1991; Chen *et al.*, 1992). Interestingly, taxol does not induce similar elongated processes though it does induce extensive microtubule bundling within these cells (Knops *et al.*, 1991). Thus, neurite extension is not a direct consequence of microtubule bundling. The microtubule–microtubule spacing in the neurite-like processes induced by MAP2 expression is similar to the spacing found in dendrites, where MAP2 is normally localized, while the spacing in those processes induced by tau is similar to that found in axons, where tau is normally localized (Chen *et al.*, 1992). Thus, in cells expressing these proteins, each of these MAPs is capable of inducing microtubule bundles with characteristics similar to those in cells in which these MAPs normally occur.

3. Effect of MAPs on Microtubule Stability and Dynamics

The expression of tau and other MAPs in a variety of cell types that normally do not produce these proteins generally results in a shift in the assembly equilibrium toward more microtubules and extensive microtubule bundling (Lewis *et al.*, 1989; Lee and Rook, 1992; Takemura *et al.*, 1992; Weisshaar *et al.*, 1992). Moreover, these bundles remain intact following treatment with microtubule depolymerizing drugs such as colchicine or nocodazole, suggesting that the microtubules become stabilized as a result of the MAP expression (Lewis *et al.*, 1989; Takemura *et al.*, 1992). Indeed, attempts to obtain stable MAP2 transformants of cultured cells have been unsuccessful. Stable transformants would require that the transformed cells be capable of dividing. The expression of MAP2, if it bundles and stabilizes microtubules, might be incompatible with the process of mitosis, in which dynamic microtubules are required (Lewis *et al.*, 1989). Interestingly, inhibition of MAP2 expression by antisense oligonucleotides allows nonmitotic, embryonic carcinoma cells, induced with retinoic acid, to reenter mitosis (Dinsmore and Solomon, 1991). Thus, it is possible that MAPs, such as MAP2 and tau, which bundle and stabilize microtubules, are only expressed in cells that have left the cell cycle. Their normal location, in neurons, which are nonmitotic, is consistent with this hypothesis.

While most microtubules in cells are dynamic and rapidly turn over, some microtubules persist over time. The existence of a stable class of microtubules was first recognized by the lack of incorporation of microinjected labeled tubulin into a subset of a cell's microtubules (Schulze and Kirschner, 1986). These stable microtubules often contain post-translationally modified tubulin, either detyrosinated or acetylated α-tubulin (Gundersen et al., 1984; Piperno and Fuller, 1985). While the tubulin modifications mark stable microtubules, the modifications do not account for their stability (Khawaja et al., 1988; Webster et al., 1990). Various studies suggest that the presence or absence of MAPs in cells results in differences in the extent of tubulin modification; these modifications may thereby reveal the extent to which the MAPs have stabilized the endogenous microtubules. Cells transfected with cDNA clones encoding MAP2, tau, or MAP1B exhibit microtubules which are stable to microtubule depolymerizing drugs and which have an increased level of acetylated tubulin (Takemura et al., 1992). Conversely, when tau synthesis is blocked by antisense oligonucleotides, there is a corresponding block in the normal generation of acetylated microtubules (Caceres et al., 1992). Thus, these tubulin modifications provide additional evidence that MAPs stabilize microtubules in vivo.

The ability to stabilize microtubules may depend on the phosphorylation state of MAPs. MAP2, when microinjected into rat fibroblasts, binds to endogenous microtubules only when this MAP has been phosphorylated on a subset of its total phosphorylatable sites (Brugg and Matus, 1991). MAP2 binds to the regulatory subunit of PKA (cAMP-dependent protein kinase) (Theurkauf and Vallee, 1982) and can be phosphorylated by this and a number of other kinases (Sloboda et al., 1975; Vallano et al., 1985; Tsuyama et al., 1986). Similarly, tau can be phosphorylated at multiple sites, and this modification affects tau's ability to stimulate microtubule assembly (Lindwall and Cole, 1984). A 220-kDa MAP is in an unphosphorylated state when isolated from interphase Xenopus eggs and can bind to microtubules. Isolated from mitotic cells, this same MAP is phosphorylated and cannot bind to microtubules. Further, when isolated from interphase, but not mitotic cells, this MAP can be phosphorylated in vitro by either maturation-promoting factor (MPF) or by MAP kinase, suggesting that the interphase form of this MAP can be phosphorylated by kinases that become active as the cell progresses into the M-phase (Shiina et al., 1992a). Thus, the ability of MAPs to influence microtubule behavior may depend on the state of phosphorylation of the MAPs. This modification, in turn, can be controlled by the cell, either via the cell cycle or cell signaling. The cell cycle-dependent behavior of microtubules is discussed in greater depth in Section II,C.

4. Effect of MAPs on Microtubule–Motor Interactions

In addition to affecting the overall architecture of microtubules in cells, MAPs might also affect cell motility by competing with motor molecules for microtubule binding. MAP2, tau, kinesin, and cytoplasmic dynein bind to overlapping or closely adjacent sites on the microtubule (Paschal *et al.*, 1989; Rodionov *et al.*, 1990; Hagiwara *et al.*, 1994). Both tau and MAP2 compete with kinesin and cytoplasmic dynein for binding to microtubules (Hagiwara *et al.*, 1994) and can inhibit dynein- or kinesin-driven microtubule gliding *in vitro* (von Massow *et al.*, 1989; Lopez and Sheetz, 1993; Hagiwara *et al.*, 1994). Recent studies have shown that MAP4 can be localized to a subset of microtubules *in vivo* (Chapin and Bulinski, 1994). Interestingly, in axons, there appear to be preferred microtubules along which vesicles are transported *in vivo* (Miller *et al.*, 1987). Perhaps the MAP composition of those microtubules which serve as tracks for vesicle movement is different from that of those microtubules along which vesicle transport is not observed.

C. Cell Cycle Control of Microtubule Behavior

1. Control of Microtubule Dynamics and Assembly

Microtubules are dynamic structures, alternating between phases of polymerization and depolymerization, with abrupt transitions occurring between the two (Mitchison and Kirschner, 1984a,b; Horio and Hotani, 1986). Measurements of microtubule turnover rates in cells have revealed that interphase microtubules have longer half lives and are therefore less dynamic than mitotic microtubules (Saxton *et al.*, 1984; Salmon *et al.*, 1984). This difference in microtubule behavior can be mimicked in *in vitro* systems. In *Xenopus*, microtubule assembly and disassembly rates are similar in mitotic and interphase extracts, but the frequency of catastrophes is greatly increased in the former (Belmont *et al.*, 1990). Changes in microtubule dynamics can be correlated with changes in the activity of cell cycle-dependent protein kinases. Treatment of interphase extracts of *Xenopus* eggs with p34^{cdc2} kinase results in a stimulation of microtubule dynamic behavior (Verde *et al.*, 1990). Similarly, inhibition of phosphatase activity in sea urchin extracts converts the dynamic behavior of the microtubules in an interphase extract to that of microtubule behavior in mitotic extracts, i.e., the dynamic behavior is enhanced as the frequency of rescue decreases (Glicksman *et al.*, 1992). That the state of microtubule assembly is directly affected by cell cycle-dependent changes is strongly suggested in studies showing that the p34^{cdc2} kinase induces microtubule disassembly (Faruki *et al.*, 1992; Lieuvin *et al.*, 1994). Addition of purified p34^{cdc2} kinase to interphase *Xenopus* egg extracts containing centrosomes with nucleated microtubules results in rapid disassembly of the centrosomal microtubules. Although MAP2 can stabilize these microtubules from depolymerization by drugs such as nocodazole, MAP2 cannot protect these microtubules from depolymerization induced by adding p34^{cdc2} (Faruki *et al.*, 1992). Furthermore, addition of this kinase to lysed mammalian cultured cells results in an immediate and complete destabilization of the microtubule network, while cAMP-dependent protein kinase, MAP kinase, casein kinase II, and Ca^{2+} calmodulin-dependent protein kinase are without effect (Lieuvin *et al.*, 1994). Thus, the p34^{cdc2} kinase specifically might be responsible for converting the interphase microtubule network into the much more dynamic and unstable mitotic microtubule network.

In mammalian cells, the cell cycle changes in microtubule behavior may be regulated by the phosphorylation state of MAP4. In non-neuronal cells, MAP4 is the primary microtubule-associated protein. MAP4 is a heterogeneous protein, consisting of multiple species of a 210-kDa polypeptide, in which heterogeneity is generated both by alternate splicing of a single gene and by various phosphorylations of the gene products (Chapin and Bulinski, 1991). MAP4 stimulates microtubule assembly *in vitro* and is found along microtubules *in vivo* (Chapin and Bulinski, 1994). Recent studies demonstrate that MAP4 binds to cyclin and is phosphorylated by p34^{cdc2} kinase. Because both phosphorylated and unphosphorylated MAP4 bind to microtubules but only unphosphorylated MAP4 stabilizes microtubules, it is possible that the increased dynamic behavior of microtubules during mitosis can be accounted for by the phosphorylation state of MAP4 (Ookata *et al.*, 1995). During interphase, unphosphorylated MAP4 would stabilize microtubules. During mitosis, when cyclin B levels are high and the p34^{cdc2} kinase is activated, MAP4 would become phosphorylated and could no longer stabilize microtubules. These would, as a consequence, become more dynamic. Interestingly, protein phosphatase 2A is also localized on microtubules and is present on them throughout the cell cycle. However, the activity of this phosphatase is much higher in S-phase than it is during mitosis (Sontag *et al.*, 1995), and might, during interphase, reverse the mitotic phosphorylations conferred onto MAP4 by p34^{cdc2} kinase.

2. Microtubule-Severing Proteins

The assembled state of microtubules can be dramatically altered by microtubule severing. A microtubule-severing activity was first demonstrated in mitotic extracts of *Xenopus* eggs (Vale, 1991). These studies revealed that extracts from mitotic but not interphase cells were capable of causing taxol-stabilized microtubules to become severed into short pieces which were then rapidly depolymerized. Treatment of interphase extracts with cyclin B conferred upon them the ability to sever microtubules (Vale, 1991), suggesting that the severing protein may be activated via a cell cycle-dependent protein phosphorylation event. Additional studies assessed the effects of both cyclin A and cyclin B on microtubule integrity in *Xenopus* egg extracts (Verde *et al.*, 1992). Addition of cyclin B to interphase extracts inhibited spontaneous microtubule assembly and instead resulted in the formation of very short astral microtubules. On the other hand, addition of cyclin A, which also blocked spontaneous microtubule assembly, resulted in the formation of long astral microtubules. These studies suggest that the microtubule array in cells is under direct control of the cell cycle. Cyclins, through their activation of cell cycle-dependent protein kinases, may regulate microtubule-severing proteins to control microtubule lengths (Verde *et al.*, 1992).

Two proteins have been isolated from *Xenopus* egg extracts and shown to have microtubule-severing activity (Shiina *et al.*, 1992b, 1994). Treatment of *Xenopus* egg interphase extracts with maturation-promoting factor results in a rapid increase in the ability of this extract to severe microtubules. An MPF-dependent severing factor has been isolated and purified and shown to be a single polypeptide of 56 kDa which may form a homopentamer. This protein severs microtubules in the absence of ATP (Shiina *et al.*, 1992b). Most intriguing are the recent findings that one of the *Xenopus* microtubule-severing proteins is identical by sequence analysis and antigenicity to the essential protein synthesis factor, EF-1α (Shiina *et al.*, 1994). Moreover, purified EF-1α exhibits very high microtubule-severing activity. At present, it is not known if EF-1α functions *in vivo* to sever microtubules in addition to its role in protein synthesis. Cells with normal, endogenous levels of EF-1α have a microtubule array; in cells induced to overexpress EF-1α, these microtubule arrays are depolymerized (Siina *et al.*, 1994). If EF-1α functions *in vivo* to sever microtubules, then this activity must be regulated so that severing occurs only as the cell transits from interphase into mitosis.

A protein in sea urchin egg extracts that is capable of severing microtubules has been isolated and purified (McNally and Vale, 1993). This protein, named katanin, forms a heterodimer containing two subunits of 81 kDa and 60 kDa. Katanin requires ATP to sever and depolymerize microtubules, and thus is unlikely to be a homolog of the *Xenopus* severing proteins discussed earlier. The resulting tubulin is capable of repolymerizing into microtubules when the ATP is depleted, thereby indicating that katanin does not irreversibly alter the tubulin by virtue of its severing capabilities (McNally and Vale, 1993). It is not yet known if katanin is cell cycle regulated.

Microtubule severing followed by rapid depolymerization has been observed *in vivo* in the extensive microtubule networks of the giant protistan *Reticulomyxa* (Chen and Schliwa, 1990). An even more dramatic and remarkable example of this phenomenon occurs in the heliozoan, *Actincoryne contractilis*. This single-celled organism is characterized by a microtubule-filled stalk by which it attaches to a substratum. In its fully extended state, the stalk can be up to 800 μm long. When the cells are stimulated, with Ca^{2+}, for example, a rapid contraction occurs, and, within about 20 msec, the stalk is usually less than 90% of its original length (Febvre-Chevalier and Febvre, 1992). This remarkable shortening is accompanied by a rapid disassembly of microtubules which itself may be preceded by rapid and numerous breaks in the microtubules which would serve to increase the number of microtubule ends. Only by such a severing mechanism is it possible to explain the rapid disassembly that occurs in these cells. In this case, a severing and/or depolymerizing factor would presumably be activated by Ca^{2+} via a signal-transduction event.

As we turn our attention now to microtubule motors and their regulation, it is useful to keep in mind that motors can act only if there are microtubules present on which to move. As detailed earlier, the assembled state of microtubules in cells is controlled by nucleation via γ tubulin and by stabilization and bundling via MAPs. Microtubule length and disassembly may be controlled by severing proteins. The interaction of these various proteins with microtubules may be regulated by changes in the phosphorylation state of key components of this system. Regulation by phosphorylation occurs either as the cell normally progresses through the cell cycle or, instead, as the cell's surface receptors are stimulated by specific agonists. Table I summarizes the various mechanisms by which microtubules may be regulated *in vivo*. The mechanisms discussed earlier and summarized in the table may not represent the entire repertoire available to a cell. Currently, the *in vivo* functions of various MAPs are not well understood, and the behaviors of γ-tubulin and of severing proteins, relative newcomers to the field, have also not been thoroughly elucidated. Future studies should provide additional information about the mechanisms by which these proteins interact with microtubules and are regulated. While the organized array of microtubules will dictate where in the cell microtubule-based motility can occur, it is likely that it is the regulation of microtubule motors that

TABLE I
Regulation of Microtubules

Effector	Effect on Microtubules	Phosphorylation event implicated?	Kinase or phosphatase identified?	Signal	References
γ-Tubulin	Enhances nucleation	Phosphorylated epitope (MPM-2) and γ-tubulin appear at centrosome coincident with nucleating activity	Cell cycle kinase implicated	Cell cycle progression	Joshi et al. (1992), Centonze and Borisy (1990), Félix et al. (1994), Stearns and Kirschner (1994)
MAPs	Induce bundling, increase MT stability; dampen MT dynamic instability	Phosphorylation of MAPs affects their interaction with MTs	PKA, p34^{cdc2}	Signal transduction, cell cycle progression	Caceres and Kosik (1990), Dinsmore and Solomon (1991), Faruki et al. (1992), Shiina et al. (1992a), Lieuvin et al. (1994), Ookata et al. (1995)
Severing proteins	Sever and depolymerize MTs	Activity of some severing proteins is cell cycle dependent	p34^{cdc2}	Cell cycle progression, signal transduction	Vale (1991), Verde et al. (1992), Shiina et al. (1992b, 1994), McNally and Vale (1993), Febvre-Chevalier and Febvre (1992)

activates and controls the direction of movements of organelles along microtubules.

III. Regulation of Microtubule Motors

Actin and myosin interact with each other in only one polarity, and bidirectional movement in actin-based systems can occur because the actin filaments are oriented in opposite polarities, as, for example, in skeletal muscle
on the opposite sides of the sarcomere (Huxley, 1969) or in characean
algae, on opposite sides of the clear zone (Kersey et al., 1976). Thus, in actin-
myosin based motility, regulation is required to initiate and halt motility, but
not to control direction. In contrast, microtubules are oriented predominantly with a uniform polarity, in which the minus ends are located at the
MTOC and the plus ends at the cell periphery (Euteneuer and McIntosh,
1981). Moreover, bidirectional transport of organelles along a single microtubule has been shown in axons (Hayden and Allen, 1984), extruded axoplasm (Schnapp et al., 1985), and the protistan Reticulomyxa (Koonce and
Schliwa, 1985). Because two different families of microtubule motors exist—the dynein gene family, whose members drive transport to the minus
ends of microtubules, and the kinesin gene family, which contains members
that are plus end motors and others that are minus end motors—regulation
of microtubule motors may be required not only to initiate motility but
also to control its direction.

A. Dynein-Based Transport

1. Dynein Family Characteristics

The dynein gene family is composed of two major classes: axonemal and
cytoplasmic dyneins. There are several genes for the different isoforms of
axonemal dynein and perhaps as few as one gene for cytoplasmic dynein
(Witman, 1992; Vallee, 1993; Holzbaur and Vallee, 1994). Dyneins are
microtubule binding proteins having a native mass greater than 1.2×10^6 Da and characterized by very large ($\sim$500 kDa) heavy chains which
can be photolysed into two fragments in the presence of vanadate and ATP
(Lee-Eiford et al., 1986; Paschal et al., 1987a). The ATPase and motor
activity of dyneins are sensitive to inhibition by low concentrations of
vanadate (Gibbons et al., 1978; Johnson, 1985). The dynein ATP hydrolysis
cycle displays kinetics similar to that of myosin, suggesting that the mechanism by which mechanical force is generated by these two motors is similar

(Johnson, 1983). Dynein's ATPase activity is stimulated by microtubules (Omoto and Johnson, 1986; Shpetner *et al.*, 1988). All dyneins characterized to date induce movement toward the minus ends of microtubules (Sale and Satir, 1977; Fox and Sale, 1987; Paschal and Vallee, 1987; Paschal *et al.*, 1987b; Vale and Toyoshima, 1988; Lye *et al.*, 1989; Kagami and Kamiya, 1992).

The β-chain of the sea urchin flagellar outer arm dynein was the first dynein subunit to be sequenced (Gibbons *et al.*, 1991; Ogawa, 1991), and contains several consensus phosphate-binding domains, P loops, which are frequently found in ATP-binding proteins. The most NH$_2$ terminally located of these P loops may be the catalytic site for ATP hydrolysis, as photocleavage in the presence of ATP and vanadate has been mapped to this site (the V1 site) (Gibbons *et al.*, 1991). A number of dynein heavy chains have now been partially or completely sequenced, and the presence of the four P loops is a feature common to them all (Koonce *et al.*, 1992; Mikami *et al.*, 1993; Eschel *et al.*, 1993; Li *et al.*, 1993; Zhang *et al.*, 1993; Asai *et al.*, 1994; Mitchell and Brown, 1994; Rasmusson *et al.*, 1994; Wilkerson *et al.*, 1994; Xiang *et al.*, 1994). Cytoplasmic dynein heavy chains from different species exhibit a relatively high degree of conservation, while cytoplasmic and axonemal dynein heavy chains vary significantly throughout their amino terminal third but exhibit conservation within the remaining carboxy-terminal two thirds (Vallee, 1993). The different amino terminals may specify dynein function. Axonemal dyneins are responsible for force production in cilia and flagella that leads to microtubule sliding. Cytoplasmic dynein is involved in organelle transport to the minus ends of microtubules. In addition, cytoplasmic dynein may function in mitotic movements.

2. Axonemal Dynein-Driven Movements

a. Properties of Axonemal Dynein The now-classic studies of Gibbons and co-workers have revealed the basic underlying mechanism generating cilia and flagellar beating. Motility is generated by the molecular motor, axonemal dynein, which cyclically cross-bridges and induces sliding between adjacent outer doublet microtubules. Dynein forms two rows of arms, the inner and outer arms, that run the length of each outer doublet microtubule (I. R. Gibbons, 1963). The sliding forces exerted by dynein are resisted by structures within the axoneme that constrain the extent of relative motion between the doublet microtubules, causing the sliding to be converted into bending. When these constraints are removed by proteolysis, the doublet microtubules slide past each other, causing the axoneme to telescope apart into individual doublet microtubules (Summers and Gibbons, 1971). Sliding between different groups of doublet microtubules must occur in a regulated manner in intact axonemes if linear sliding is to be converted into coordinated bending.

Flagellar outer arm dyneins are composed of two or three heavy chains and, correspondingly, have two or three globular head domains (Sale *et al.*, 1985). Each heavy chain contains microtubule binding, ATPase domains (Pfister *et al.*, 1984, 1985), and a part of the stalk. In addition, each heavy chain is associated with a set of intermediate and/or light chains. In *Chlamydomonas,* the intermediate chain IC70 has been localized to the base of the dynein particle, and its close association with IC78 and several light chains suggests that these will be localized to the base also (King and Witman, 1990). IC78 interacts with α-tubulin in an ATP-insensitive manner, again suggesting its localization to the base of the dynein particle and further implicating IC78 in dynein microtubule binding to the A-fiber of the axonemal doublet microtubules (King *et al.*, 1991). While there appears to be a single molecular species of outer arm dynein, there are at least three different sets of inner arms in *Chlamydomonas,* and each is composed of a distinct set of polypeptides (Piperno *et al.*, 1990; reviewed in Witman, 1992). Moreover, the dynein isoforms comprising the inner arms may vary along the length of the axoneme, with one class of inner-arm dyneins localizing to the proximal portion of the axoneme (Piperno and Ramanis, 1991). Recent studies in sea urchin embryos have identified 14 dynein heavy-chain mRNA transcipts, each of which is derived from a unique dynein gene. Eleven of these increase in abundance when the embryos are deciliated and are therefore likely to be axonemal dyneins (Gibbons *et al.*, 1994). The need to coordinate motility not only among the nine groups of doublets, but also along the length of the axoneme, may require that there be multiple species of dynein which can be differentially activated.

The precise contributions of specific classes of dynein molecules within the axoneme to motility have been the subject of analysis in a number of studies. Removal of the outer row of sea urchin sperm flagellar dynein arms reduces axonemal beating to half its original frequency without affecting the wave form (Gibbons and Gibbons, 1973). Similarly, in *Chlamydomonas,* the outer dynein arm, *oda,* mutants exhibit a decrease in swimming speed and beat frequency (Kamiya and Okamoto, 1985; Mitchell and Rosenbaum, 1985), while mutants of *Chlamydomonas* lacking a subset of the inner arms exhibit a sharp reduction in the amplitude of the bend and little change in the beat frequency (Brokaw and Kamiya, 1987). Thus, the inner and outer rows of dynein arms may have distinct functions in which the outer dynein arms function to overcome viscous resistances during beating, and thereby affect speed, while different subsets of the inner arms may be responsible for bend initiation and propagation, which may affect wave form (Brokaw, 1994).

The outer dynein arms in *Chlamydomonas* are composed of three heavy-chain subunits, the α-, β- and γ-chains. The contribution that each of these makes toward force generation has been elucidated using outer dynein arm mutants. One of these, *oda 1,* lacks the α-chain and one of the light chains. This mutant swims more slowly than the wild type but significantly faster than *oda4,* which lacks the entire outer arm (Sakakibara *et al.,* 1991). Mutants carrying an allele of *oda4, oda4-s7,* possess outer dynein arms that contain a truncated β heavy chain and lack only a small part of the structure of the outer arm. Interestingly, these mutants exhibit the same reduced swimming speed as the *oda4* mutants lacking the entire outer arm (Sakakibara *et al.,* 1993). Thus, the swimming speed function of the outer dynein arm appears to be a property principally of the β heavy chain.

The contribution to motility of each dynein arm and each heavy chain within an arm can be assessed by microtubule gliding assays *in vitro* in which dynein, absorbed to a glass slide, causes microtubules to glide relative to the surface of the slide (Paschal *et al.,* 1987b; Vale and Toyoshima, 1988). Such analyses have shown that both the inner arm (Fox and Sale, 1987; Kagami and Kamiya, 1992) and outer arm (Paschal *et al.,* 1987b) dyneins generate force in the same direction, toward the minus end of the microtubule. The velocity of translocation, however, differs among isoforms. The inner-arm isotypes I2 and I3 and the outer arm dyneins translocate at much greater velocities than the inner arm isotype I1 (Smith and Sale, 1991), potentially reflecting functional differences among the structurally different inner arm dyneins (Smith and Sale, 1992b). Moreover, the isolated β complex, which is derived from the outer dynein arm in sea urchin flagella, is itself sufficient to generate force (Sale and Fox, 1988; Moss *et al.,* 1992), and does so at a faster rate than intact 21S outer-arm dynein (Sale and Fox, 1988). Similarly, the isolated γ-chain of *Tetrahymena* 22S outer-arm dynein induces microtubule gliding (Mimori and Miki-Noumura, 1995). It has not yet been demonstrated that the α heavy chain of the outer arm possesses microtubule motor activity although it has ATPase activity as well as the invariant ATP binding domain (the P1 loop) present in all dyneins sequenced to date (Pfister and Witman, 1984; Mitchell and Brown, 1994). It is somewhat daunting to consider how these various dyneins, which generate different rates of microtubule sliding under similar *in vitro* conditions and which are all present in the same axoneme, can be regulated *in vivo* to generate coordinated motility.

b. Regulation of Ciliary and Flagellar Beating

If all nine doublet microtubules in axonemes were to slide simultaneously, there would be no net movement. An analysis of doublet sliding in sea urchin sperm axonemes demonstrates clearly that when active sliding in the principal bend occurs between doublets 6–9, doublets 1–4 remain quiescent (Sale, 1986). Numerous studies have demonstrated that ciliary and flagellar beating is affected by the second messengers, Ca^{2+} and cAMP, though their effects on motility differ among different organisms. Nonetheless, recent studies suggest that dynein may be one target of the regulatory system, and these data may provide the first step toward understanding potentially multiple levels of regulation in this system.

   i. Role of cAMP in Regulating Motility. Biochemical Studies.   In many cilia and flagella, cAMP activates motility (Opresko and Brokaw, 1983; Stommel and Stephens, 1985; Bonini *et al.*, 1986; Murakami, 1987). Beating in *Chlamydomonas* flagella is inhibited rather than activated by cAMP and will be discussed separately later. In *Paramecium* cilia, membrane-permeable analogs of cAMP or phosphodiesterase inhibitors increase the swimming speed of live cells (Bonini *et al.*, 1986) while cAMP has the same effect on the swimming speed of lysed models (Bonini and Nelson, 1988; Hamasaki *et al.*, 1991). The cAMP mediates its effect via PKA-induced phosphorylation of specific proteins in the axoneme, and a number of studies now indicate that subunits of dynein may be major targets of this regulatory system.

   In *Paramecium,* stimulation of PKA activity in lysed cells results in an increase in phosphorylation of a number of the ciliary axonemal proteins. Extraction of the outer dynein arms from these axonemes results in an enrichment of two phosphoproteins, one at 65 kDa and the other at 29 kDa (Hamasaki *et al.*, 1989, 1991; Bonini and Nelson, 1990). The 29-kDa polypeptide is phosphorylated in the presence of cAMP; this phosphorylation is inhibited by micromolar levels of Ca^{2+} (Hamasaki *et al.*, 1989). The rate of dynein-mediated *in vitro* gliding of microtubules is significantly faster when the 29-kDa polypeptide in the dynein preparation is phosphorylated (Hamasaki *et al.*, 1991; Barkalow *et al.*, 1994). Because the 29-kDa polypeptide copurifies stoichiometrically with dynein heavy chain, this polypeptide may be an axonemal dynein regulatory light chain (Barkalow *et al.*, 1994). Dynein subunits of molecular weight similar to the *Paramecium* 29-kDa light chain become phosphorylated in a cAMP-dependent manner in cilia of *Tetrahymena* (Chilcote and Johnson, 1990), *Spisula solidissima,* and *Mytilus edulis* (Stephens and Prior, 1992), thus indicating that phosphorylation of a dynein light chain may be a general feature of ciliary regulation.

   The phosphorylation of a dynein light chain by a cAMP-mediated process may characterize ciliary but not flagellar regulation. Motility is stimulated in both molluscan gill cilia and in sperm flagella by a cAMP-dependent pathway. In the cilia of the two species analyzed, phosphorylation occurs primarily on one or two dynein light chains (Stephens and Prior, 1992). Most interestingly, while motility in the sperm of these species is also activated by cAMP, it is the α heavy chain of the outer dynein arms, rather than a light chain, that is the principal target of the regulatory system.

In *Ciona* sperm, both a dynein heavy chain and a light chain become phosphorylated in a cAMP-dependent manner during activation of motility both *in vivo* and *in vitro* (Dey and Brokaw, 1991). These studies suggest that, while targeting dynein in both cilia and flagella, the regulatory pathways controlling activation of motility may vary.

Genetic Studies. Motility in *Chlamydomonas* flagella can be examined genetically, and paralyzed mutants have been very useful for understanding how regulation may be mediated. In *Chlamydomonas,* unlike in most cilia and flagella, cAMP inhibits rather than activates flagellar motility (Hasegawa *et al.,* 1987). *Chlamydomonas* mutants lacking the radial spokes or central pair complex are paralyzed, but can undergo microtubule sliding if the axonemes are briefly digested with trypsin (Witman *et al.,* 1978; Smith and Sale, 1992a). Extragenic suppressor mutations have been isolated which restore motility but do not restore the disrupted radial spokes or central pair microtubules. In each case, these suppressors are second-site mutations in either the outer or inner dynein arms (Brokaw *et al.,* 1982, Huang *et al.,* 1982; Porter *et al.,* 1992, 1994) or in the dynein regulatory complex, discussed later (Piperno *et al.,* 1992, 1994). Two extragenic suppressor mutations of radial spoke and central pair defects have been mapped to a locus within the β heavy chain of the outer dynein arm. The two mutations are distinct, with the loss of 7 or of a different 10 amino acids in a region adjacent to the nucleotide binding domain (Porter *et al.,* 1994). These studies suggest that the radial spoke–central pair complex normally activates dynein, but, in the absence of this complex, the requirement for this activation can be circumvented by a change in dynein itself. This region may act like a regulatory inhibitory domain, as occurs in other enzymes such as myosin light chain kinase or calcineurin, and deletion of this domain may release dynein from inhibition. In wild-type cells, this inhibition would be released by the radial spoke–central pair complex, potentially by inducing dephosphorylation of the dynein, as suggested by the following studies.

Activation, or release from inhibition, of dynein by the central pair–spoke complex is implicated by an analysis of sliding velocities in *Chlamydomonas* axonemes lacking radial spokes (Smith and Sale, 1992a). Sliding in the mutant axonemes is significantly slower than that in wild-type controls. Inner-arm dynein, extracted from the mutants and reconstituted onto wild-type axonemes which had been depleted of dynein, induces sliding at wild-type rates. Moreover, dynein from the wild-type axonemes induces sliding at wild-type rates when reconstituted onto the spokeless axonemes. These observations are consistent with the hypothesis that radial spokes are required to activate dynein, and that, once activated by exposure to radial spokes, the dynein can function in their absence (Smith and Sale, 1992a). Because inhibitors of cAMP-dependent protein kinase activity increase the velocity of sliding in spokeless mutants, but have no effect on wild-type sliding rate, the radial spokes may affect dynein phosphorylation. These and additional analyses suggest that the radial spokes may activate inner dynein arms by inhibiting dynein phosphorylation (Howard *et al.*, 1994). However, it has not yet been directly demonstrated that phosphorylation of inner-arm dynein occurs, nor has it been shown that this phosphorylation inhibits the motor activity of inner-arm dynein.

Phosphorylation of an outer dynein arm polypeptide in *Chlamydomonas* does occur, although it is not yet known if this phosphorylation is correlated with activity. Among the outer dynein arm polypeptides in *Chlamydomonas*, only the α-subunit becomes phosphorylated *in vivo;* the same six sites that are phosphorylated *in vivo* are also phosphorylated in isolated axonemes, suggesting that the kinase is tightly associated with its substrate (King and Witman, 1994). Interestingly, the α-subunit is the only one in the outer arm that has not yet been shown to possess motor activity, although this activity might be predicted to occur since this subunit contains the conserved P loops found in other dynein heavy chains.

The dynein regulatory complex in axonemes of *Chlamydomonas* was identified as a complex of six proteins in which mutations either cause motility defects or, instead, suppress radial spoke or central pair defects without repairing them (Piperno *et al.*, 1992, 1994). This complex appears to serve as a binding site or to be an essential component of the binding site for the inner dynein arms (Piperno *et al.*, 1994). Consistent with this biochemical evidence are structural studies which show, using computer averaging of electron microscopic images, that the dynein regulatory complex is localized to the base of the second of the radial spokes in *Chlamydomonas*, in close association with the inner dynein arms (Gardner *et al.*, 1994). Because defects in both the central pair and radial spokes can be suppressed by mutations in dynein or in the dynein regulatory complex, it has been proposed that regulation in the axonemes of *Chlamydomonas* occurs by a cascade in which the central pair apparatus may control the radial spokes, which, in turn, control the dynein regulatory complex, which finally controls the dynein arms (Piperno *et al.*, 1994). The final sequence of this cascade may be to control the phosphorylation state of dynein, as discussed earlier. Thus, the dynein regulatory complex may possess or control kinase or phosphatase activity.

ii. Role of Ca^{2+} in Regulating Axonemal Motility.   In addition to cAMP, Ca^{2+} also acts as a second messenger in controlling ciliary and flagellar motility and may reverse the effects of cAMP. In *Paramecium*, Ca^{2+}, at a concentration of 10^{-4} M, inhibits the cAMP-dependent phosphorylation of the 29-kDa dynein regulatory light chain and inhibits the cAMP-mediated stimulation in swimming speed (Hamasaki *et al.*, 1991). Low Ca^{2+} produces symmetrical beating in reactivated sperm while higher concentrations induce an asymmetrical waveform (Brokaw *et al.*, 1974). Millimolar levels of

Ca^{2+} completely inhibit motility in sea urchin sperm axonemes (Gibbons and Gibbons, 1980), and the effect appears to be mediated by calmodulin (Brokaw and Nagayama, 1985). Similarly, in *Chlamydomonas* axonemes, Ca^{2+} concentration controls waveform (Bessen *et al.*, 1980) and may act via calmodulin (Gitelman and Witman, 1980). The Ca^{2+}/calmodulin-dependent phosphatase, calcineurin, has been localized to sperm axonemes and has been shown to modulate sperm motility (Tash *et al.*, 1988). Ca^{2+} can stimulate both kinases and phosphatases, and it is possible that this second messenger mediates control via multiple targets in the axoneme.

Because the waveform of beating flagella is a three-dimensional event, and the wave can take on more than one form, regulation of beating may be more complex than a simple activation or inhibition of specific dynein arms. The relatively simpler two-dimensional motility of organelles translocating along microtubules may yield its regulatory secrets more readily than will the beating flagellum.

3. Cytoplasmic Dynein-Driven Movements

a. Properties of Cytoplasmic Dynein   The discovery that cytoplasmic dynein existed as an isoform distinct from axonemal forms of dynein came from studies using microtubule gliding assays, such as those that had been developed with the discovery of kinesin, which demonstrated that molecules with properties very similar to those of axonemal dynein could be purified from the cytoplasm of cells that lack cilia or flagella (Lye *et al.*, 1987; Paschal and Vallee, 1987; Paschal *et al.*, 1987a; Vallee *et al.*, 1988). Cytoplasmic dynein consists of two globular heads connected to a stalk domain which terminates in a globular base. The molecule contains two identical heavy chains, two or three intermediate chains of about 74 kDa, and several light chains of about 50 kDa (Paschal *et al.*, 1987a; Vallee *et al.*, 1988). The heavy chain of cytoplasmic dynein has been sequenced in rat (Mikami *et al.*, 1993; Zhang *et al.*, 1993), *Dictyostelium* (Koonce *et al.*, 1992), *Drosophila* (Li *et al.*, 1994), *S. cerevisiae* (Eschel *et al.*, 1993; Li *et al.*, 1993), and *Aspergillus* (Xiang *et al.*, 1994) and exhibits extensive homology among species.

The 74-kDa subunit has also been sequenced and exhibits some homology to the *Chlamydomonas* axonemal dynein intermediate chains IC70 (Mitchell and Kang, 1991; Paschal *et al.*, 1992). Cytoplasmic dynein has microtubule-activated ATPase activity (Shpetner *et al.*, 1988) and moves microtubules *in vitro* as a minus end-directed motor (Paschal and Vallee, 1987; Lye *et al.*, 1989). This dynein has been localized to membranous organelles (Hirokawa *et al.*, 1990; Koonce and McIntosh, 1990; Lacey and Haimo, 1992; Lin and Collins, 1992; Yu *et al.*, 1992) and to kinetochores both at the light microscopic (Pfarr *et al.*, 1990; Steuer *et al.*, 1990) and electron microscopic (Wordeman *et al.*, 1991) levels. It is interesting to consider what the functions of cytoplasmic dynein might be *in vivo,* given that cells may contain only one cytoplasmic dynein heavy chain. If it were to function both in membranous organelle transport and in chromosome movement, does the identical molecule function in each or are there differences among the intermediate or light chains or post-translational modifications of the heavy chain that define localization and thus function?

b. Function of Cytoplasmic Dynein   A variety of studies implicate cytoplasmic dynein in organelle transport. Isolated organelles from squid axoplasm (Schnapp and Reese, 1989) or from chick embryonic fibroblast cells (Schroer *et al.,* 1989) can move along microtubules toward their minus ends in the presence of soluble cytoplasmic extracts. If these extracts are treated with UV light in the presence of ATP and vanadate to photocleave the dynein, the extracts can then support minus-end transport only when they are supplemented with exogenous cytoplasmic dynein. Similarly, the ability of isolated endosomes to move along microtubules *in vitro* depends on the presence of a soluble cell extract containing intact cytoplasmic dynein (Bomsel *et al.,* 1990; Aniento *et al.,* 1993). Cytoplasmic dynein has also been implicated in the capture and minus end-directed transport of Golgi (Corthésy-Theulaz *et al.,* 1992). In none of these studies, however, could organelle movements occur in the presence of cytoplasmic dynein alone. Other, unidentified factors in the cell extracts were also required.

In *S. cerevisiae* (Eschel *et al.,* 1993; Li *et al.,* 1993), *Aspergillus* (Xiang *et al.,* 1994), and *Neurospora* (Plamann *et al.,* 1994), mutations in the gene encoding cytoplasmic dynein heavy chains result in defects in cell growth and dramatic defects in nuclear migration and spindle orientation. In yeast, the dividing nucleus is not faithfully oriented toward the bud, resulting in a high frequency of anucleate buds and binucleate mother cells. In *Aspergillus* and *Neurospora,* mutations in the cytoplasmic dynein heavy chain gene result in lack of transport of the nuclei along the elongating hypha. Both nuclear positioning and nuclear transport require the interaction of the nuclear envelope with microtubules, and these studies provide reasonably compelling evidence that at least one function of cytoplasmic dynein is to transport membrane-bounded organelles along microtubules.

Biochemical studies provide evidence that cytoplasmic dynein binds directly to membrane phospholipids. Cytoplasmic dynein has been characterized as a peripheral membrane protein of synaptic vesicles. Exogenous cytoplasmic dynein can bind to these membranes after they have been alkali stripped of all peripheral membrane proteins, suggesting that cytoplasmic dynein requires no other peripheral membrane proteins for its membrane binding (Lacey and Haimo, 1992). While one report suggests that membrane binding of cytoplasmic dynein is reduced after trypsin treatment (Yu *et al.,* 1992), another suggests that dynein will bind to trypsinized, alkali-treated membranes or to pure phospholipid vesicles composed of acidic phospholipids with a binding affinity not significantly different than that of binding to native membranes (Lacey and Haimo, 1994). These studies suggest that cytoplasmic dynein binding to membranes can occur in the absence of a membrane receptor. If, indeed, there are no membrane receptors for these motors, it will be an interesting problem to understand how specificity of binding to particular organelles might be established. However, because the dynactin complex, discussed later, is required for dynein to mediate vesicle transport (Gill *et al.,* 1991), a physiological interaction of dynein with membranes may be more complex than a simple motor–phospholipid interaction.

A functional role for dynein *in vivo* in mitotic movements is suggested from studies using antibodies directed against the catalytic domain of cytoplasmic dynein (Vaisberg *et al.,* 1993). When injected into mitotic cells in prophase, these antibodies blocked spindle formation. When injected into mitotic cells after prophase, the antibodies had no detectable effect on spindle morphology, chromosome movements, or spindle elongation. Thus, dynein function might be implicated in the formation of the spindle, but not in any of the movements normally associated with anaphase. Chromosomes can move *in vitro* toward the minus ends of microtubules, and the set of ATP analogs that supports this movement is similar to that which supports cytoplasmic dynein-driven movements (Hyman and Mitchison, 1991). However, because *Drosophila* Ncd protein, a minus end-directed kinesin motor, has a nucleotide fingerprint similar to cytoplasmic dynein (Shimizu *et al.,* 1995), it is not possible to identify with certainty the motor involved in these chromosomal minus-end movements on the basis of the nucleotide usage profile alone. While deletion of dynein alone is not lethal in *S. cerevisiae* or in *A. nidulans* (Eschel *et al.,* 1993; Li *et al.,* 1993; Xiang *et al.,* 1994), simultaneous defects in cytoplasmic dynein and in two kinesin-like proteins do block chromosome segregation (Saunders *et al.,* 1995). Thus, cytoplasmic dynein may function during mitosis, but its functions may overlap with those of other motors. Given the importance of correct chromosome segregation to cell viability, these genetic studies might be interpreted to indicate that cells have evolved redundant motors to ensure that this process occurs with fidelity.

c. Regulation of Cytoplasmic Dynein   Recent studies suggest that the behavior of cytoplasmic dynein may be regulated, mediated by an interaction of dynein with an activator, the dynactin complex (Gill *et al.,* 1991; Schroer and Sheetz, 1991), or by modification of dynein itself via changes in its state of phosphorylation (Dillman and Pfister, 1994; Lin *et al.,* 1994).

i. Regulation via the Dynactin Complex.   The dynactin complex was originally identified as a 20S particle that copurified through sucrose gradient sedimentation with cytoplasmic dynein but could be subsequently separated from the dynein by ion exchange chromatography. In the absence of dynactin, dynein promotes microtubule gliding along glass but only poorly supports vesicle transport along microtubules. Readdition of the dynactin complex results in a significant increase in the number of vesicles observed to move along microtubules (Gill *et al.*, 1991; Schroer and Sheetz, 1991). The dynactin complex contains a 150–160-kDa dynactin polypeptide (p150^{glued}) that is homologous throughout its length to the *glued* gene in *Drosophila* (Gill *et al.*, 1991; Holzbaur *et al.*, 1991). *Glued* heterozygotes have defects in development of the nervous system, particularly in the visual system, while *glued* homozygous mutations are lethal (Swaroop *et al.*, 1987). The dynactin complex contains several proteins in addition to the p150^{glued}. Interestingly, the most abundant of these is an actin-related protein, Arp1 (Lees-Miller *et al.*, 1992; Paschal *et al.*, 1993). The dynactin complex has recently been visualized by deep-etch electron microscopy and appears as a short, 37-nm, filamentous structure with a thin projection. Antibodies, used to map the location of some of these polypeptides within the dynactin complex, reveal that Arp1 may comprise the 37-nm long filament which appears to be the major backbone of the structure while the dynactin p150^{glued} polypeptide may form the thin projection (Schafer *et al.*, 1994).

Recent studies using genetic systems in which mutations occur in homologs of the Arp1 gene, provide the most direct evidence that cytoplasmic dynein and dynactin interact *in vivo*. As discussed earlier, disruption of the single cytoplasmic dynein gene in *S. cerevisiae* (Eschel *et al.*, 1993; Li *et al.*, 1993) and mutations in the cognate genes in *A. nidulans* (Xiang *et al.*, 1994) and *Neurospora crassa* (Plamann *et al.*, 1994) result in defects in spindle positioning or in nuclear migration, respectively. Most intriguingly, a phenotype indistinguishable from these is observed when mutations instead are in the Arp1 gene homolog. In *S. cerevisiae* (Clark and Meyer, 1994; Muhua *et al.*, 1994), mutations in the Arp1 gene result in a defect in nuclear positioning into the bud. In *N. crassa,* defects in either the cytoplasmic dynein gene or the Arp1 gene were isolated as identical phenotypic *ropy* mutants; these have curled hyphae and abnormal nuclear distribution (Plamann *et al.*, 1994). Further, in yeast, double mutants in the dynein heavy chain and Arp1 genes have the same phenotype as the single mutants (Muhua *et al.*, 1994). These observations strongly suggest that dynein and Arp1 function in the same processes *in vivo*. Further, because in vertebrate cells, all Arp1 is present in the dynactin complex (Paschal *et al.*, 1993), Arp1 can serve as a marker for the entire complex. Thus, defects in Arp1 should be reflected by defects in dynactin function. Additional evidence for the interaction of dynein and the dynactin complex *in vivo* comes from the observations that the *glued* mutation in *Drosophila* can be suppressed by an extragenic mutation that has been mapped to the cytoplasmic dynein heavy chain (T. Hays, personal communication). Together, these studies provide strong support for the dependence of dynein on dynactin function. What is not clear from these studies, however, is why defects in cytoplasmic dynein or Arp1 in haploid cells are not lethal while *glued* homozygous mutations in *Drosophila* are.

In situ localization studies are consistent with the genetic evidence that dynein and dynactin interact to promote membranous organelle movements *in vivo*. In vertebrate cells, antibodies directed against components of the dynactin complex exhibit punctate staining, which may correspond to vesicles, and localize to centrosomes (Gill *et al.,* 1991; Clark and Meyer, 1992; Paschal *et al.,* 1993). Given that centrosomes are located at the minus ends of microtubules, it is possible that dynactin, if necessary for dynein-mediated organelle transport, will be associated with dynein on these organelles and will accumulate with them at the minus ends of microtubules. In addition, a recent biochemical analysis of Golgi-derived small vesicles reveals that both dynein and dynactin are present in the same membrane preparations (Fath *et al.,* 1994).

An analysis of microtubule-dependent transport in cell extracts of *Xenopus* eggs has revealed that the extent of vesicle movements toward the minus ends of microtubules is greatly reduced in mitotic versus interphase extracts. However, the ability of microtubules to exhibit minus end-directed transport is comparable in both extracts (Allan and Vale, 1991). These findings suggest that dynein is active in all parts of the cell cycle but that its ability to interact functionally with vesicles may be cell-cycle regulated. Therefore, it will be of considerable interest to determine if it is dynein–dynactin interactions that are affected in a cell cycle-dependent manner or if the cell cycle exercises its effects directly on dynein. Because cell cycle regulation involves changes in protein phosphorylation, it will be important to determine if dynactin is the target of regulation. While such regulation may occur, recent studies correlate changes in cytoplasmic dynein activity with protein phosphorylation of dynein itself.

ii. Regulation of Cytoplasmic Dynein via Phosphorylation. The interaction of cytoplasmic dynein with membranes may be regulated by protein phosphorylation. Recent studies show a correlation between the level of dynein heavy chain phosphorylation and the presence of this motor on membranous organelles (Lin *et al.,* 1994). Serum starvation of cultured fibroblasts induces a dissociation of cytoplasmic dynein from lysosomes, as determined by immunofluorescent staining (Lin and Collins, 1993). The extent of heavy chain phosphorylation increases during starvation while the 74-kDa intermediate chain is unaffected. Further, treatment of the cells with okadaic acid to inhibit protein phosphatase activity mimics the effects of serum starvation, causing a redistribution of dynein from membranes to the soluble pool with a concomitant increase in phosphorylation of the dynein heavy chain (Lin *et al.,* 1994). These studies demonstrate that cytoplasmic dynein is phosphorylated *in vivo,* and that its phosphorylation state may be regulated to control the distribution of dynein between the membrane and soluble compartments.

Given that dynein binds to acidic phospholipids and that this binding can be reversed by high salt (Lacey and Haimo, 1994), dynein may bind electrostatically to membranes, an interaction that may be inhibited by the introduction of a highly negatively charged moiety onto the dynein. If phosphorylation of dynein induces it to dissociate from organelles, organelle transport toward the minus ends of microtubules would be precluded. However, localization of cytoplasmic dynein on vesicles in ligated neurons indicates that those vesicles accumulated on either side of the ligation, and thus moving in opposing directions, exhibit equivalent amounts of cytoplasmic dynein staining (Hirokawa *et al.,* 1990). Thus, dynein dissociation from vesicles is not a necessary prerequisite for organelle transport to the plus ends of microtubules.

Other studies suggest that the state of dynein phosphorylation may vary with its state of activity (Dillman and Pfister, 1994). Introduction of labeled [^{32}P]orthophosphate into the optic lobe can be used to trace phosphorylated molecules that are being transported in the anterograde direction toward the synapse. Following *in vivo* ^{32}P-labeling of the optic lobe, the heavy chain of cytoplasmic dynein was found to incorporate less phosphate and the intermediate chain more phosphate than the dynein in the general dynein pool, i.e., not undergoing net anterograde transport. Because dynein should be inactive in anterograde transport, it is possible that this differential phosphorylation is correlated with dynein activity (Dillman and Pfister, 1994). However, these studies did not show that dynein activity differed in these two populations, so, at present, it is only possible to conclude that different populations of dynein may be differentially phosphorylated. An analysis of endoplasmic reticulum (ER) movements along microtubules driven by cytoplasmic dynein in *Xenopus* extracts suggests that phosphorylation regulates dynein activity, although these studies did not identify the target of the regulatory system (Allan, 1995). Inhibition of phosphatase activity by okadaic acid greatly enhanced the amount of dynein-driven ER membrane movements along microtubules *in vitro* but did not affect the total amount of dynein bound to these membranes. Moreover, the activity of the soluble dynein in these preparations was not enhanced by okadaic acid treatment. These observations suggest that only the membrane-associated dynein is affected by the inhibition of phosphatase activity, and that inhibition activates membrane-bound but not soluble dynein (Allan, 1995).

An intriguing relationship between dynein activity and phosphorylation is suggested in an analysis of the ropy mutants of *N. crassa* (Plamann *et al.,* 1994). Mutations in either the dynein heavy chain gene or in the Arp1 gene can suppress a defect in a gene, *cot1,* that encodes a serine–threonine kinase. In the absence of suppression, *cot1* mutants do not extend hypae. One possible explanation for this suppression is that a change in the gene for either cytoplasmic dynein or Arp1 might alleviate the requirement for dynein to be activated by this kinase, similar to the behavior of *Chlamydomonas* inner arm dynein, discussed earlier.

The studies to date on cytoplasmic dynein regulation are intriguing but have not yet provided a thorough picture of how the activity of this motor is controlled. If the dynactin complex activates dynein interaction with membranes, does it enhance dynein binding to the membranes or does it enhance the ability of bound dynein to interact with microtubules? Is the interaction of dynactin with dynein cell cycle regulated, and, if so, are any proteins in the dynactin complex subject to cell cycle-dependent phosphorylation? Is cytoplasmic dynein itself subject to cell cycle-dependent phosphorylation? Does the phosphorylation of dynein affect dynein's active site, its interaction with dynactin, with the membrane itself, or with microtubules? Because there are multiple phosphorylation sites on dynein, does phosphorylation of different sites have distinct effects on dynein activity? The resolution of many of these questions will allow a clearer picture to emerge of the mechanism by which cytoplasmic dynein-dependent movements are controlled.

Table II summarizes our current understanding of both axonemal and cytoplasmic dynein regulation. Both classes of dynein interact with very large protein complexes, which are distinct from the motor molecules themselves, and which are implicated in the regulatory pathway. The central pair complex, radial spokes, and dynein regulatory complex may produce a regulatory cascade that culminates in activation of axonemal dynein arms. Similarly, the dynactin complex may activate the ability of cytoplasmic dynein to induce vesicle transport. Both axonemal and cytoplasmic dynein can be phosphorylated and, in both cases, there is a correlation between this phosphorylation and motor activity. While the details of the regulatory systems controlling this motor family are not yet fully elucidated, rapid progress has been made in this area in the past year or two, and it is likely that considerable advances in understanding in this field will be shortly forthcoming.

B. Kinesin-Based Transport

1. Kinesin Family Characteristics

Kinesin was discovered in 1985 based on its ability to form a rigor-like complex between microtubules and organelles in squid axoplasm and on

TABLE II

Regulation of Dynein Activity

Target	Effect on motility	Phosphorylation event implicated?	Regulatory molecule identified?	Signal	References
1. Axonemal dynein					
Outer dynein arm (cilia)	Increases swimming speed, increases rate of MT gliding	DLC 29-kDa subunit phosphorylated	PKA	cAMP activates, Ca^{2+} inhibits	Bonini and Nelson (1990), Hamasaki et al. (1991), Barkalow et al. (1994)
Outer dynein arm (flagella)	Activates motility	DHC and DLC phosphorylated	PKA	cAMP	Stephens and Prior (1992), Dey and Brokaw (1991)
Inner dynein arm	Inhibits sliding	Yes, but exact target unknown	Regulatory cascade involving central pair/radial spokes and dynein regulatory complex and PKA	cAMP	Smith and Sale (1992a), Howard et al. (1994), Piperno et al. (1994)
2. Cytoplasmic dynein					
Dynein	Enhances dynein-driven organelle motility in vitro	Not known	Dynactin	Not known	Gill et al. (1991), Schroer and Sheetz (1991)
DHC	Dynein dissociates from lysosomes in vivo	DHC phosphorylated	Protein phosphatase implicated, since okadaic acid treatment mimics effect of serum starvation	Ca^{2+} or serum starvation	Lin and Collins (1993), Lin et al. (1994)
Likely dynein	Enhanced dynein-driven transport of ER membranes along MTs in vitro	Not known, but likely since enhanced transport occurs when phosphatase activity is inhibited	Protein phosphatase 1 implicated	Not known	Allan (1995)
DHC, DIC	Not known, but phosphorylation may be correlated with activity	DHC and DIC are differentially phosphorylated during anterograde transport relative to dynein in the general pool	Not known	Not known	Dillman and Pfister (1994)

its ability to promote microtubule gliding in an *in vitro* motility assay (Brady, 1985; Vale *et al.*, 1985b). Molecules related to but distinct from kinesin were identified in 1990 (Goldstein, 1993; Skoufias and Scholey, 1993; Walker and Sheetz, 1993), and it has subsequently become apparent that kinesin is but one of a very large gene family. Kinesin and kinesin-like proteins exhibit substantial homology in their motor domain but differ extensively in their tail domains. While conventional kinesin and several kinesin-like proteins are plus end-directed motors (Vale *et al.*, 1985c; Nislow *et al.*, 1992; Sawin *et al.*, 1992; Cole *et al.*, 1993, 1994; Kondo *et al.*, 1994; Nangaku *et al.*, 1994; Sekine *et al.*, 1994; Noda *et al.*, 1995), kinesin-like proteins generating force in the opposite direction have been identified in *Drosophila* and in *S. cerevisiae* (McDonald *et al.*, 1990; Walker *et al.*, 1990; Endow *et al.*, 1994; Middleton and Carbon, 1994). Sequence analysis of plus-end directed kinesins places their motor domain at the amino terminal of the protein, with the exception of two mammalian kinesin-like proteins whose motor domains reside in the middle of the molecule (Aizawa *et al.*, 1992; Wordeman and Mitchison, 1995). Conversely, the motor domains of the minus end-directed kinesins described to date are at their carboxy ends (Endow *et al.*, 1990; Meluh and Rose, 1990; McDonald and Goldstein, 1990). Nonetheless, minus end-directed motor activity is an inherent property of the motor domain and not a function of its placement within the molecule (Stewart *et al.*, 1993).

2. Conventional Kinesin-Driven Movements

a. Properties of Conventional Kinesin   Kinesin is a heterotetramer consisting of two heavy chains of ~110–130 kDa and two light chains of ~60–70 kDa. An amino-terminal domain in the heavy chains forms globular heads which contain the microtubule binding and ATPase sites (Bloom *et al.*, 1988; Kuznetsov *et al.*, 1989; Scholey *et al.*, 1989). The central rod-like stalk domain is predicted to be an alpha helical coiled-coil (Yang *et al.*, 1989), and the two light chains are associated with the carboxy-terminal tail domain (Hirokawa *et al.*, 1989). Expression of kinesin heavy chain, or heavy chain fragments, in fibroblasts has indicated that both the amino terminal motor domain and the carboxy-terminal tail domain interact with microtubules, but that the stalk does not (Navone *et al.*, 1992).

Kinesin's ATPase activity is significantly stimulated by microtubules (Kuznetsov and Gelfand, 1986), a property that would keep hydrolysis of ATP by kinesin to a minimum when kinesin is not producing force. Indeed, kinesin's microtubule-activated ATPase activity is directly correlated with its ability to translocate microtubules (Cohn *et al.*, 1987). Kinetic studies of ATP hydrolysis by kinesin indicate that ADP release is rate-limiting; microtubules stimulate this release by 10,000-fold (Hackney, 1988, 1994a).

Presteady-state kinetic analysis of the kinesin motor activity using chemical quench flow confirms that ADP release is rate-limiting and that microtubules activate kinesin ATPase activity by increasing the rate of ADP dissociation from kinesin from 0.01/sec to 10 to 20/sec (Gilbert and Johnson, 1994).

The cross-bridge cycle of kinesin differs significantly from that of myosin and dynein since ADP causes kinesin to dissociate from microtubules (Romberg and Vale, 1993), while for myosin (Lymn and Taylor, 1971) and for axonemal dynein (Johnson, 1983, 1985) phosphate release is the rate-limiting step and ATP binding induces dissociation from the filament. A single kinesin molecule per vesicle is sufficient for translocation *in vitro* (Howard *et al.*, 1989; Block *et al.*, 1990), a property which also distinguishes it from myosin (Uyeda *et al.*, 1991). This difference may reflect the duty cycle of kinesin, which remains attached to the cross-bridged microtubule during the greater part of the cross-bridge cycle (Hackney, 1988). Recent studies provide evidence that the two motor domains of a kinesin molecule undergo alternate cycles of ATP hydrolysis; while one head attaches to a microtubule and releases bound ADP, the other head does not. Thus, kinesin might translocate along a microtubule by sequentially alternating cross-bridge activity between its two heads. Such a cycle would provide a mechanism for the kinesin to remain attached to the microtubule throughout most of its cross-bridge cycle (Hackney, 1994b; Gilbert *et al.*, 1995). Two heads, with one or the other always attached, not only keep kinesin from diffusing away from the microtubule, but also confer on the motor the ability to track along a single protofilament (Berliner *et al.*, 1995).

The size of the kinesin step has recently been measured using laser trapping; this step, 8 nm (Svoboda *et al.*, 1993), is the size of the tubulin dimer. Similarly, the force exerted by a single kinesin molecule at each cross-bridge cycle has been measured by several methods which have yielded somewhat conflicting results. The isometric force of a trapped latex bead attached to a microtubule being driven by kinesin was measured at 1.9 pN at each ATP cycle (Kuo and Sheetz, 1993). In an entirely different method, the ability of kinesin to translocate demembranated sea urchin sperm against a centrifugal force yielded an isometric force value for kinesin of 0.12 pN (Hall *et al.*, 1993), a value more than an order of magnitude less than that of the former study. Two recent studies, using different methods, obtained higher values; the ability of kinesin attached to silica beads to move along microtubules in the face of an applied optical trap resulted in a calculation of a force of 5–6 pN exerted by kinesin (Svoboda and Block, 1994). The drag forces sufficient to slow microtubules gliding on kinesin resulted in a calculation of 4.2 pN (Hunt *et al.*, 1994). Knowledge of the magnitude of this force, the step size, and an understanding of the duty cycle, for both dynein and kinesin would permit hypothesis to be made as to whether directional transport is possible when both motors are present and active.

b. Function of Conventional Kinesin A number of *in vitro* studies implicate kinesin in vesicle transport. Cytosolic fractions containing kinesin promote vesicle transport along microtubules (Vale *et al.*, 1985a); purified kinesin promotes the movement of chromaffin granule membranes along microtubules (Urrutia *et al.*, 1991); and loss of vesicle transport is correlated with immunodepletion of kinesin from cytosolic fractions that otherwise support transport of organelles along microtubules *in vitro* (Schroer *et al.*, 1988; Bomsel *et al.*, 1990). In addition, the formation of an ER-like network *in vitro* depends on the presence of kinesin (Dabora and Sheetz, 1988; Vale and Hotani, 1988). Antibodies directed against kinesin inhibit vesicle transport in extruded squid axoplasm, but, perplexingly, transport is inhibited in both directions (Brady *et al.*, 1990). These latter studies might indicate that inhibition of one motor can affect the activity of another.

Although a large fraction of kinesin exists as a soluble component in cells (Murofushi *et al.*, 1988; Neighbors *et al.*, 1988; Hollenbeck, 1989; Schmitz *et al.*, 1994), kinesin has been localized to a variety of membranous organelles (Neighbors *et al.*, 1988; Hollenbeck, 1989; Pfister *et al.*, 1989; Leopold *et al.*, 1992; Yu *et al.*, 1992) and can bind to membrane fractions *in vitro*. This binding does not require the kinesin light chains and is mediated by the tail domain of the kinesin heavy chain (Skoufias *et al.*, 1994). A kinesin membrane receptor, kinectin, has been identified in detergent extracts of microsomal membranes and isolated based on its ability to bind to an immunoaffinity column containing immobilized kinesin (Toyoshima *et al.*, 1992). Antibodies directed against kinectin block both kinesin binding to membranes and organelle transport (Kumar *et al.*, 1995). At present, it is not known if kinectin is present in the membrane of all organelles that bind kinesin.

That kinesin indeed drives organelle transport *in vivo* is supported by a variety of studies. Extension of lysosomes to form a tubular network is inhibited by microinjection of kinesin antibodies (Hollenbeck and Swanson, 1990). Pigment granule dispersion, but not aggregation, is inhibited in melanophores also following microinjection of kinesin antibodies. Because dispersion is toward the plus ends of the melanophore microtubules, these studies demonstrate that the inhibition is specific for the direction of transport predicted to be kinesin driven (Rodionov *et al.*, 1991). This same technique has also shown that movement of material from the Golgi to the ER, but not from the ER to Golgi, also appears to be mediated by kinesin (Lippincott-Schwartz *et al.*, 1995). Using antisense oligonucleotides to kinesin heavy chain, recent studies have shown that ER extension, transport through the "late" Golgi, and endosomal dispersion all require kinesin, while mitochondrial movements do not (Feiguin *et al.*, 1994). This last result is particularly interesting in light of other findings that the function of one kinesin-like protein may be to transport mitochondria (Nangaku *et al.*, 1994).

Kinesin has also been implicated in the distribution of vimentin filaments in cells. Microinjection of fibroblasts with kinesin antibodies causes the vimentin filaments to collapse, despite the continued persistence of the microtubules (Gyoeva and Gelfand, 1991). Thus, kinesin may be involved not only in organelle transport but also in the active organization of the cytoskeleton.

c. Regulation of Conventional Kinesin

i. Regulation of Kinesin-Organelle Interactions. Both the heavy and light chains of kinesin can be phosphorylated *in vivo* (Hollenbeck, 1993; Matthies *et al.*, 1993), and this phosphorylation may affect either kinesin's ability to bind to vesicles and/or kinesin's motor activity. Localization of kinesin *in situ* in ligated neurons indicates there is a statistically greater amount of kinesin on vesicles being transported in the anterograde than in the retrograde direction (Hirokawa *et al.*, 1991). Moreover, phosphorylation of kinesin *in vitro* inhibits its ability to bind synaptic vesicles and translocate vesicles along microtubules (Sato-Yoshitake *et al.*, 1992). Thus, regulation of the direction of organelle transport could be mediated by controlling the ability of kinesin to bind to organelles. If kinesin were to become phosphorylated when a vesicle reached the synapse and subsequently be dissociated from the vesicle, then, in the absence of kinesin, the vesicle could be transported back to the cell body, potentially by dynein, which may be present on the vesicles during both directions of transport (Hirokawa *et al.*, 1990). However, in contrast to these findings, which suggest that phosphorylation of kinesin inhibits membrane binding, other studies suggest that increased phosphorylation of kinesin is correlated with its enhanced membrane association and with increased levels of organelle transport (Lee and Hollenbeck, 1995). Because there are multiple phosphoisoforms of kinesin, it is possible that phosphorylation at particular sites has a specific effect on membrane binding; phosphorylation at one site may enhance membrane binding; phosphorylation at another site may decrease this binding.

Kinectin, the membrane receptor for kinesin, can also be phosphorylated (Hollenbeck, 1993), but there are no data to date that correlate changes in its phosphorylation state with kinesin binding. An analysis of both kinesin and dynein binding to brain microsomal membranes has indicated that they compete and may therefore bind to the same or neighboring sites (Yu *et al.*, 1992), although binding of dynein to kinectin has not been demonstrated directly. If they do bind to the same receptor, a regulatory mechanism might be required to control which motor is bound at a given time (Yu *et al.*, 1992). In ligated neurons, vesicles accumulating on either site of the ligation contain equivalent amounts of dynein, but differ in their amount of kinesin, suggesting that only binding of kinesin to membranes is regulated (Hirokawa *et al.*, 1991). However, an analysis of kinesin on intermediate compartment vesicles, which are thought to cycle continually between the ER and Golgi, indicates that kinesin is present on these vesicles regardless of whether they are moving toward the Golgi or toward the ER (Lippincott-Schwartz *et al.*, 1995). Thus, for organelles that are continually undergoing bidirectional transport, the attachment of motor molecules to the membrane may not be regulated.

   ii. Regulation of Kinesin Motor Activity. Phosphorylation of kinesin may control its activity. Calmodulin binds to kinesin light chain and reduces by half kinesin's microtubule-stimulated ATPase activity. Phosphorylation of kinesin by PKA reverses the calmodulin inhibition of ATPase activity (Matthies *et al.*, 1993). It is the light chains of kinesin that most heavily incorporate phosphate when treated with PKA, but the role of these chains is not known. In rats, there are three light chain isoforms; these are derived from a single gene which undergoes alternate splicing to yield distinct carboxy-terminals (Cyr *et al.*, 1991). The light chains are not required for kinesin to exhibit motor activity (Yang *et al.*, 1990), but these observations do not preclude a role in regulation. Kinesin can undergo a change in conformation from a folded 9S particle to an extended 6S particle, the latter having a higher ATPase activity. Because in the folded conformation, the kinesin light chains are in close proximity to the motor domain and because kinesin heavy chains lacking light chains have higher ATPase activity than the native molecule, the light chains may function to inhibit kinesin activity (Hackney *et al.*, 1991, 1992). Potentially, phosphorylation of the light chains affects their interaction with the heavy chains.

   Phosphorylation of proteins that interact with kinesin may control kinesin's activity. Three polypeptides—150 kDa, 78 kDa, and 73 kDa—copurify with kinesin and are highly phosphorylated when the kinesin preparation is subjected to *in vitro* phosphorylation in the presence of okadaic acid, a protein phosphatase inhibitor. This phosphorylation is correlated both with an increase in kinesin ATPase activity and with a slight increase in the rate of microtubule gliding *in vitro*. The same proteins are also phosphorylated *in vivo* when cells are incubated in okadaic acid (McIlvain *et al.*, 1994). Moreover, treatment of cells with okadaic acid results in an overall increase in organelle transport (Hamm-Alvarez *et al.*, 1993). Thus, it is possible that organelle transport is regulated by controlling the activity of kinesin via a phosphorylation of kinesin-associated proteins. However, it should be noted that treatment of PC12 cells with nerve growth factor to induce kinesin-dependent neurite outgrowth resulted in an increase in the extent of kinesin phosphorylation but no change in the phosphorylation level of the kinesin-associated proteins (Lee and Hollenbeck, 1995).

3. Kinesin-like Protein-Driven Transport

a. Properties of Kinesin-like Proteins Kinesin-like proteins (KLPs, also called kinesin-related proteins) have been identified in a wide range of cell types using biochemical, immunological, genetic, and molecular methods (Endow and Titus, 1992; Goldstein, 1993; Skoufias and Scholey, 1993; Walker and Sheetz, 1993; Hoyt, 1994). The kinesin superfamily contains widely divergent members that have in common a relatively well-conserved motor domain. These KLPs participate in a variety of functions *in vivo*.

Two KLPs, the sea urchin egg $KRP_{85/95}$ (Cole *et al.*, 1993) and *Drosophila* embryo KRP_{130} (Cole *et al.*, 1994), have been biochemically isolated from their native cells and shown to possess motor activity. The $KRP_{85/95}$ has two subunits of 85 and 95 kDa, and, because each is recognized by a peptide antibody generated against a conserved sequence in the motor domain, each may have motor activity. The *Drosophila* KRP_{130} appears to be a homotetramer composed of 4 subunits of the 130-kDa polypeptide, each with motor activity. Both of these KLPs are plus end motors. $KRP_{85/95}$ drives microtubules at ~0.4 μm/sec while KRP_{130} drives microtubules at only 0.4 μm/sec, which is significantly slower than the 0.6- to 0.8-μm/sec rate of microtubule gliding induced by conventional kinesin (Cole *et al.*, 1993, 1994).

The *Drosophila* KRP_{130} may form a bipolar motor, with heads located at either end. If so, then this kinesin might be capable of cross-linking microtubules. An expressed, mammalian kinesin-like plus end motor, MKLP1, cross-bridges microtubules of opposite polarity and localizes to the spindle midzone (Nislow *et al.*, 1992). Such a kinesin might be capable of inducing spindle elongation, which could be brought about by the sliding apart of interdigitating microtubules arising from opposite spindle poles. A number of other KLPs, including Eg5, KIF1B, KIF2, KIF3A, and KIF4, have also been purified following expression of their genes in foreign cells and are plus end-directed motors (Sawin *et al.*, 1992; Kondo *et al.*, 1994; Nangaku *et al.*, 1994; Sekine *et al.*, 1994; Noda *et al.*, 1995). In contrast, Kar3p and ncd protein, KLPs from *S. cerevisiae* and *Drosophila*, respectively, are minus end-directed motors (McDonald *et al.*, 1990; Walker *et al.*, 1990; Endow *et al.*, 1994; Middleton and Carbon, 1994).

b. Function of Kinesin-like Proteins KLPs have been implicated in organelle transport and in the various motile events associated with mitosis and meiosis. In addition, KLPs have recently been identified in the axoneme of *Chlamydomonas;* their functions in this organelle are not yet known.

i. Kinesin-like Proteins in Organelle Transport. Although conventional kinesin seems clearly to be involved in organelle transport, some KLPs also function to move organelles. The product of the *unc 104* gene in *C.*

elegans is required for fast axonal transport of synaptic vesicles (Hall and Hedgecock, 1991). Five different classes of the kinesin family have been identified in embryonic mouse brain (KIF1-5) and characterized (Aizawa *et al.*, 1992). KIF5 encodes conventional kinesin; KIF1A is homologous to the *unc104* gene product and is therefore also most likely involved in synaptic vesicle transport. KIF2, KIF3A, and KIF4 have each been expressed in S9 cells and encode three different plus end-directed kinesin-like proteins that promote microtubule gliding *in vitro* at about 0.47, 0.6, and 0.2 μm/sec, respectively. Because antibodies directed against these kinesin-like proteins stain vesicular structures in neurons, it has been proposed that they too may drive anterograde fast axonal transport (Kondo *et al.*, 1994; Sekine *et al.*, 1994). KIF2 localizes to vesicles that are distinct from synaptic vesicles and thus, may have a specific set of vesicles which it is responsible for transporting. KIF1B localizes specifically to mitochondria and drives mitochondrial but not synaptic vesicle transport along microtubules *in vitro* (Nangaku *et al.*, 1994).

An unusual KLP may interact with an actin-based organelle transport system. In *S. cerevisiae*, mutations in the *myo2* gene, a member of the myosin superfamily, can be suppressed by the product of a gene that belongs to the kinesin superfamily. Yeast normally grow by budding, but in *myo2* temperature-sensitive mutants, the cells at the restrictive temperature increase in mass but do not form buds because vesicles are not transported to a potential bud site. Extragenic suppression can occur by overexpression of the *SMY1* (suppression of myosin) gene which bears sequence homology to the kinesin superfamily (Lillie and Brown, 1992), and antibodies directed against both Myo2p and Smy1p colocalize to regions of incipient bud formation, small buds, and to the mother-daughter neck before cell separation (Lillie and Brown, 1994). Interestingly, the product of the *SMY1* gene is not essential because its deletion from otherwise wild-type cells results in a wild-type phenotype. However, deletion of this gene from *myo2* temperature-sensitive mutants is lethal (Lillie and Brown, 1992). Moreover, actual deletion of the *MYO2* gene, rather than introduction of a temperature-sensitive mutation into it, cannot be suppressed by overexpression of *SMY1*, indicating that Smy1p does not perform a completely redundant function with Myo2p and that only some loss of myosin function can be accommodated by an enhancement in kinesin-like motor activity. In the *myo2* temperature-sensitive mutant, at the restrictive temperature, normal actin and myo2p localization is lost but can be restored by overexpression of *SMY1* (Lillie and Brown, 1994). It will be most informative to understand the mechanism by which this suppression of myosin function is mediated. Does Smy1p interact with microtubules? Is the motor domain of Smy1p essential for *myo2* suppression? What would result if site-specific mutagenesis were directed against the ATP binding domain of *SMY1* and this mutated gene were then used for overexpression and attempted suppression?

ii. Kinesin-like Proteins in Mitosis and Meiosis. Many of the KLPs, for which a function has been ascribed, may be involved in various motile events occurring during meiosis or mitosis. Indeed, the wide variety of kinesin-like proteins in cells may have evolved to function in very limited and specific cell division events. KLPs may be involved in chromosome movements during prometaphase and in anaphase, in spindle formation, and in spindle elongation during anaphase.

Some KLPs are located on chromosomes and are implicated in their movements. Localization of kinesin-like proteins by immunofluorescence has shown at least two to be present at the centromere. In mammalian cells, CENP-E, a kinesin-like protein, is localized at the kinetochore until some time in anaphase (Yen *et al.,* 1992; Lombillo *et al.,* 1995). An unrelated kinesin-like protein, MCAK, also localizes to the centromeres throughout mitosis (Wordeman and Mitchison, 1995). Although KLPs at the centromere might be predicted, recent studies show that some of these kinesins not only have a motor domain but also contain a DNA-binding domain and localize along the chromosomal arms. These kinesin-like proteins might be required for correct alignment of chromosomes at the metaphase plate or for their normal anaphase separation (Afshar *et al.,* 1995; Murphy and Karpen, 1995; Vernos *et al.,* 1995; Wang and Adler, 1995).

Genetic studies also provide evidence for a role for KLPs in chromosome movements. Mutations in either the *ncd* or *nod* genes, both members of the kinesin superfamily, result in chromosomal nondisjunction of nonjunctional chromosomes during meiosis in *Drosophila* (Endow *et al.,* 1990; Zhang *et al.,* 1990). Indeed, Nod is one of the kinesins that has recently been localized to the arms of the chromosomes (Afshar *et al.,* 1995), and genetic analysis reveals that it is functional along these arms (Murphy and Karpen, 1995). The *S. cerevisiae* kinesin-like protein, Kar3p, may be involved in driving chromosomes to the minus ends of microtubules. In yeast, a protein complex, CBF3, binds specifically to centromeric DNA and transports centromeric DNA-coated beads to the minus ends of microtubules *in vitro* (Hyman *et al.,* 1992). Analysis of the motor activity in this preparation has revealed that Kar3p, a minus end motor, provides the motive force. Interestingly, Kar3p itself appears to bind to DNA in general, but seems capable of driving DNA-coated beads along microtubules only in the presence of the entire CBF3 complex which binds specifically to centromeric DNA (Middleton and Carbon, 1994). Two other yeast KLPs, Kip1p and Cin8p, have also been implicated in chromosome movements because simultaneous defects in their genes, and in the cytoplasmic dynein gene, result in a block in anaphase chromosome separation (Saunders *et al.,* 1995).

Many of the kinesin-like proteins function in spindle assembly. The products of *cut* 7 in *S. pombe* (Hagan and Yanagida, 1990), *bimC* in *Aspergillus* (Enos and Morris, 1990), *CIN8* and *KIP1* in *S. cerevisiae* (Hoyt *et al.*, 1992; Roof *et al.*, 1992), *Eg5* in *Xenopus* (Sawin *et al.*, 1992), and *ncd* and *KLP61F* in *Drosophila* (Hatsumi and Endow, 1992; Heck *et al.*, 1993) genes have each been implicated in this process. Some kinesins may have multiple roles in mitotic movements; for example, Cin8p, Kip1p, and ncd protein have also been implicated in chromosomal movements, as discussed earlier.

Some of these kinesin-like proteins may have overlapping functions because mitotic defects are sometimes observed only when more than one is mutated. For example, mutations in either the yeast proteins Kip1p or Cin8p alone are not lethal, but conditional double mutants do not assemble bipolar spindles or maintain bipolar structures at a restrictive temperature (Hoyt *et al.*, 1992; Roof *et al.*, 1992; Saunders and Hoyt, 1992). Moreover, defects in both of these genes and in cytoplasmic dynein are required to inhibit chromosome segregation (Saunders *et al.*, 1995). Interestingly, in addition to this functional redundancy, opposing forces in the spindle may be generated by multiple motors that generate counteracting forces. A mutation in the motor encoding domain of the *kar3* gene can suppress *kip1* and *cin8* defects (Hoyt *et al.*, 1993). Because Kar3p is a minus end kinesin-like protein (Endow *et al.*, 1994; Middleton and Carbon, 1994), it is possible that its motor activity normally opposes plus end-directed motor activity of Cin8p and Kip1p. Similarly, in *Aspergillus*, loss of a functional *bimC* gene, which encodes a KLP and is normally required for spindle pole body separation, can be suppressed by the simultaneous loss of a minus-end KLP, the product of the *klpA* gene (O'Connell *et al.*, 1993). Thus, the product of the *bimC* and *klpA* genes may produce two KLPs whose activities oppose each other.

KLPs might function during spindle elongation. In mammalian cells, a KLP, MKLP1, has been identified that induces sliding between microtubules of opposite polarity *in vitro* and localizes to the spindle midbody. This protein may have the appropriate characteristics to drive anaphase B, in which it would be predicted that sliding would occur between microtubules of opposite polarity (Nislow *et al.*, 1992). Moreover, in diatom spindles, spindle elongation is inhibited *in vitro* by antibodies that recognize a conserved peptide within kinesin's motor domain (Hogan *et al.*, 1993). CENP-E, another mammalian KLP, localizes to the kinetochores during metaphase, but then moves to the midzone during anaphase (Yen *et al.*, 1992) and thus might function during anaphase to generate spindle elongation (Brown *et al.*, 1994).

iii. Kinesin-like Proteins in Flagella. Several KLPs have been identified in axonemes of *Chlamydomonas* (Bernstein *et al.*, 1994; Fox *et al.*, 1994; Johnson *et al.*, 1994; Walther *et al.*, 1994). Antibodies against conserved domains in the kinesin superfamily reveal at least seven reactive polypeptides on immunoblots of axonemal preparations (Fox *et al.*, 1994). Three KLPs have been localized to the central pair microtubules (Bernstein *et al.*, 1994; Fox *et al.*, 1994; Johnson *et al.*, 1994), and one of these, Klp1, has been localized to a single microtubule, the C2 microtubule of the central pair (Bernstein *et al.*, 1994). Studies in a variety of organisms, including *Chlamydomonas*, have shown that the central pair complex of microtubules rotates during ciliary beating (Omoto and Kung, 1979; Omoto and Witman, 1981; Kamiya, 1982). KLPs localized to the central pair might induce this rotation. Rotation, in turn, might result in the sequential activation of different groups of doublets, possibly by affecting the phosphorylation state of dynein via an interaction of the central arm projections with the radial spokes. If this speculation were to be true, then kinesin could be responsible for dynein regulation.

One of the KLPs in the flagellum is the product of the *FLA10* gene, a gene which is required for normal flagellar assembly (Walther *et al.*, 1994). Given that assembly occurs primarily at the flagellar tip (Johnson and Rosenbaum, 1992), movement of material toward the plus ends of microtubules might be necessary for assembly. Thus, one intriguing possibility for the function of the FLA10 protein is that it is needed to transport material required for flagellar elongation down to the growing tip of the flagellum.

c. Regulation of Kinesin-like Proteins   Recent studies are beginning to provide evidence that KLPs, like conventional kinesin and dyneins, are regulated. In particular, the kinesin-like CENP-E protein exhibits an extremely interesting, cell cycle-regulated behavior (Yen *et al.*, 1992; Brown *et al.*, 1994; Liao *et al.*, 1994). CENP-E was first identified as a kinetochore-associated protein which was absent from interphase cells, appeared at the centromeres of chromosomes during prometaphase, and relocalized to the midzone and midbody during anaphase and telophase (Yen *et al.*, 1992). Pulse-chase experiments have demonstrated that CENP-E accumulates during the cell cycle, peaking during early mitosis. At the end of mitosis, there is a ten-fold loss in CENP-E levels (Yen *et al.*, 1992; Brown *et al.*, 1994). The cells must progress through mitosis in order for CENP-E to be degraded; if cells are blocked in mitosis with colcemid, CENP-E levels remain high (Yen *et al.*, 1992). In contrast, if cytokinesis is blocked with cytochalasin, CENP-E is destroyed normally. CENP-E destruction is precisely timed and occurs only after that of cyclin B's (Brown *et al.*, 1994). These studies indicate that CENP-E's function in the cell is tightly coupled to the cell cycle and is regulated by its synthesis and destruction.

Most interestingly, the activity of CENP-E may also be regulated by a phosphorylation event. CENP-E exhibits microtubule binding, as do other kinesins, but only about 50% of the microtubule-bound CENP-E can be extracted with ATP from the microtubules (Liao *et al.*, 1994). The amino terminal of CENP-E, which contains the kinesin-like motor domain, binds to microtubules and completely dissociates from them in the presence of ATP. In contrast, the carboxy-terminal domain also binds to microtubules, but is not dissociated with ATP. This domain is rich in basic amino acids and proline, characteristics similar to microtubule binding sites on other microtubule binding proteins, and contains a consensus phosphorylation site for MPF (Yen *et al.*, 1992). When fragments containing the carboxy domain of CENP-E are phosphorylated by MPF *in vitro*, these phosphorylated fragments can no longer bind to microtubules. Moreover, CENP-E becomes specifically phosphorylated *in vivo* during mitosis (Liao *et al.*, 1994). Because CENP-E accumulates at mitosis and binds to the kinetochores, it may be involved in driving chromosome movements prior to the onset of anaphase. During this period, MPF activity would be high, CENP-E should be phosphorylated, and its carboxy-terminal should be unable to bind to microtubules. Prior to anaphase, CENP-E might serve as a motor between the chromosome and microtubules, but, because it is phosphorylated, it would be unable to cross-link microtubules.

At the onset of anaphase, cyclin B is destroyed and MPF activity falls exponentially. In the absence of this kinase activity, CENP-E should become dephosphorylated, and its carboxy-terminal would then exhibit high-affinity microtubule binding. Dephosphorylated CENP-E might dissociate from kinetochores and cross-bridge microtubules. The observed redistribution of CENP-E from the chromosome to the midzone may indicate that CENP-E has two distinct functions during mitosis and that it is a phosphorylation event that determines which function is exhibited. Other studies suggest that CENP-E may stay associated with the kinetochore until late anaphase, and, if so, might be involved in chromosome segregation (Lombillo *et al.*, 1995). Chromosomes can remain attached to the ends of depolymerizing microtubules *in vitro*, thereby moving with the depolymerizing microtubule plus ends toward the minus ends in the absence of ATP (Coue *et al.*, 1991). Interestingly, antibodies against CENP-E significantly reduce the rate of this movement but do not decouple the chromosomes from the microtubules (Lombillo *et al.*, 1995). It is not yet clear why the rate would be affected if the role of this KLP is to couple chromosomes to the depolymerizing ends of microtubules.

Our current understanding of kinesin regulation is summarized in Table III. Kinesin or kinesin-like proteins can be phosphorylated, a modification that may affect motor activity or intracellular binding sites. Alternatively, kinesin may interact with proteins whose phosphorylation state regulates kinesin behavior. Finally, synthesis and destruction of kinesin-like proteins may dictate when these motors are present and thus able to participate in intracellular movements. We turn our attention now to actual microtubule-

TABLE III
Regulation of Kinesin and KLP Activity

Target	Effect on motility	Phosphorylation event implicated?	Regulatory molecule identified?	Signal	References
Conventional kinesin	Inhibits kinesin-membrane binding	KHC, KLC phosphorylated	PKA	cAMP	Sato-Yoshitake et al. (1992)
Conventional kinesin	Reverses calmodulin inhibition of kinesin ATPase activity	KLC phosphorylated	Calmodulin and PKA	Ca^{2+} and cAMP	Matthies et al. (1993)
Conventional kinesin	Enhances kinesin-membrane binding	KHC, KLC phosphorylated	Not known	Nerve growth factor	Lee and Hollenbeck (1995)
Kinesin-associated phosphoproteins (KAPPS)	Stimulates interaction of kinesin with microtubules	KAPPS become highly phosphorylated when cells are treated with okadaic acid	Protein phosphatase implicated	Not known	McIlvain et al. (1994)
CENP-E	MT binding site on carboxy-terminal capable of MT binding only when it is dephosphorylated	Carboxy terminal phosphorylated prior to anaphase, dephosphorylated at anaphase	MPF	Cell cycle	Liao et al. (1994)

dependent motile events in cells to consider what is understood about the regulation of these movements.

IV. Regulation of Bidirectional Transport

Although it is useful to discuss dynein and kinesin as independent motor families, many movements that occur *in vivo* require the participation of more than one motor. If these motors are to generate transport in opposing directions, then a regulatory system must be established to control the direction of transport.

How do organelles know which way to go or on which filaments to travel if they contain more than one motor molecule? When microtubules are challenged *in vitro* to slide along a glass surface coated with both dynein and kinesin, the microtubules move short distances in one direction and then reverse and move in the opposite direction (Vale *et al.,* 1992). This oscillatory behavior mimics the saltatory motions that characterize many organelles, but does little to explain how net translocations over long distances can be achieved in the presence of more than one active motor. How is the direction of vesicle and chromosome transport regulated?

Chromatophores, pigment-bearing cells common among invertebrates and lower vertebrates, have been widely used as model systems for studying regulation of organelle transport. The types of chromatophores, iridophores, xanthophores, erythrophores, and melanophores, named for the type of pigment they bear, contain pigment in discrete organelles which are cyclically transported to and from the cell center following stimulation of cell surface receptors (Bagnara and Hadley, 1973). In melanophores and squirrel fish erythrophores, transport appears to be microtubule dependent (Murphy and Tilney, 1974; Beckerle and Porter, 1983), and dynein and kinesin may drive movements in opposing directions (Rodionov *et al.,* 1991; Lazzaro and Haimo, 1996). Because the two directions of transport are temporally distinct, there must be a regulatory system to control the direction.

In xanthophores (Lynch *et al.,* 1986a,b) and melanophores (Grundström *et al.,* 1985; Rozdzial and Haimo, 1986a,b), cAMP-dependent protein phosphorylation is required for dispersion of pigment toward the cell periphery. Conversely, in melanophores, protein dephosphorylation, mediated by the Ca^{2+}-dependent protein phosphatase calcineurin, is required for pigment transport toward the cell center (Rozdzial and Haimo, 1986b; Thaler and Haimo, 1990, 1992). A potential target protein has been identified and shown to be specifically phosphorylated and dephosphorylated during dispersion and aggregation, respectively. A 57-kDa protein in xanthophores of *Carrasius* is phosphorylated when live dispersing xanthophores are incubated in [^{32}P]orthophosphate (Lynch *et al.*, 1986a). When purified pigment granules are added to a xanthophore cell-free extract containing cAMP and [^{32}P]ATP, the p57 becomes phosphorylated. These data suggest that the p57 is pigment granule associated and that its phosphorylation is correlated with pigment dispersion in xanthophores (Lynch *et al.*, 1986b). In melanophores, a protein of similar molecular weight is phosphorylated and dephosphorylated during opposite directions of transport (Rozdzial and Haimo, 1986b). Phosphorylation of the xanthophore p57 appears to result in its dissociation from the cytoskeleton, suggesting that the phosphorylation state of this protein may affect its interaction with the cytoskeleton (Palazzo *et al.*, 1989). The relationship of this phosphoprotein to the microtubule motors driving pigment granule transport has not yet been elucidated.

Chromosomes *in vivo* move toward the equator during prometaphase and away from it during anaphase. Similarly, isolated mammalian chromosomes possess two motor activities; they can move toward the plus ends of microtubules under conditions in which they have been exposed to an irreversible phosphorylation event via incubation in ATPγS or they can move toward the minus ends of microtubules when they have not been thiophosphorylated (Hyman and Mitchison, 1991). Changes in the phosphorylation state of the kinetochore, correlated with kinetochore activity, have been observed *in situ*. An antibody that recognizes phosphorylated epitopes stains kinetochores of chromosomes that have not yet arrived at the equator in prometaphase cells. As a chromosome begins a directed path toward the equator, its leading kinetochore stains more intensely with this antibody than does its trailing kinetochore. Once the chromosome arrives at the equator and is stationary, neither kinetochore exhibits staining (Gorbsky and Ricketts, 1993). These studies demonstrate that transport of chromosomes to the plus ends of the mitotic microtubules is coincident with kinetochore phosphorylation and that a loss or masking of this epitope is coincident with the termination of plus end-directed movement. Phosphorylation of this epitope may be required for plus end transport. In addition, or alternatively, dephosphorylation of this epitope may be required for minus end chromosome transport or may instead serve as a cell cycle checkpoint in which entry into anaphase occurs only after all kinetochores have lost this phosphorylated epitope. The target of this phosphorylation event in the kinetochore is not known. Because motor proteins are necessary for chromosome movements and because protein phosphorylation affects the direction of chromosome movements *in vitro* and is correlated with kinetochore activity, regulation of the direction of chromosome movements may be controlled by phosphorylation of motor proteins in the kinetochore or of kinetochore proteins that interact with these motors.

V. Summary

Changes in the microtubule array and regulation of microtubule-based motile events in cells appear to be controlled, in cases where regulation is understood, either by changes associated with the cell cycle or following signal-transduction events upon stimulation of cell surface receptors. Protein phosphorylation and dephosphorylation appear to be the primary regulatory mechanism used to control microtubules and their motors, as has been shown by correlating changes in phosphorylation with changes in the microtubule cytoskeleton or changes in motility. In many cases, the target proteins of these regulatory cascades are not known. Changes in the state of phosphorylation can be very rapid and are reversible, properties that would suit the needs of a cell attempting to respond to its environment by altering its cytoskeleton and/or motility, or progressing in an orderly fashion through the motile events of mitosis.

Future studies should elucidate the entire regulatory pathway, from signaling to the final target proteins in the regulatory cascade and provide information about how changes in phosphorylation can lead to changes in motility. Not all organelles move solely on microtubules or on actin filaments. Organelles in the extruded axoplasm of the squid giant axon have been observed to jump from microtubules to actin filaments (Kuznetsov et al., 1992). Similarly, biochemical analysis of Golgi-derived vesicles from intestinal epithelial cells reveals that a single organelle can possess both myosin I and cytoplasmic dynein (Fath et al., 1994). Thus, in addition to an understanding of the mechanisms involved in regulating microtubules and their motors, it will be necessary to understand the mechanism by which the cell controls which class of filaments a given organelle translocates on in a given set of circumstances.

References

Afshar, K., Barton, N. R., Hawley, R. S., and Goldstein, L. S. B. (1995). DNA binding and meiotic chromosomal localization of the *Drosophila* nod kinesin-like protein. *Cell (Cambridge, Mass.)* **81,** 129–138.

Aizawa, H., Sekine, Y., Takemura, R., Zhang, Z., Nangaku, M., and Hirokawa, N. (1992). Kinesin family in murine central nervous system. *J. Cell Biol.* **119,** 1287–1296.

Allan, V. (1995). Protein phosphatase 1 regulates the cytoplasmic dynein-driven formation of endoplasmic reticulum networks *in vitro. J. Cell Biol.* **128,** 879–891.

Allan, V. J., and Vale, R. D. (1991). Cell cycle control of microtubule-based membrane transport and tubule formation *in vitro. J. Cell Biol.* **113,** 347–359.

Aniento, F., Emans, N., Griffiths, G., and Gruenberg, J. (1993). Cytoplasmic dynein-dependent vesicular transport from early to late endosomes. *J. Cell Biol.* **123,** 1373–1387.

Asai, D. J., Beckwith, S. M., Kandl, K. A., Keating, H. H., Tjandra, H., and Forney, J. D. (1994). The dynein genes of *Paramecium tetraurelia*. Sequences adjacent to the catalytic P-loop identify cytoplasmic and axonemal heavy chain isoforms. *J. Cell Sci.* **107,** 859–867.

Baas, P. W., Pienkowsk, T. P., and Kosik, K. S. (1991). Processes induced by tau expression in Sf9 cells have an axonlike microtubule organization. *J. Cell Biol.* **115,** 1333–1344.

Bagnara, J. T., and Hadley, M. E. (1973). "Chromatophores and Color Change: The Comparative Physiology of Animal Pigmentation." Prentice-Hall, Englewood Cliffs, NJ.

Barkalow, K., Hamasaki, T., and Satir, P. (1994). Regulation of 22S dynein by a 29-kD light chain. *J. Cell Biol.* **126,** 727–735.

Beckerle, M. C., and Porter, K. R. (1983). Analysis of the role of microtubules and actin in erythrophore intracellular motility. *J. Cell Biol.* **96,** 354–362.

Belmont, L. D., Hyman, A. A., Sawin, K. E., and Mitchison, T. J. (1990). Real-time visualization of cell cycle-dependent changes in microtubule dynamics in cytoplasmic extracts. *Cell (Cambridge, Mass.)* **62,** 579–589.

Berliner, E., Young, E. C., Anderson, K., Mahtani, H. K., and Gelles, J. (1995). Failure of a single-headed kinesin to track parallel to microtubule protofilaments. *Nature (London)* **373,** 718–721.

Bernhardt, R., and Matus, A. (1984). Light and electron microscopic studies of the distribution of microtubule-associated protein 2 in rat brain: A difference between the dendritic and axonal cytoskeletons. *J. Comp. Neurol.* **226,** 203–221.

Bernstein, M., Beech, P. L., Katz, S. G., and Rosenbaum, J. L. (1994). A new kinesin-like protein (Klp1) localized to a single microtubule of the *Chlamydomonas* flagellum. *J. Cell Biol.* **125,** 1313–1326.

Bessen, M., Fay, R. B., and Witman, G. B. (1980). Calcium control of waveform in isolated flagellar axonemes of *Chlaymdomonas*. *J. Cell Biol.* **86,** 446–455.

Binder, L. I., Frankfurter, A., and Rehbun, L. I. (1985). The distribution of tau in the mammalian nervous system. *J. Cell Biol.* **101,** 1371–1378.

Block, S. M., Goldstein, L. S. B., and Schnapp, B. J. (1990). Bead movement by single kinesin molecules studied with optical tweezers. *Nature (London)* **348,** 348–352.

Bloom, G. S., Schoenfeld, T. A., and Vallee, R. B. (1984). Widespread distribution of the major polypeptide component of MAP1 (microtubule-associated protein 1) in the nervous system. *J. Cell Biol.* **98,** 320–330.

Bloom, G. S., Wagner, M. C., Pfister, K. K., and Brady, S. T. (1988). Native structure and physical properties of bovine brain kinesin and identification of the ATP binding subunit polypeptide. *Biochemistry* **27,** 3409–3416.

Bomsel, M., Parton, R., Kuznetsov, S. A., Schroer, T. A., and Gruenberg, J. (1990). Microtubule- and motor-dependent fusion *in vitro* between apical and basolateral endocytic vesicels from MDCK cells. *Cell (Cambridge, Mass.)* **62,** 719–731.

Bonini, N. M., and Nelson, D. L. (1988). Differential regulation of *Paramecium* ciliary motility by cAMP and cGMP. *J. Cell Biol.* **106,** 1615–1624.

Bonini, N. M., and Nelson, D. L. (1990). Phosphoproteins associated with cyclic nuceotide stimulation of ciliary motility in *Paramecium*. *J. Cell Sci.* **95,** 219–230.

Bonini, N. M., Gustin, M. C., and Nelson, D. L. (1986). Regulation of ciliary motility by membrane potential in *Paramecium*. *Cell Motil. Cytoskel.* **6,** 256–272.

Brady, S. T. (1985). A novel brain ATPase with properties expected for the fast axonal motor. *Nature (London)* **317,** 73–75.

Brady, S. T., Pfister, K. K., and Bloom, G. S. (1990). A monoclonal antibody against kinesin inhibits both anterograde and retrograde fast axonal transport in squid axoplasm. *Proc. Natl. Acad. Sci. U.S.A.* **87,** 1061–1065.

Bré, M.-H., Pepperkok, R., Hill, A. M., Levilliers, N., Ansorge, W., Stelzer, E. H. K., and Karsenti, E. (1990). Regulation of microtubule dynamics and nucleation during polarization in MDCK II cells. *J. Cell Biol.* **111,** 3013–3021.

Brokaw, C. J. (1994). Control of flagellar bending: A new agenda based on dynein diversity. *Cell Motil. Cytoskel.* **28,** 199–204.

Brokaw, C. J., and Kamiya, R. (1987). Bending patterns of *Chlamydomonas* flagella: IV. Mutants with defects in inner and outer dynein arms indicate differences in arm function. *Cell Motil. Cytoskel.* **8,** 68–75.

Brokaw, C. J., and Nagayama, S. M. (1985). Modulation of the asymmetry of sea urchin sperm falgellar bending by calmodulin. *J. Cell Biol.* **100,** 1875–1883.

Brokaw, C. J., Josslin, R., and Bobrow, L. (1974). Calcium ion regulation of flagellar beat symmetry in reactivated sea urchin spermatozoa. *Biochem. Biophys. Res. Commun.* **58,** 795–800.

Brokaw, C. J., Luck, D. J. L., and Huang, B. (1982). Analysis of the movement of *Chlamydomons* flagella: The function of the radial-spoke system is revealed by comparison of wild-type and mutant flagella. *J. Cell Biol.* **92,** 722–732.

Brown, K. D., Coulson, R. M. R., Yen, T. J., and Cleveland, D. W. (1994). Cyclin-like accumulation and loss of the putative kinetochore motor CENP-E results from coupling continuous synthesis with specific degradation at the end of mitosis. *J. Cell Biol.* **125,** 1303–1312.

Brugg, B., and Matus, A. (1991). Phosphorylation determines the binding of microtubule associated protein 2 (MAP2) to microtubules in living cells. *J. Cell Biol.* **114,** 735–743.

Buendia, B., Draetta, G., and Karsenti, E. (1992). Regulation of the microtubule nucleating activity of centrosomes in *Xenopus* extracts: Role of cyclin A-associated protein kinase. *J. Cell Biol.* **116,** 1431–1442.

Bulinski, J. C., and Borisy, G. G. (1980). Widespread distribution of a 210,000 mol wt microtubule-associated protein in cells and tissues of primates. *J. Cell Biol.* **87,** 802–808.

Burgoyne, R. D., and Cumming, R. (1984). Ontogeny of microtubule-associate protein 2 in rat cerebellum: differential expression of the doublet polypeptides. *Neuroscience* **11,** 157–167.

Caceres, A., and Kosik, K. S. (1990). Inhibition of neurite polarity by tau antisense oligonucleotides in primary cerebellar neurons. *Nature (London)* **343,** 461–463.

Caceres, A., Binder, L. I., Payne, M. R., Bender, P., Rehbun, L., and Steward, O. (1984). Differential subcellular localizations of tubulin and the microtubule-associated protein MAP2 in brain tissue as revealed by immunohistochemistry with monoclonal hybridoma antibodies. *J. Neurosci.* **4,** 394–410.

Caceres, A., Mautino, J., and Kosik, K. S. (1992). Suppression of MAP2 in cultured cerebellar macroneurons inhibits minor neurite formation. *Neuron* **9,** 607–618.

Centonze, V. A., and Borisy, G. G. (1990). Nucleation of microtubules from mitotic centrosomes is modulated by a phosphorylated epitope. *J. Cell Sci.* **95,** 405–411.

Chapin, S. J., and Bulinski, J. C. (1991). Non-neuronal 210 X 10^3 M_r microtubule-associated protein (MAP4) contains a domain homologus to the microtubule-binding domains of neuronal MAP2 and tau. *J. Cell Sci.* **98,** 27–36.

Chapin, S. J., and Bulinski, J. C. (1994). Cellular microtubules heterogeneous in their content of microtubule-associated protein 4 (MAP4). *Cell Motil. Cytoskel.* **27,** 133–149.

Chen, J., Kanai, Y., Cowan, N. J., and Hirokawa, N. (1992). Projection domains of MAP2 and tau determine spacings between microtubules in dendrites and axons. *Nature (London)* **360,** 674–677.

Chen, Y.-T., and Schliwa, M. (1990). Direct observation of microtubule dynamics in *Reticulomyxa:* Unusual rapid length changes and microtubule sliding. *Cell Motil. Cytoskel.* **17,** 214–226.

Cheney, R. E., Riley, M. A., and Mooseker, M. S. (1993). Phylogenetic analysis of the myosin superfamily. *Cell Motil. Cytoskel.* **24,** 215–223.

Chilcote, T. J., and Johnson, K. A. (1990). Phosphorylation of *Tetrahymena* 22S dynein. *J. Biol. Chem.* **265,** 17257–17266.

Clark, S. W., and Meyer, D. I. (1992). Centractin is an actin homologue associated with the centrosome. *Nature (London)* **359**, 246–250.

Clark, S. W., and Meyer, D. I. (1994). ACT3: A putative centractin homologue in *S. cerevisiae* is required for proper orientation of the mitotic spindle. *J. Cell Biol.* **127**, 129–138.

Cohn, S. A., Ingold, A. L., and Scholey, J. M. (1987). Correlation between the ATPase and microtubule translocating activities of sea urchin egg kinesin. *Nature (London)* **328**, 160–163.

Cole, D. G., Chinn, S. W., Wedaman, K. P., Hall, K., Vuong, T., and Scholey, J. M. (1993). Novel heterotrimeric kinesin-related protein purified from sea urchin eggs. *Nature (London)* **366**, 268–270.

Cole, D. G., Saxton, W. M., Sheehan, K. B., and Scholey, J. M. (1994). A "slow" homotetrameric kinesin-related motor protein purified from *Drosophila* embryos. *J. Biol. Chem.* **269**, 22913–22916.

Condeelis, J. (1993). Life at the leading edge: The formation of cell protrusions. *Annu. Rev. Cell Biol.* **9**, 411–444.

Corthésy-Theulaz, I., Pauloin, A., and Pfeffer, S. R. (1992). Cytoplasmic dynein participates in the centrosomal localization of the Golgi complex. *J. Cell Biol.* **118**, 1333–1345.

Coue, M., Lombillo, V. A., and McIntosh, J. R. (1991). Microtubule depolymerization promotes particle and chromosome movement *in vitro*. *J. Cell Biol.* **112**, 1165–1175.

Cyr, J. L., Pfister, K. K., Bloom, G. S., Slaughter, C. A., and Brady, S. T. (1991). Molecular genetics of kinesin light chains: Generation of isoforms by alternative splicing. *Proc. Natl. Acad. Sci. U.S.A.* **88**, 10114–10118.

Dabora, S. L., and Sheetz, M. P. (1988). The microtubule-dependent formation of a tubulovesicular network with characteristics of the ER from cultured cell extracts. *Cell (Cambridge, Mass.)* **54**, 27–35.

Davis, F. M., Tsao, T. Y., Fowler, S. K., and Rao, P. N. (1983). Monoclonal antibodies to mitotic cells. *Proc. Natl. Acad. Sci. U.S.A.* **80**, 2926–2930.

DeCamilli, P., Miller, P. E., Navone, F., Theurkauf, W. E., and Vallee, R. B. (1984). Distribution of microtubule-associated protein 2 in the nervous system of the rat studied by immunofluorescence. *Neuroscience* **11**, 817–846.

Dey, C. S., and Brokaw, C. J. (1991). Activation of *Ciona* sperm motility: Phosphorylation of dynein polypeptides and effects of a tyrosine kinase inhibitor. *J. Cell Sci.* **100**, 815–824.

Dillman, J. F., and Pfister, K. K. (1994). Differential phosphorylation *in vivo* of cytoplasmic dynein association with anterogradely moving organelles. *J. Cell Biol.* **127**, 1671–1681.

Dinsmore, J. H., and Solomon, F. (1991). Inhibition of MAP2 expression affects both morphological and cell division phenotypes of neuronal differentiation. *Cell* **64**, 817–826.

Endow, S. A., and Titus, M. A. (1992). Genetic approaches to molecular motors. *Annu. Rev. Cell Biol.* **8**, 29–66.

Endow, S. A., Henikoff, S., and Niedziela, L. S. (1990). Mediation of meiotic and early mitotic chromosome segregation in *Drosophila* by a protein related to kinesin. *Nature (London)* **345**, 81–83.

Endow, S. A., Kang, S. J., Satterwhite, L. L., Rose, M. D., Skeen, V. P., and Salmon, E. D. (1994). Yeast Kar3 is a minus-end microtubule motor protein that destabilizes microtubules preferentially at the minus ends. *EMBO J.* **13**, 2708–2713.

Enos, A. P., and Morris, N. R. (1990). Mutation of a gene that encodes a kinesin-like protein blocks nuclear division in *A. nidulans*. *Cell (Cambridge, Mass.)* **60**, 1019–1027.

Eschel, D., Urrestarazu, L. A., Vissers, S., Jauniaux, J.-C., van Vliet-Reedijk, J. C., Planta, R. J., and Gibbons, I. R. (1993). Cytoplasmic dynein is required for normal nuclear segregation in yeast. *Proc. Natl. Acad. Sci. U.S.A.* **90**, 11172–11176.

Euteneuer, U., and McIntosh, J. R. (1981). Polarity of some motility-related microtubules. *Proc. Natl. Acad. Sci. U.S.A.* **78**, 372–376.

Faruki, S., Dorée, M., and Karsenti, E. (1992). cdc2 kinase-induced destabilization of MAP2-coated microtubules in *Xenopus* egg extracts. *J. Cell Sci.* **101**, 69–78.

Fath, K. R., Trimbur, G. M., and Burgess, D. R. (1994). Molecular motors are differentially distributed on Golgi membranes from polarized epithelial cells. *J. Cell Biol.* 126, 661–675.

Febvre-Chevalier, C., and Febvre, J. (1992). Microtubule disassembly *in vivo:* Intercalary destabilization and breakdown of microtubules in the Heliozoan *Actinocoryne contractilis*. *J. Cell Biol.* 118, 585–594.

Feiguin, F., Ferreira, A., Kosik, K. S., and Caceres, A. (1994). Kinesin-mediated organelle translocation revealed by specific cellular manipulations. *J. Cell Biol.* 127, 1021–1039.

Félix, M.-A., Antony, C., Wright, M., and Maro, B. (1994). Centrosome assembly *in vitro:* Role of gamma-tubulin recruitment in *Xenopus* sperm aster formation. *J. Cell Biol.* 124, 19–31.

Fox, L. A., and Sale, W. S. (1987). Direction of force generated by the inner row of dynein arms on flagellar microtubules. *J. Cell Biol.* 105, 1781–1787.

Fox, L. A., Sawin, K. E., and Sale, W. S. (1994). Kinesin-related proteins in eukaryotic flagella. *J. Cell Sci.* 107, 1545–1550.

Gardner, L. C., O'Toole, E., Perrone, C. A., Giddings, T., and Porter, M. E. (1994). Components of a "dynein regulatory complex" are located at the junction between the radial spokes and the dynein arms in *Chlamydomonas* flagella. *J. Cell Biol.* 127, 1311–1325.

Gelfand, V. I., and Bershadsky, A. D. (1991). Microtubule dynamics: Mechanism, regulation, and function. *Annu. Rev. Cell Biol.* 7, 93–116.

Gibbons, B. H., and Gibbons, I. R. (1973). The effect of partial extraction of dynein arms on the movement of reactivated-sea urchin sperm. *J. Cell Sci.* 13, 337–357.

Gibbons, B. H., and Gibbons, I. R. (1980). Calcium-induced quiescence in reactivated sea urchin sperm. *J. Cell Biol.* 84, 13–27.

Gibbons, B. H., Asai, D. J., Tang, W. J., Hays, T. S., and Gibbons, I. R. (1994). Phylogeny and expression of axonemal and cytoplasmic dynein genes in sea urchins. *Mol. Cell. Biol.* 5, 57–70.

Gibbons, I. R. (1963). Studies on the protein components of cilia from *Tetrahymena pyriformis*. *Proc. Natl. Acad. Sci. U.S.A.* 50, 1002–1010.

Gibbons, I. R., Cosson, M. P., Evans, J. A., Gibbons, B. H., Houck, B. B., Martinson, K. H., Sale, W. S., and Tang, W. J. (1978). Potent inhibition of dynein adenosine-triphosphatase and of motility of cilia and sperm flagella by vanadate. *Proc. Natl. Acad. Sci. U.S.A.* 75, 2220–2224.

Gibbons, I. R., Gibbons, B. H., Mocz, G., and Asai, D. J. (1991). Multiple nucleotide-binding sites in the sequence of dynein beta heavy chain. *Nature (London)* 352, 640–643.

Gilbert, S. P., and Johnson, K. A. (1994). Pre-steady-state kinetics of the microtubule-kinesin ATPase. *Biochemistry* 33, 1951–1960.

Gilbert, S. P., Webb, M. R., Brune, M., and Johnson, K. A. (1995). Pathway of processive ATP hydrolysis by kinesin. *Nature (London)* 373, 671–676.

Gill, S. R., Schroer, T. A., Szilak, I., Steuer, E. R., Sheetz, M. P., and Cleveland, D. W. (1991). Dynactin, a conserved, ubiquitously expressed component of an activator of vesicle motility mediated by cytoplasmic dynein. *J. Cell Biol.* 115, 1639–1650.

Gitelman, S. E., and Witman, G. B. (1980). Purification of calmodulin from *Chlamydomonas:* Calmodulin occurs in cell bodies and flagella. *J. Cell Biol.* 98, 764–770.

Glicksman, N. R., Parsons, S. F., and Salmon, E. D. (1992). Okadaic acid induces interphase to mitotic-like microtubule dynamic instability by inactivating rescue. *J. Cell Biol.* 119, 1271–1276.

Goldstein, L. S. B. (1993). With apologies to Scheherazade: Tails of 1001 kinesin motors. *Annu. Rev. Genet.* 27, 319–351.

Gorbsky, G. J., and Ricketts, W. A. (1993). Differential expression of a phosphoepitope at the kinetochore of moving chromosomes. *J. Cell Biol.* 122, 1311–1321.

Grundström, N., Karlsson, J. O. G., and Andersson, R. G. G. (1985). The control of granule movement in fish melanophores. *Acta Physiol. Scand.* 125, 415–422.

Gundersen, G. G., Kalnoski, M. H., and Bulinski, J. C. (1984). Distinct populations of microtubules: Tyrosinated and nontyrosinated alpha tubulin are distributed differently *in vivo. Cell (Cambridge, Mass.)* **38,** 779–789.

Gyoeva, F. K., and Gelfand, V. I. (1991). Coalignment of vimentin intermediate filaments with microtubules depends on kinesin. *Nature (London)* **353,** 445–448.

Hackney, D. D. (1988). Kinesin ATPase: Rate-limiting ADP release. *Proc. Natl. Acad. Sci. U.S.A.* **85,** 6314–6318.

Hackney, D. D. (1994a). The rate-limiting step in microtubule-stimulated ATP hydrolysis by dimeric kinesin head domains occurs while bound to the microtubule. *J. Biol. Chem.* **269,** 16508–16511.

Hackney, D. D. (1994b). Evidence for alternating head catalysis by kinesin during microtubule-stimulated ATP hydrolysis. *Proc. Natl. Acad. Sci. U.S.A.* **91,** 6865–6869.

Hackney, D. D., Levitt, J. D., and Wagner, D. D. (1991). Characterization of $\alpha_2\beta_2$ and α_2 forms of kinesin. *Biochem. Biophys. Res. Commun.* **174,** 810–815.

Hackney, D. D., Levitt, J. D., and Suhan, J. (1992). Kinesin undergoes a 9S to 6S conformational transition. *J. Biol. Chem.* **267,** 8696–8701.

Hagan, I., and Yanagida, M. (1990). Novel potential mitotic motor protein encoded by the fission yeast cut7⁺ gene. *Nature (London)* **347,** 563–566.

Hagiwara, H., Yorifuji, H., Sato-Yoshitake, R., and Hirokawa, N. (1994). Competition between motor molecules (kinesin and cytoplasmic dynein) and fibrous microtubule-associated proteins in binding to microtubules. *J. Biol. Chem.* **269,** 3581–3589.

Hall, D. H., and Hedgecock, E. M. (1991). Kinesin-related gene unc-104 is required for axonal transport of synaptic vesicles in *C. elegans. Cell (Cambridge, Mass.)* **65,** 837–847.

Hall, K., Cole, D. G., Yeh, Y., Scholey, J. M., and Baskin, R. J. (1993). Force-velocity relationships in kinesin-driven motility. *Nature (London)* **364,** 457–459.

Hamasaki, T., Murtaugh, T. J., Satir, B. H., and Satir, P. (1989). *In vitro* phosphorylation of *Paramecium* axonemes and permeabilized cells. *Cell Motil. Cytoskel.* **12,** 1–11.

Hamasaki, T., Barkalow, K., Richmond, J., and Satir, P. (1991). cAMP-stimulated phosphorylation of an axonemal polypeptide that copurifies with the 22S dynein arm regulates microtubule translocation velocity and swimming speed in *Paramecium. Proc. Natl. Acad. Sci. U.S.A.* **88,** 7918–7922.

Hamm-Alvarez, S. F., Kim, P. Y., and Sheetz, M. P. (1993). Regulation of vesicle transport in CV-1 cells and extracts. *J. Cell Sci.* **106,** 955–966.

Harada, A., Oguchi, K., Okabe, S., Kuno, J., Terada, S., Ohshima, T., Sato-Yoshitake, R., Takel, Y., Noda, T., and Hirokawa, N. (1994). Altered microtubule organization in small-calibre axons of mice lacking tau protein. *Nature (London)* **369,** 488–491.

Hasegawa, E., Hayashi, H., Asakura, S., and Kamiya, R. (1987). Stimulation of *in vitro* motility of *Chlamydomonas* axonemes by inhibition of cAMP-dependent phosphorylation. *Cell Motil. Cytoskel.* **8,** 302–311.

Hatsumi, M., and Endow, S. A. (1992). Mutants of the microtubule motor protein, nonclaret disjunctional, affect spindle structure and chromosome movement in meiosis and mitosis. *J. Cell Sci.* **103,** 1013–1020.

Hayden, J. H., and Allen, R. D. (1984). Detection of single microtubules in living cells: Particle transport can occur in both directions along the same microtubule. *J. Cell Biol.* **99,** 1785–1793.

Heck, M. M. S., Pereira, A., Pesavento, P., Yannoni, Y., Spradling, A. C., and Goldstein, L. S. B. (1993). The kinesin-like protein KLP61F is essential for mitosis in *Drosophila. J. Cell Biol.* **123,** 665–679.

Hirokawa, N. (1994). Microtubule organization and dynamics dependent on microubule-associated proteins. *Curr. Opin. Cell Biol.* **6,** 74–81.

Hirokawa, N., Pfister, K. K., Yorifuji, H., Wagner, M. C., Brady, S. T., and Bloom, G. S. (1989). Submolecular domains of bovine brain kinesin identified by electron microscopy and monoclonal antibody decoration. *Cell (Cambridge, Mass.)* **56,** 867–878.

Hirokawa, N., Sato-Yoshitake, R., Yoshida, T., and Kawashima, T. (1990). Brain dynein (MAP1C) localizes on both anterogradely and retrogradely transported membranous organelles in vivo. J. Cell Biol. 111, 1027–1037.

Hirokawa, N., Sato-Hoshitake, R., Kobayashi, N., Pfister, K. K., Bloom, G. S., and Brady, S. T. (1991). Kinesin associates with anterogradely transported membranous organelles in vivo. J. Cell Biol. 114, 295–302.

Hogan, C. J., Wein, H., Wordeman, L., Scholey, J. M., Sawin, K. E., and Cande, W. Z. (1993). Inhibition of anaphase spindle elongation in vitro by a peptide antibody that recognizes kinesin motor domain. Proc. Natl. Acad. Sci. U.S.A. 90, 6611–6615.

Hollenbeck, P. J. (1989). The distribution, abundance and subcellular localization of kinesin. J. Cell Biol. 108, 2335–2342.

Hollenbeck, P. J. (1993). Phosphorylation of neuronal kinesin heavy and light chains in vivo. J. Neurochem. 60, 2265–2275.

Hollenbeck, P. J., and Swanson, J. A. (1990). Radial extension of macrophage tubular lysosomes supported by kinesin. Nature (London) 346, 864–866.

Holzbaur, E. L. F., and Vallee, R. B. (1994). Dyneins: Molecular structure and cellular function. Annu. Rev. Cell. Biol. 10, 339–372.

Holzbaur, E. L. F., Hammarback, J. A., Paschal, B. M., Kravit, N. G., Pfister, K. K., and Vallee R. B. (1991). Homology of a 150kD cytoplasmic dynein-associated polypeptide with the Drosophila gene Glued. Nature (London) 351, 579–583.

Horio, T., and Hotani, H. (1986). Visualization of the dynamic instability of individual microtubules by dark-field microscopy. Nature (London) 321, 605–607.

Horio, T., and Oakley, B. R. (1994). Human γ-tubulin functions in fission yeast. J. Cell Biol. 126, 1465–1473.

Howard, D. R., Habermacher, G., Glass, D. B., Smith, E. F., and Sale, W. S. (1994). Regulation of Chlamydomonas flagellar dynein by an axonemal protein kinase. J. Cell Biol. 127, 1683–1692.

Howard, J., Hudspeth, A. J., and Vale, R. D. (1989). Movement of microtubules by single kinesin molecules. Nature (London) 342, 154–158.

Hoyt, M. A. (1994). Cellular roles of kinesin and related proteins. Curr. Opin. Cell Biol. 6, 63–68.

Hoyt, M. A., He, L., Loo, K. K., and Saunders, W. S. (1992). Two Saccharomyces cerevisiae kinesin-related gene products required for mitotic spindle assembly. J. Cell Biol. 118, 109–120.

Hoyt, M. A., He, L., Totis, L., and Saunders, W. S. (1993). Loss of function of Saccharomyces cerevisiae kinesin-related CIN8 and KIP1 is suppressed by KAR3 motor mutations. Genetics 135, 35–44.

Huang, B., Ramanis, Z., and Luck, D. J. L. (1982). Suppressor mutations in Chlamydomonas reveal a regulatory mechanism for flagellar function. Cell (Cambridge, Mass.) 28, 115–124.

Hunt, A. J., Gittes, F., and Howard, J. (1994). The force exerted by a single kinesin molecule against a viscous load. Biophys. J. 67, 766–781.

Huxley, H. E. (1969). The mechanism of muscular contraction. Science 164, 1356–1366.

Hyman, A. A., and Mitchison, T. J. (1991). Two different microtubule-based motor activities with opposite polarities in kinetochores. Nature (London) 351, 206–211.

Hyman, A. A., Middleton, K., Centola, M., Mitchison, T. J., and Carbon, J. (1992). Microtubule-motor activity of a yeast centromere-binding protein complex. Nature (London) 359, 533–536.

Johnson, K. A. (1983). The pathway of ATP hydrolysis by dynein. Kinetics of a presteady state phosphate burst. J. Biol. Chem. 258, 13825–13832.

Johnson, K. A. (1985). Pathway of the microtubule-dynein ATPase and structure of dynein: A comparison with actomyosin. Annu. Rev. Biophys. Biophys. Chem. 14, 161–188.

Johnson, K. A., and Rosenbaum, J. L. (1992). Polarity of flagellar assembly in *Chlamydomonas*. *J. Cell Biol.* **119,** 1605–1611.

Johnson, K. A., Haas, M. A., and Rosenbaum, J. L. (1994). Localization of a kinesin-related protein to the central pair apparatus of the *Chlamydomonas reinhardtii* flagellum. *J. Cell Sci.* **107,** 1551–1556.

Joshi, H. C., Palacios, M. J., McNamara, L., and Cleveland, D. (1992). γ-tubulin is a centrosomal protein required for cell cycle-dependent microtubule nucleation. *Nature (London)* **356,** 80–82.

Kagami, O., and Kamiya, R. (1992). Translocation and rotation of microtubules caused by multiple species of *Chlamydomonas* inner-arm dynein. *J. Cell Sci.* **103,** 653–664.

Kamiya, R. (1982). Extrusion and rotation of the central-pair microtubules in detergent-treated *Chlamydomonas* flagella. *Cell Motil., Suppl.* **1,** 169–173.

Kamiya, R., and Okamoto, M. (1985). A mutant of *Chlamydomonas reinhardtii* that lacks the flagellar outer dynein arm but can swim. *J. Cell Sci.* **74,** 181–191.

Kersey, Y. M., Hepler, P. K., Palevitz, C. A., and Wessells, N. K. (1976). Polarity of actin filaments in *Characean* algae. *Proc. Natl. Acad. Sci. U.S.A.* **73,** 165–168.

Khawaja, S., Gundersen, G. G., and Bulinski, J. C. (1988). Enhanced stability of microtubules enriched in detyrosinated tubulin is not a direct function of detyrosination level. *J. Cell Biol.* **106,** 141–150.

King, S. M., and Witman, G. B. (1990). Immunolocalization of an intermediate chain of outer arm dynein by immunoelectron microscopy. *J. Biol. Chem.* **265,** 19807–19811.

King, S. M., and Witman, G. B. (1994). Multiple sites of phosphorylation within the heavy chain of *Chlamyodmonas* outer arm dynein. *J. Biol. Chem.* **269,** 5452–5457.

King, S. M., Wilkerson, C. G., and Witman, G. B. (1991). The M_r 78,000 intermediate chain of *Chlamydomonas* outer arm dynein interacts with α-tubulin *in situ. J. Biol. Chem.* **266,** 8401–8407.

Knops, J., Kosik, K. S., Lee, G., Pardee, J. D., Cohen-Gould, L., and McConlogue, L. (1991). Overexpression of tau in a nonneuronal cell induces long cellular processes. *J. Cell Biol.* **114,** 725–733.

Kondo, S., Sato-Yoshitake, R., Noda, Y., Aizawa, H., Nakata, T., Matsuura, Y., and Hirokawa, N. (1994). KIF3A is a new microtubule-based anterograde motor in the nerve axon. *J. Cell Biol.* **125,** 1109–1117.

Koonce, M. P., and McIntosh, J. R. (1990). Identification and immunolocalization of cytoplasmic dynein in *Dictyostelium. Cell Motil. Cytoskel.* **15,** 51–62.

Koonce, M. P., and Schliwa, M. (1985). Bidirectional organelle transport can occur in cell processes that contain single microtubules. *J. Cell Biol.* **100,** 322–326.

Koonce, M. P., Grissom, P. M., and McIntosh, J. R. (1992). Dynein from *Dictyostelium:* Primary structure comparisons between a cytoplasmic motor enzyme and flagellar dynein. *J. Cell Biol.* **119,** 1597–1604.

Kumar, J., Yu, H., and Sheetz, M. P. (1995). Kinectin, an essential anchor for kinesin-driven vesicle motility. *Science* **267,** 1834–1837.

Kuo, S. C., and Sheetz, M. P. (1993). Force of single kinesin molecules measured with optical tweezers. *Science* **260,** 232–234.

Kuriyama, R., and Borisy, G. G. (1981). Microtubule-nucleating activity of centrosomes in Chinese hamster ovary cells is independent of the centriole cycle but coupled to the mitotic cycle. *J. Cell Biol.* **91,** 822–826.

Kuznetsov, S. A., and Gelfand, V. I. (1986). Bovine brain kinesin is a microtubule-activated ATPase. *Proc. Natl. Acad. Sci. U.S.A.* **83,** 8530–8534.

Kuznetsov, S. A., Vaisberg, Y. A., Rothwell, S. W., Murphy, D. B., and Gelfand, V. I. (1989). Isolation of a 45-kDa fragment from the kinesin heavy chain with enhanced ATPase and microtubule-binding activity. *J. Biol. Chem.* **264,** 589–595.

Kuznetsov, S. A., Langford, G. M., and Weiss, D. G. (1992). Actin-dependent organelle movement in squid axoplasm. *Nature (London)* **356,** 722–725.

Lacey, M. L., and Haimo, L. T. (1992). Cytoplasmic dynein is a vesicle protein. *J. Biol. Chem.* **267,** 4793–4798.

Lacey, M. L., and Haimo, L. T. (1994). Cytoplasmic dynein binds to phospholipid vesicles. *Cell Motil. Cytoskel.* **28,** 205–212.

Lazzaro, M., and Haimo, L. T. (1996). In preparation.

Lee, G., and Rook, S. L. (1992). Expression of tau protein in non-neuronal cells: Microtubule binding and stabilization. *J. Cell Sci.* **102,** 227–237.

Lee, K.-D., and Hollenbeck, P. J. (1995). Phosphorylation of kinesin *in vivo* correlates with organelle association and neurite outgrowth. *J. Biol. Chem.* **270,** 5600–5605.

Lee-Eiford, A., Ow, R. A., and Gibbons, I. R. (1986). Specific cleavage of dynein heavy chains by ultraviolet irradiation in the presence of ATP and vanadate. *J. Biol. Chem.* **261,** 2337–2342.

Lees-Miller, J. P., Helfman, D. M., and Schroer, T. A. (1992). A vertebrate actin-related protein is a component of a multisubunit complex involved in microtubule-based vesicle motility. *Nature (London)* **359,** 244–246.

Leopold, P. L., McDowall, A. W., Pfister, K. K., Bloom, G. S., and Brady, S. T. (1992). Association of kinesin with characterized membrane-bounded organelles. *Cell Motil. Cytoskel.* **23,** 19–33.

Lewis, S. A., Wang, D., and Cowan, N. J. (1988). Microtubule-associated protein MAP2 shares a microtubule binding motif with tau protein. *Science* **242,** 936–939.

Lewis, S. A., Ivanov, I. E., Lee, G.-H., and Cowan, N. J. (1989). Organization of microtubules in dendrites and axons is determined by a short hydrophobic zipper in microtubule-associated proteins MAP2 and tau. *Nature (London)* **342,** 498–505.

Li, M., McGrail, M., Serr, M., and Hays, T. S. (1994). *Drosophila* cytoplasmic dynein, a microtubule motor that is asymmetrically localized in the oocyte. *J. Cell Biol.* **126,** 1475–1494.

Li, Y.-Y., Yeh, E., Hays, T., and Bloom, K. (1993). Disruption of mitotic spindle orientation in a yeast dynein mutant. *Proc. Natl. Acad. Sci. U.S.A.* **90,** 10096–10100.

Liao, H., Li, G., and Yen, T. J. (1994). Mitotic regulation of microtubule cross-linking activity of CENP-E kinetochore protein. *Science* **265,** 394–398.

Lieuvin, A., Labbé, J., Dorée, M., and Job, D. (1994). Intrinsic microtubule stability in interphase cells. *J. Cell Biol.* **124,** 985–996.

Lillie, S. H., and Brown, S. S. (1992). Suppression of a myosin defect by a kinesin-related gene. *Nature (London)* **356,** 358–361.

Lillie, S. H., and Brown, S. S. (1994). Immunofluorescence localization of the unconventional myosin, Myo2p, and the putative kinesin-related protein, Smy1p, to the same regions of polarized growth in *Saccharomyces cerevisiae. J. Cell Biol.* **125,** 825–842.

Lin, S. X. H., and Collins, C. A. (1992). Immunolocalization of cytoplasmic dynein to lysosomes in cultured cells. *J. Cell Sci.* **101,** 125–137.

Lin, S. X. H., and Collins, C. A. (1993). Regulation of the intracellular distribution of cytoplasmic dynein by serum factors and calcium. *J. Cell Sci.* **105,** 579–588.

Lin, S. X. H., Ferro, K. L., and Collins, C. A. (1994). Cytoplasmic dynein undergoes intracellular redistribution concomitant with phosphorylation of the heavy chain in response to serum starvation and okadaic acid. *J. Cell Biol.* **127,** 1009–1019.

Lindwall, G., and Cole, R. D. (1984). Phosphorylation affects the ability of tau protein to promote microtubule assembly. *J. Biol. Chem.* **259,** 5301–5305.

Lippincott-Schwartz, J., Cole, N. B., Marotta, A., Conrad, P. A., and Bloom, G. S. (1995). Kinesin is the motor for microtubule-mediated Golgi-to-ER membrane traffic. *J. Cell Biol.* **128,** 293–306.

Lombillo, V. A., Nislow, C., Yen, T. J., Gelfand, V. I., and McIntosh, J. R. (1995). Antibodies to the kinesin motor domain and CENP-E inhibit microtubule depolymerization-dependent motion of chromosomes in vitro. *J. Cell Biol.* **128,** 107–115.

Lopez, L. A., and Sheetz, M. P. (1993). Steric inhibition of cytoplasmic dynein and kinesin motility by MAP2. *Cell Motil. Cytoskel.* **24,** 1–16.

Lye, R. J., Porter, M. E., Scholey, J. M., and McIntosh, J. R. (1987). Identification of a microtubule-based cytoplasmic motor in the nematode *C. elegans. Cell (Cambridge, Mass.)* **51,** 309–318.

Lye, R. J., Pfarr, C. M., and Porter, M. E. (1989). Cytoplasmic dynein and microtubule translocators. *In* "Cell Movements" (F. D. Warner and J. R. McIntosh, eds.), Vol. 2, pp. 141–154. Alan R. Liss, New York.

Lymn, R. W., and Taylor, E. W. (1971). Mechanism of adenosine triphosphate hydrolysis by actomyosin. *Biochemistry* **10,** 4617–4624.

Lynch, T. J., Taylor, J. D., and Tchen, T. T. (1986a). Regulation of pigment organelle translocation. I: Phosphorylation of the organelle associated protein p57. *J. Biol. Chem.* **261,** 4204–4211.

Lynch, T. J., Wu, B.-Y., Taylor, J. D., and Tchen, T. T. (1986b). Regulation of pigment organelle translocation. II: Participation of a cAMP dependent protein kinase. *J. Biol. Chem.* **261,** 4212–4216.

Matthies, H. J. G., Miller, R. J., and Palfrey, H. C. (1993). Calmodulin binding to and cAMP-dependent phosphorylation of kinesin light chains modulate kinesin ATPase activity. *J. Biol. Chem.* **268,** 11176–11187.

McDonald, H. B., and Goldstein, L. S. B. (1990). Identification and characterization of a gene encoding a kinesin-like protein in *Drosophila. Cell (Cambridge, Mass.)* **61,** 991–1000.

McDonald, H. B., Stewart, R. J., and Goldstein, L. S. B. (1990). The kinesin-like ncd protein of *Drosophila* is a minus end-directed microtubule motor. *Cell (Cambridge, Mass.)* **63,** 1159–1165.

McIlvain, J. M., Burkhardt, J. K., Hamm-Alvarez, S., Argon, Y., and Sheetz, M. P. (1994). Regulation of kinesin activity by phosphorylation of kinesin-associated proteins. *J. Biol. Chem.* **269,** 19176–19182.

McNally, F. J., and Vale, R. D. (1993). Identification of katinin, an ATPase that severs and disassembles stable microtubules. *Cell (Cambridge, Mass.)* **75,** 419–429.

Meluh, P. B., and Rose, M. D. (1990). KAR3, a kinesin-related gene required for nuclear fusion. *Cell (Cambridge, Mass.)* **60,** 1029–1041.

Middleton, K., and Carbon, J. (1994). KAR3-encoded kinesin is a minus-end-directed motor that functions with centromere binding proteins (CBF3) on an *in vitro* yeast kinetochore. *Proc. Natl. Acad. Sci. U.S.A.* **91,** 7212–7216.

Mikami, A., Paschal, B. M., Mazumdar, M., and Vallee, R. B. (1993). Molecular cloning of the retrograde transport motor cytoplasmic dynein (MAP1C). *Neuron* **10,** 787–796.

Miller, R. H., Lasek, R. J., and Katz, M. J. (1987). Preferred microtubules for vesicle transport in lobster axons. *Science* **235,** 220–222.

Mimori, Y., and Miki-Noumura, T. (1995). Extrusion of rotating microtubules on the dynein-track from a microtubule-dynein γ-complex. *Cell Motil. Cytoskel.* **30,** 17–25.

Mitchell, D. R. (1994). Cell and molecular biology of flagellar dyneins. *Int. Rev. Cytol.* **155,** 141–180.

Mitchell, D. R., and Brown, K. S. (1994). Sequence analysis of the *Chlamydomonas* alpha and beta dynein heavy chain genes. *J. Cell Sci.* **107,** 635–644.

Mitchell, D. R., and Kang, Y. (1991). Reversion analysis of dynein intermediate chain function. *J. Cell Biol.* **113,** 835–842.

Mitchell, D. R., and Rosenbaun, J. L. (1985). A motile *Chlamydomonas* flagellar mutant that lacks outer dynein arms. *J. Cell Biol.* **100,** 1228–1234.

Mitchison, T., and Kirschner, M. (1984a). Dynamic instability of microtubule growth. *Nature (London)* **312,** 232–237.

Mitchison, T., and Kirschner, M. (1984b). Microtubule assembly nucleated by isolated centrosomes. *Nature (London)* **312,** 237–242.

Mooseker, M. (1993). A multitude of myosins. *Curr. Biol.* **3**, 245–248.

Moss, A. G., Gatti, J.-L., and Witman, G. B. (1992). The motile β/IC1 subunit of sea urchin sperm outer arm dynein does not form a rigor bond. *J. Cell Biol.* **118**, 1177–1188.

Muhua, L., Karpova, T. S., and Cooper, J. A. (1994). A yeast actin-related protein homologus to that in vertebarate dynactin complex is important for spindle orientation and nuclear migration. *Cell (Cambridge, Mass.)* **78**, 669–679.

Murakami, A. (1987). Control of ciliary beat frequency in the gill of *Mytilus*-I. Activation of the lateral cilia by cyclic AMP. *Comp. Biochem. Physiol. C* **86C**, 273–279.

Murofushi, H., Ikai, A., Okuhara, K., Kotani, S., Aizawa, H., Kamakura, K., and Sakai, H. (1988). Purification and characterization of kinesin from bovine adrenal medulla. *J. Biol. Chem.* **263**, 12744–12750.

Murphy, D. B., and Tilney, L. G. (1974). The role of microtubules in the movement of pigment granules in teleost melanophores. *J. Cell Biol.* **61**, 757–779.

Murphy, T. D., and Karpen, G. H. (1995). Interactions between the nod⁺ kinesin-like gene and extracentromeric sequences are required for transmission of a *Drosophila* minichromosome. *Cell Cambridge, Mass.)* **81**, 139–148.

Nangaku, M., Sato-Yoshitake, R., Okada, Y., Noda, Y., Takemura, R., Yamazaki, H., and Hirokawa, N. (1994). KIF1B, a novel microtubule plus end-directed monomeric motor protein for transport of mitochondria. *Cell (Cambridge, Mass.)* **79**, 1209–1220.

Navone, F., Niclas, J., Hom-Booher, N., Sparks, L., Bernstein, H. D., McCaffrey, G., and Vale, R. D. (1992). Cloning and expression of a human kinesin heavy chain gene: Interaction of the COOH-terminal domain with cytoplasmic microtubules in transfected CV-1 cells. *J. Cell Biol.* **117**, 1263–1275.

Neighbors, B. W., Williams, R. C., Jr., and McIntosh, J. R. (1988). Localization of kinesin in cultured cells. *J. Cell Biol.* **106**, 1193–1204.

Nislow, C., Lombillo, V. A., Kuriyama, R., and McIntosh, J. R. (1992). A plus-end-directed motor enzyme that moves antiparallel microtubules *in vitro* localizes to the interzone of mitotic spindles. *Nature (London)* **359**, 543–547.

Noble, M., Lewis, S. A., and Cowan, N. J. (1989). The microtubule binding domain of microtubule-associated protein MAP1B contains a repeated sequence motif unrelated to that of MAP2 and tau. *J. Cell Biol.* **109**, 3367–3376.

Noda, Y., Sato-Yoshitake, R., Kondo, S., Nangaku, M., and Hirokawa, N. (1995). KIF2 is a new microtubule-based anterograde motor that transport membranous organelles distinct from those carried by kinesin heavy chain or KIP3A/B. *J. Cell Biol.* **129**, 157–167.

Oakley, B. R., Oakley, C. E., Yoon, Y., and Jung, M. K. (1990). γ-tubulin is a component of the spindle pole body that is essential for microtubule function in *Aspergillus nidulans. Cell (Cambridge, Mass.)* **61**, 1289–1301.

Oakley, C. E., and Oakley, B. R. (1989). Identification of γ tubulin, a new member of the tubulin superfamily encoded by mipA gene of *Aspergillus nidulans. Nature (London)* **338**, 662–664.

O'Connell, M. J., Meluh, P. B., Rose, M. D., and Morris, N. R. (1993). Suppression of the bimC4 mitotic spindle defect by deletion of klpA, a gene encoding a KAR3-related kinesin-like protein in *Aspergillus nidulans. J. Cell Biol.* **120**, 153–162.

Ogawa, K. (1991). Four ATP-binding sites in the midregion of the beta heavy chain of dynein. *Nature (London)* **352**, 643–645.

Omoto, C. K., and Johnson, K. A. (1986). Activation of the dynein adenosine-triphosphatase by microtubules. *Biochemistry* **25**, 419–427.

Omoto, C. K., and Kung, C. (1979). The pair of central tubules rotates during ciliary beat in *Paramecium. Nature (London)* **279**, 532–534.

Omoto, C. K., and Witman, G. W. B. (1981). Functionally significant central-pair rotation in a primitive eukaryotic flagellum. *Nature (London)* **290**, 708–710.

Ookata, K., Hisanaga, S., Bulinski, J. C., Murofushi, H., Aizawa, H., Itoh, T. J., Hotani, H., Okumura, E., Tachibana, K., and Kishimoto, T. (1995). Cyclin B interaction with microtubule-associated protein 4 (MAP4) targets p34^{cdc2} kinase to microtubules and is a potential regulator of M-phase microtubule dynamics. *J. Cell Biol.* **128,** 849–862.

Opresko, L. K., and Brokaw, C. J. (1983). cAMP-dependent phosphorylation associated with activation of motility of *Ciona* sperm flagella. *Gamete Res.* **8,** 201–218.

Palazzo, R. E., Lynch, T. J., Taylor, J. D., and Tchen, T. T. (1989). cAMP-independent and cAMP-dependent protein phosphorylations by isolated goldfish xanthophore cytoskeletons: Evidence for the association of cytoskeleton with a carotenoid droplet protein. *Cell Motil. Cytoskel.* **13,** 21–29.

Paschal, B. M., and Vallee R. B. (1987). Retrograde transport by the microtubule-associated protein MAP 1C. *Nature (London)* **330,** 181–183.

Paschal, B. M., Shpetner, H. S., and Vallee, R. B. (1987a). MAP1C is a microtubule-activated ATPase which translocates microtubules in vitro and has dynein-like properties. *J. Cell Biol.* **105,** 1273–1282.

Paschal, B. M., King, S. M., Moss, A. G., Collins, C. A., Vallee, R. B., and Witman, G. B. (1987b). Isolated flagellar outer arm dynein translocates brain microtubules *in vitro. Nature (London)* **330,** 672–674.

Paschal, B. M., Obar, R. A., and Vallee, R. B. (1989). Interaction of brain cytoplasmic dynein and MAP2 with a common sequence at the C terminus of tubulin. *Nature (London)* **342,** 569–572.

Paschal, B. M., Mikami, A., Pfister, K. K., and Vallee, R. B. (1992). Homology of the 74-kD cytoplasmic dynein subunit with a flagellar dynein polypeptide suggests an intracellular targeting function. *J. Cell Biol.* **118,** 1133–1143.

Paschal, B. M., Holzbaur, E. L., Pfister, K. K., Clark, S., Meyer, D. I., and Vallee, R. B. (1993). Characterization of a 50-kDa polypeptide in cytoplasmic dynein preparations reveals a complex with p150^{glued} and a novel actin. *J. Biol. Chem.* **268,** 15318–15323.

Pfarr, C. M., Coue, M., Grissom, P. M., Hays, T. S., Porter, M. E., and McIntosh, J. R. (1990). Cytoplasmic dynein is localized to kinetochores during mitosis. *Nature (London)* **345,** 263–265.

Pfister, K. K., and Witman, G. B. (1984). Subfractionation of *Chlamydomonas* 18S dynein into two unique subunits containing ATPase activity. *J. Biol. Chem.* **259,** 12072–12080.

Pfister, K. K., Haley, B. E., and Witman, G. B. (1984). The photoaffinity probe 8-azidoadenosine 5′-triphosphate selectively labels the heavy chain of *Chlamydomonas* 12 S dynein. *J. Biol. Chem.* **259,** 8499–8504.

Pfister, K. K., Haley, B. E., and Witman, G. B. (1985). Labeling of *Chlamydomonas* 18S dynein polypeptides by 8-azidoadenosine 5′-triphosphate, a photoaffinity analog of ATP. *J. Biol. Chem.* **260,** 12844–12850.

Pfister, K. K., Wagner, M. C., Stenoien, D. L., Brady, S. T., and Bloom, G. S. (1989). Monoclonal antibodies to kinesin heavy and light chains stain vesicle-like structures, but not microtubules, in cultured cells. *J. Cell Biol.* **108,** 1453–1464.

Piperno, G., and Fuller, M. T. (1985). Monoclonal antibodies specific for an acetylated form of α-tubulin recognize the antigen in cilia and flagella from a variety of organisms. *J. Cell Biol.* **101,** 2085–2094.

Piperno, G., and Ramanis, Z. (1991). The proximal portion of *Chlamydomonas* flagella contains a distinct set of inner dynein arms. *J. Cell Biol.* **112,** 701–709.

Piperno, G., Ramanis, Z., Smith, E. F., and Sale, W. S. (1990). Three distinct inner dynein arms in *Chlamydomonas* flagella: Molecular composition and location in the axoneme. *J. Cell Biol.* **110,** 379–389.

Piperno, G., Mead, K., and Shestak, W. (1992). The inner dynein arms I2 interact with a "dynein regulatory complex" in *Chlamydomonas* flagella. *J. Cell Biol.* **118,** 1455–1463.

Piperno, G., Mead, K., LeDizet, M., and Moscatelli, A. (1994). Mutations in the "dynein regulatory complex" alter the ATP-insensitive binding sites for inner arm dyneins in *Chlamydomonas* axonemes. *J. Cell Biol.* **125**, 1109–1117.

Plamann, M., Minke, P. F., Tinsley, J. H., and Bruno, K. S. (1994). Cytoplasmic dynein and actin-related Arp1 are required for normal nuclear distribution in filamentous fungi. *J. Cell Biol.* **127**, 139–149.

Porter, M. E., Power, J., and Dutcher, S. K. (1992). Extragenic suppressors of paralyzed flagellar mutations in *Chlamydomonas reinhardtii* identify loci that alter the inner dynein arms. *J. Cell Biol.* **118**, 1163–1176.

Porter, M. E., Knott, J. A., Gardner, L. C., Mitchell, D. R., and Dutcher, S. K. (1994). Mutations in the sup^{pf1} locus of *Chlamydomonas reinhardtii* identify a regulatory domain in the beta dynein heavy chain. *J. Cell Biol.* **126**, 1495–1508.

Rasmusson, K., Serr, M., Gepner, J., Gibbons, I., and Hays, T. S. (1994). A family of dynein genes in *Drosophila melanogaster. Mol. Biol. Cell* **5**, 45–55.

Rodionov, V. I., Gyoeva, F. K., Ksashina, A. S., Kuznetsov, S. A., and Gelfand, V. I. (1990). Microtubule-associated proteins and microtubule-based translocators have different binding sites on tubulin molecules. *J. Biol. Chem.* **265**, 5702–5707.

Rodionov, V. I., Gyoeva, F. K., and Gelfand, V. I. (1991). Kinesin is responsible for centrifugal movement of pigment granules in melanophores. *Proc. Natl. Acad. Sci. U.S.A.* **88**, 4956–4960.

Romberg, L., and Vale, R. D. (1993). Chemomechanical cycle of kinesin differs from that of myosin. *Nature (London)* **361**, 168–170.

Roof, D. M., Meluh, P. B., and Rose, M. D. (1992). Kinesin-related proteins required for assembly of the mitotic spindle. *J. Cell Biol.* **118**, 95–108.

Rozdzial, M. M., and Haimo, L. T. (1986a). Reactivated melanophore motility: differential regulation and nucleotide requirements of bidirectional pigment granule transport. *J. Cell Biol.* **103**, 2755–2764.

Rozdzial, M. M., and Haimo, L. T. (1986b). Bidirectional pigment granule movements of melanophores are regulated by protein phosphorylation and dephosphorylation. *Cell (Cambridge, Mass.)* **47**, 1061–1070.

Ruppel, K. M., and Spudich, J. A. (1995). Myosin motor function: Structural and mutagenic approaches. *Curr. Opin. Cell Biol.* **7**, 89–93.

Sakakibara, H., Mitchell, D. R., and Kamiya, R. (1991). A *Chlamydomonas* outer dynein mutant missing the α heavy chain. *J. Cell Biol.* **113**, 615–622.

Sakakibara, H., Takada, S., King, S. M., Witman, G. B., and Kamiya, R. (1993). A *Chlamydomonas* outer arm dynein mutant with a truncated β heavy chain. *J. Cell Biol.* **122**, 653–661.

Sale, W. S. (1986). The axonemal axis and Ca^{2+}-induced asymmetry of active microtubule sliding in sea urchin sperm tails. *J. Cell Biol.* **102**, 2042–2052.

Sale, W. S., and Fox, L. A. (1988). Isolated β-heavy chain subunit of dynein translocates microtubules *in vitro. J. Cell Biol.* **107**, 1793–1797.

Sale, W. S., and Satir, P. (1977). Direction of active sliding of microtubules in *Tetrahymena* cilia. *Proc. Natl. Acad. Sci. U.S.A.* **74**, 2045–2049.

Sale, W. S., Goodenough, U. W., and Heuser, J. E. (1985). The substructure of isolated and in situ outer dynein arms of sea urchin sperm flagella. *J. Cell Biol.* **101**, 1400–1412.

Salmon, E. D., Leslie, R. J., Saxton, W. M., Karow, M. L., and McIntosh, J. R. (1984). Spindle microtubule dynamics in sea urchin embryos. Analysis using fluorescence-labeled tubulin and measurements of fluorescence redistribution after laser photobleaching. *J. Cell Biol.* **99**, 2165–2174.

Sato-Yoshitake, R., Yorifuji, H., Inagaki, M., and Hirokawa, N. (1992). The phosphorylation of kinesin regulates its binding to synaptic vesicles. *J. Biol. Chem.* **267**, 23930–23936.

Saunders, W. S., and Hoyt, M. A. (1992). Kinesin-related proteins required for structural integrity of the mitotic spindle. *Cell (Cambridge, Mass.)* **70**, 451–458.

Saunders, W. S., Koshland, D., Eschel, D., Gibbons, I. R., and Hoyt, M. A. (1995). *Saccharomyces cerevisiae* kinesin- and dynein-related proteins required for anaphase chromosome segregation. *J. Cell Biol.* **128**, 617–624.

Sawin, K. E., Le Guellec, K. L., Philippe, M., and Mitchison, T. J. (1992). Mitotic spindle organization by a plus end-directed microtubule motor. *Nature (London)* **359**, 540–543.

Saxton, W. M., Stemple, D. L., Leslie, R. J., Salmon, E. D., Zavortink, M., and McIntosh, J. R. (1984). Tubulin dynamics in cultured mammalian cells. *J. Cell Biol.* **99**, 2175–2186.

Schafer, D. A., Gill, S. R., Cooper, J. A., Heuser, J. E., and Schroer, T. A. (1994). Ultrastructural analysis of the dynactin complex: An actin-related protein is a component of a filament that resembles F-actin. *J. Cell Biol.* **126**, 403–412.

Schmitz, F., Wallis, K. T., Rho, M., Drenckhahn, D., and Murphy, D. B. (1994). Intracellular distribution of kinesin in chromaffin cells. *Eur. J. Cell Biol.* **63**, 77–83.

Schnapp, B. J., and Reese, T. S. (1989). Dynein is the motor for retrograde axonal transport of organelles. *Proc. Natl. Acad. Sci. U.S.A.* **86**, 1548–1552.

Schnapp, B. J., Vale, R. D., Sheetz, M. P., and Reese, T. S. (1985). Single microtubules from squid axoplasm support bidirectional movement of organelles. *Cell (Cambridge, Mass.)* **40**, 455–462.

Scholey, J. M., Heuser, J., Yang, J. T., and Goldstein, L. S. B. (1989). Identification of globular mechanochemical heads of kinesin. *Nature (London)* **38**, 335–357.

Schroer, T. A. (1994). Structure, function and regulation of cytoplasmic dynein. *Curr. Opin. Cell Biol.* **6**, 69–73.

Schroer, T. A., and Sheetz, M. P. (1991). Two activators of microtubule-based vesicle transport. *J. Cell Biol.* **115**, 1309–1318.

Schroer, T. A., Schnapp, B. J., Reese, T. S., and Sheetz, M. P. (1988). The role of kinesin and other soluble factors in organelle movements along microtubules. *J. Cell Biol.* **107**, 1785–1792.

Schroer, T. A., Steuer, E. R., and Sheetz, M. P. (1989). Cytoplasmic dynein is a minus end-directed motor for membranous organelles. *Cell (Cambridge, Mass.)* **56**, 937–946.

Schulze, E., and Kirschner, M. (1986). Microtubule dynamics in interphase cells. *J. Cell Biol.* **102**, 1020–1031.

Sekine, Y., Okada, Y., Noda, Y., Kondo, S., Aizawa, H., Takemura, R., and Hirokawa, N. (1994). A novel microtubule-based motor protein (KIF4) for organelle transport, whose expression is regulated developmentally. *J. Cell Biol.* **127**, 187–201.

Shiina, N., Moriguchi, T., Ohta, K., Gotoh, Y., and Nishida, E. (1992a). Regulation of a major microtubule associated protein by MPF and MAP kinase. *EMBO J.* **11**, 3977–3984.

Shiina, N., Gotoh, Y., and Nishida, E. (1992b). A novel homo-oligomeric protein responsible for an MPF-dependent microtubule-severing activity. *EMBO J.* **11**, 4723–4731.

Shiina, N., Gotoh, Y., Kubomura, N., Iwamatsu, A., and Nishida, E. (1994). Microtubule severing by elongation factor 1α. *Science* **266**, 282–285.

Shimizu, T., Toyoshima, Y. Y., Edamatsu, M., and Vale, R. D. (1995). Comparison of the motile and enzymatic properties of two microtubule minus-end-directed motors, ncd and cytoplasmic dynein. *Biochemistry* **34**, 1575–1582.

Shpetner, H. S., Paschal, B. M., and Vallee, R. B. (1988). Characterization of the microtubule-activated ATPase of brain cytoplasmic dynein (MAP1C). *J. Cell Biol.* **107**, 1001–1009.

Skoufias, D. A., and Scholey, J. M. (1993). Cytoplasmic microtubule-based motor proteins. *Curr. Opin. Cell Biol.* **5**, 95–104.

Skoufias, D. A., Cole, D. G., Wedaman, K. P., and Scholey, J. M. (1994). The carboxyl-terminal domain of kinesin heavy chain is important for membrane binding. *J. Biol. Chem.* **269**, 1477–1485.

Sloboda, R. D., Rudolph, S. A., Rosenbaum, J. L., and Greengard, P. (1975). Cyclic AMP-dependent endogenous phosphorylation of a microtubule-associated protein. *Proc. Natl. Acad. Sci. U.S.A.* **72**, 177–181.

Smith, E. F., and Sale, W. S. (1991). Microtubule binding and translocation by inner dynein arm subtype I1. *Cell Motil. Cytoskel.* **18**, 258–268.

Smith, E. F., and Sale, W. S. (1992a). Regulation of dynein-driven microtubule sliding by the radial spokes in flagella. *Science* **257**, 1557–1559.

Smith, E. F., and Sale, W. S. (1992b). Structural and functional reconstitution of inner dynein arms in *Chlamydomonas* flagellar axonemes. *J. Cell Biol.* **117**, 573–581.

Sontag, E., Nunbhakdi-Craig, V., Bloom, G. S., and Mumby, M. C. (1995). A novel pool of protein phosphatase 2A is associated with microtubules and is regulated during the cell cycle. *J. Cell Biol.* **128**, 1131–1144.

Stearns, T., and Kirschner, M. (1994). *In vitro* reconstitution of centrosome assembly and function: The central role of gamma-tubulin. *Cell (Cambridge, Mass.)* **76**, 623–637.

Stearns, T., Evans, L., and Kirschner, M. (1991). γ-tubulin is a highly conserved component of the centrosome. *Cell (Cambridge, Mass.)* **65**, 825–836.

Stephens, R. E., and Prior, G. (1992). Dynein from serotonin-activated cilia and flagella: Extraction characteristics and distinct sites of cAMP-dependent protein phosphorylation. *J. Cell Sci.* **103**, 999–1012.

Steuer, E. R., Wordeman, L., Schroer, T. A., and Sheetz, M. P. (1990). Localization of cytoplasmic dynein to mitotic spindles and kinetochores. *Nature (London)* **345**, 266–268.

Stewart, R. J., Thaler, J. P., and Goldstein, L. S. (1993). Direction of microtubule movement is an intrinsic property of the motor domains of kinesin heavy chain and *Drosophila* ncd protein. *Proc. Natl. Acad. Sci. U.S.A.* **90**, 5209–5213.

Stommel, E. W., and Stephens, R. E. (1985). Calcium-dependent phosphatidylinositol phosphorylation in lamellibranch gill lateral cilia. *J. Comp. Physiol. A* **157A**, 451–459.

Summers, K. E., and Gibbons, I. R. (1971). Adenosine triphosphate-induced sliding of tubules in trypsin-treated flagella of sea-urchin sperm. *Proc. Natl. Acad. Sci. U.S.A.* **68**, 3092–3096.

Svoboda, K., and Block, S. M. (1994). Force and velocity measured for single kinesin molecules. *Cell (Cambridge, Mass.)* **77**, 773–784.

Svoboda, K., Schmidt, C. F., Schnapp, B. J., and Block, S. M. (1993). Direct observation of kinesin stepping by optical trapping interferometry. *Nature (London)* **365**, 721–727.

Swaroop, A., Swaroop, M., and Garen, A. (1987). Sequence analysis of the complete cDNA and encoded polypeptide for the glued gene of *Drosophila melanogaster. Proc. Natl. Acad. Sci. U.S.A.* **84**, 6501–6505.

Takemura, R., Okabe, S., Umeyama, T., Kanai, Y., Cowan, N. J., and Hirokawa, N. (1992). Increased microtubule stability and alpha tubulin acetylation in cells transfected with microtubule-associated proteins, MAP1B, MAP2 or tau. *J. Cell Sci.* **103**, 953–964.

Tan, J. L., Ravid, S., and Spudich, J. A. (1992). Control of nonmuscle myosins by phosphorylation. *Annu. Rev. Biochem.* **61**, 721–759.

Tash, J. S., Krinks, M., Patel, J., Means, R. L., Klee, C. B., and Means, A. R. (1988). Identification, characterization, and functional correlation of calmodulin-dependent protein phosphatase in sperm. *J. Cell Biol.* **106**, 1625–1633.

Thaler, C. D., and Haimo, L. T. (1990). Regulation of organelle transport in melanophores by calcineurin. *J. Cell Biol.* **111**, 1939–1948.

Thaler, C. D., and Haimo, L. T. (1992). Control of organelle transport in melanophores: regulation of Ca^{2+} and cAMP levels. *Cell Motil. Cytoskel.* **22**, 175–184.

Theurkauf, W. E., and Vallee, R. B. (1982). Molecular characterization of the cAMP-dependent protein kinase bound to microtubule-associated protein 2. *J. Biol. Chem.* **257**, 3284–3290.

Titus, M. A. (1993). Multitasking with myosin. *Trends Genet.* **9**, 187–188.

Toyoshima, I., Yu, H., Steuer, E. R., and Sheetz, M. P. (1992). Kinectin, a major kinesin binding protein. *J. Cell Biol.* **118**, 1121–1131.

Tsuyama, S., Bramblett, G. T., Huang, K. P., and Flavin, M. (1986). Calcium/phospholipid-dependent kinase recognizes sites in microtubule-associated protein 2 which are phosphorylated in living brain and are not accessible to other kinases. *J. Biol. Chem.* **261**, 4110–4116.

Urrutia, R., McNiven, M. A., Albanesi, J. P., Murphy, D. B., and Kachar, B. (1991). Purified kinesin promotes vesicle motility and induces active sliding between microtubules *in vitro*. *Proc. Natl. Acad. Sci. U.S.A.* **88**, 6701–6705.

Uyeda, T. Q., Warrick, H. M., Kron, S. J., and Spudich, J. A. (1991). Quantized velocities at low myosin densities in an *in vitro* motility assay. *Nature (London)* **352**, 307–311.

Vaisberg, E. A., Koonce, M. P., and McIntosh, J. R. (1993). Cytoplasmic dynein plays a role in mammalian mitotic spindle formation. *J. Cell Biol.* **123**, 849–858.

Vale, R. D. (1991). Severing of stable microtubules by a mitotically-activated protein in *Xenopus* egg extracts. *Cell (Cambridge, Mass.)* **64**, 827–839.

Vale, R. D., and Hotani, H. (1988). Formation of membrane networks *in vitro* by kinesin-driven microtubule movement. *J. Cell Biol.* **107**, 2233–2241.

Vale, R. D., and Toyoshima, Y. Y. (1988). Rotation and translocation of microtubules *in vitro* induced by dyneins from *Tetrahymena* cilia. *Cell (Cambridge, Mass.)* **52**, 459–469.

Vale, R. D., Schnapp, B. J., Reese, T. S., and Sheetz, M. P. (1985a). Organelle, bead, and microtubule translocations promoted by soluble factors from the squid giant axon. *Cell (Cambridge, Mass.)* **40**, 559–569.

Vale, R. D., Reese, T. S., and Sheetz, M. P. (1985b). Identification of a novel force-generating protein, kinesin, involved in microtubule-based motility. *Cell (Cambridge, Mass.)* **42**, 39–50.

Vale, R. D., Schnapp, B. J., Mitchison, T., Steuer, E., Reese, T. S., and Sheetz, M. P. (1985c). Different axoplasmic proteins generate movements in opposite directions along microtubules *in vitro*. *Cell (Cambridge, Mass.)* **43**, 623–632.

Vale, R. D., Malik, F., and Brown, D. (1992). Directional instability of microtubule transport in the presence of kinesin and dynein, two opposite polarity motor proteins. *J. Cell Biol.* **119**, 1589–1596.

Vallano, M. L., Goldering, J. R., Buckholz, T. M., Larson, R. E., and Delorenzo, R. J. (1985). Separation of endogenous calmodulin and cAMP dependent kinases from microtubule preparations. *Proc. Natl. Acad. Sci. U.S.A.* **83**, 3203–3208.

Vallee, R. (1993). Molecular analysis of the microtubule motor dynein. *Proc. Natl. Acad. Sci. U.S.A.* **90**, 8769–8772.

Vallee, R. B., Wall, J. S., Paschal, B. M., and Shpetner, H. S. (1988). Microtubule-associated protein 1C from brain is a two-headed cytosolic dynein. *Nature (London)* **332**, 561–563.

Verde, F., Labbé, J. C., Dorée, M., and Karsenti, E. (1990). Regulation of microtubule dynamics by cdc2 protein kinase in cell-free extracts of *Xenopus* eggs. *Nature (London)* **343**, 233–238.

Verde, R., Dogterom, M., Stelzer, E., Karsenti, E., and Leibler, S. (1992). Control of microtubule dynamics and length by cyclin A- and cyclin B-dependent kinases in *Xenopus* egg extracts. *J. Cell Biol.* **118**, 1097–1108.

Vernos, I., Raats, J., Hirano, T., Heasman, J., Karsenti, E., and Wylie, C. (1995). Xklp1, a chromosomal *Xenopus* kinesin-like protein essential for spindle organization and chromosome positioning. *Cell (Cambridge, Mass.)* **81**, 117–127.

von Massow, A., Mandelkow, E. M., and Mandelkow, E. (1989). Interaction between kinesin, microtubules and microtubule-associated protein 2. *Cell Motil. Cytoskel.* **14**, 562–571.

Walker, R. A., and Sheetz, M. P. (1993). Cytoplasmic microtubule-associated motors. *Annu. Rev. Biochem.* **62**, 429–451.

Walker, R. A., Salmon, E. D., and Endow, S. A. (1990). The *Drosophila* claret segregation protein is a minus-end directed motor molecule. *Nature (London)* **347**, 780–782.

Walther, Z., Vashishtha, M., and Hall, J. L. (1994). The *Chlamydomonas* FLA10 gene encodes a novel kinesin-homologous protein. *J. Cell Biol.* **126**, 175–188.

Wang, S.-Z., and Adler, R. (1995). Chromokinesin: A DNA-binding, kinesin-like nuclear protein. *J. Cell Biol.* **128**, 761–768.

Webster, D. R., Wehland, J., Weber, K., and Borisy, G. G. (1990). Detyrosination of α-tubulin does not stabilize microtubules *in vivo*. *J. Cell Biol.* **111**, 113–122.

Weiss, A., Mayer, D. C., and Leinwand, L. A. (1994). Diversity of myosin-based motility: multiple genes and functions. *Soc. Gen. Physiol. Ser.* **49,** 159–171.

Weisshaar, B., Doll, T., and Matus, A. (1992). Reorganization of the microtubular cytoskeleton by embryonic microtubule-associated proteins 2 (MAP2c). *Development (Cambridge, Mass.)* **116,** 1151–1161.

Wilkerson, C. G., King, S. M., and Witman, G. B. (1994). Molecular analysis of the γ heavy chain of *Chlamydomonas* flagellar outer-arm dynein. *J. Cell Sci.* **107,** 497–506.

Witman, G. B. (1992). Axonemal dyneins. *Curr. Opin. Cell Biol.* **4,** 74–79.

Witman, G. B., Plummer, J., and Sander, G. (1978). *Chlamydomonas* flagellar mutants lacking radial spokes and central tubules. *J. Cell Biol.* **76,** 729–747.

Wordeman, L., and Mitchison, T. J. (1995). Identification and partial characterization of mitotic centromere-associated kinesin, a kinesin-related protein that associates with centrosomes during mitosis. *J. Cell Biol.* **128,** 95–105.

Wordeman, L., Steuer, E. R., Sheetz, M. P., and Mitchison, T. J. (1991). Chemical subdomains within the kinetochore domain of isolated CHO mitotic chromosomes. *J. Cell Biol.* **114,** 285–294.

Xiang, X., Beckwith, S. M., and Morris, N. R. (1994). Cytoplasmic dynein is involved in nuclear migration in *Aspergillus nidulans. Proc. Natl. Acad. Sci. U.S.A.* **91,** 2100–2104.

Yang, J. T., Laymon, R. A., and Goldstein, L. S. B. (1989). A three-domain structure of kinesin heavy chain revealed by DNA sequence and microtubule binding analyses. *Cell (Cambridge, Mass.)* **56,** 879–889.

Yang, J. T., Saxton, W. M., Stewart, R. J., Raff, E. C., and Goldstein L. S. B. (1990). Evidence that the head of kinesin is sufficient for force generation and motility *in vitro. Science* **249,** 42–47.

Yen, T. J., Li, G., Schaar, B. R., Szilak, I., and Cleveland, D. W. (1992). CENP-E is a putative kinetochore motor that accumulates just before mitosis. *Nature (London)* **359,** 536–539.

Yu, H., Toyoshima, I., Steuer, E. R., and Sheetz, M. P. (1992). Kinesin and cytoplasmic dynein binding to brain microsomes. *J. Biol. Chem.* **267,** 20457–20464.

Zhang, P., Knowles, A., Goldstein, L. S. B., and Hawley, R. S. (1990). A kinesin-like protein required for distributive chromosome segregation in *Drosophila. Cell (Cambridge, Mass.)* **62,** 1053–1062.

Zhang, Z., Tanaka, Y., Nonaka, S., Aizawa, H., Kawasaki, H., Nakata, T., and Hirokawa, N. (1993). The primary structure of rat brain (cytoplasmic) dynein heavy chain, a cytoplasmic motor enzyme. *Proc. Natl. Acad. Sci. U.S.A.* **90,** 7928–7932.

Zheng, Y., Jung, M. K., and Oakley, B. R. (1991). γ tubulin is present in *Drosophila melanogaster* and *Homo sapiens* and is associated with the centromere. *Cell (Cambridge, Mass.)* **65,** 817–823.

INDEX